Nuclear Choices

New Liberal Arts Series

*This book is published as part of an Alfred P. Sloan Foundation
program.*

Nuclear Choices

A Citizen's Guide to Nuclear Technology

Richard Wolfson

The MIT Press
Cambridge, Massachusetts
London, England

Library of Congress Cataloging-in-Publication Data

Wolfson, Richard.
  Nuclear choices : a citizen's guide to nuclear technology / Richard Wolfson.
     p.   cm. — (The New liberal arts series)
  Includes bibliographical references and index.
  1. Nuclear energy—Popular works. I. Title. II. Series.
TK9145.W59   1991b
333.792′4—dc20   90-13545
ISBN: 0-262-23158-1   CIP

To the memory of Annabel Wolfson and Mary Swift, who taught me much about independent thinking.

Contents

Series Foreword

The Alfred P. Sloan Foundation's New Liberal Arts (NLA) Program stems from the belief that a liberal education for our time should involve undergraduates in meaningful experiences with technology and with quantitative approaches to problem solving in a wide range of subjects and fields. Students should understand not only the fundamental concepts of technology and how structures and machines function, but also the scientific and cultural settings within which engineers work, and the impacts (positive and negative) of technology on individuals and society. They should be much more comfortable than they are with making calculations, reasoning with numbers and symbols, and applying mathematical and physical models. These methods of learning about nature are increasingly important in more and more fields. They also underlie the process by which engineers create the technologies that exercise such vast influence over all our lives.

The program is closely associated with the names of Stephen White and James D. Koerner, both vice-presidents (retired) of the foundation. Mr. White wrote an internal memorandum in 1980 that led to the launching of the program two years later. In it he argued for quantitative reasoning and technology as "new" liberal arts, not as replacements for the liberal arts as customarily identified, but as liberating modes of thought needed for understanding the technological world in which we now live. Mr. Koerner administered the program for the foundation, successfully leading it through its crucial first four years.

The foundation's grants to 36 undergraduate colleges and 12 universities have supported a large number of seminars, workshops, and symposia on topics in technology and applied mathematics. Many new courses have been developed and existing courses modified at these colleges. Some minors or concentrations in technology studies have been organized. A Resource Center for the NLA Program, located at Stony Brook, publishes and distributes a monthly newsletter, collects and disseminates syllabi, teaching modules, and other materials prepared at the colleges and universities taking part in the program, and serves in a variety of ways to bring news of NLA activities to all who express interest and request information.

As the program progressed, faculty members who had developed successful new liberal arts courses began to prepare textbooks. Also, a number of the foundation's grants to universities were used to support writing projects of professors—often from engineering departments—who had taught well-attended courses in technology and applied mathematics that had been designed to be accessible to liberal arts undergraduates. It seemed appropriate not only to encourage the preparation of books for such courses, but also to find a way to publish and thereby make available to the widest possible audience the best products of these teaching experiences and writing projects. This is the background with which the foundation approached The MIT Press and the McGraw-Hill Publishing Company about publishing a series of books on the new liberal arts. Their enthusiastic response led to the launching of the New Liberal Arts Series.

The publishers and the Alfred P. Sloan Foundation express their appreciation to the members of the Editorial Advisory Board for the New Liberal Arts Series: John G. Truxal, Distinguished Teaching Professor, Department of Technology and Society, State University of New York, Stony Brook, Chairman; Joseph Bordogna, Alfred Fitler Moore Professor and Dean, School of Engineering and Applied Science, University of Pennsylvania; Robert W. Mann, Whitaker Professor of Biomedical Engineering, Massachusetts Institute of Technology; Merritt Roe Smith, Professor of the History of Technology, Massachusetts Institute of Technology; J. Ronald Spencer, Associate Academic Dean and Lecturer in History, Trinity College; and Allen B. Tucker, Jr., Professor of Computer Science, Bowdoin College. In developing this new

publication program, The MIT Press has been represented by Frank
P. Satlow and the McGraw-Hill Publishing Company by Eric M.
Munson.

Samuel Goldberg
Program Officer
Alfred P. Sloan Foundation

Preface

Nuclear technology is an inescapable part of our lives. Nuclear reactors provide a significant share of our electrical energy. Techniques of nuclear medicine offer new promise in the diagnosis and treatment of disease. Nuclear processes help industry produce better and safer products, help airlines detect terrorists' bombs, and help archaeologists understand our past. And through the second half of the twentieth century, nuclear weapons have purportedly kept the peace by threatening the annihilation of modern civilization.

But with nuclear technology come dangers. Nuclear war is an obvious one. So are reactor accidents like those at Chernobyl and Three Mile Island. The mining of uranium, the manufacture of nuclear weapons, and the normal operation of nuclear power plants all release radioactivity to the environment. Nuclear medicine carries risks that must be weighed against its potential benefits. Even such non-nuclear technologies as aviation and house construction have nuclear dangers associated with them.

The news media regularly bring nuclear technology and its dangers to our attention. Nuclear technology provokes vigorous debates at the local, national, and global levels. Nuclear issues force us to make nuclear choices—individually in the voting booth, through our elected representatives, and through our leaders at the highest levels of international negotiation.

I have written this book on the premise that nuclear choices are best made by citizens who know something about the underlying issues, who understand the basics of nuclear technology, and who can judge for themselves statements advocating particular positions. In that spirit, the book demands no prior knowledge of nuclear matters. It does ask that readers be open to the range of opinions, be willing to grasp some basic technological considerations, and be willing to bring informed judgment to their own choices.

Nuclear Choices arises from a course I have taught in recent years at Middlebury College. As I hope the book will do, that course has given people with no particular scientific, technological, or political background the understanding to help them make informed choices about nuclear issues. Although the book should find use in similar college courses, its intended audience is much broader. Citizens of today's industrialized societies cannot avoid nuclear technology, and the book should help them to become familiar with it and to gain confidence in making nuclear choices.

My specific goal here is to introduce readers to the ideas they will need in order to understand nuclear issues as they are presented in the contemporary news media. By covering essentially all nuclear technologies in one book, I have been able to stress the connections among them—especially the multifaceted relation between nuclear power and nuclear weapons. Readers seeking a deeper understanding of individual nuclear technologies are referred to the more thorough works listed at the ends of the chapters.

A great many people and institutions contributed to the making of this book. Corporations, government agencies, national laboratories, universities, and individuals supplied photographs. The reference staff of Middlebury College's Starr Library—especially Terry Plum—accepted enthusiastically the often obscure challenges I put to them. John Truxal and several unnamed reviewers made helpful suggestions on the entire manuscript. Geneticist George Saul kindly reviewed chapter 4, while Rush Holt of the Princeton Plasma Physics Laboratory supplied helpful comments on the treatment of nuclear fusion. Political scientists Russ Leng and Ted Rueter assessed the coverage of strategic nuclear doctrines. The support of the Alfred P. Sloan Foundation's New Liberal Arts Program has been most important. Through major grants to liberal arts colleges, the NLA program has sought to bring technological literacy to students with academic interests well removed from science and technology. Middlebury's multi-year NLA grant allowed, among other activities, the development and teaching of the course from which this book grew. And a subsequent NLA Special-Leave Grant gave me the time and support that allowed me to complete this book in a timely fashion. I am grateful to the Sloan Foundation for its generous backing of this project. It was a pleasure to work with Paul Bethge at The MIT Press as the book went through its editing, and with George Nichols, Robin Brickman, and Amanda Tate as they prepared the final artwork. Finally, I thank my family for their patience and encouragement during the time this book was in the making.

Nuclear Choices

Nuclear News, Nuclear Choices

1

In 1988, Massachusetts voters were confronted by a ballot question asking whether they approved of a proposed law providing that "there shall be no further generation of electric power by commercial nuclear power plants . . . by means which result in the production of nuclear waste." Since all nuclear power plants produce nuclear waste, the effect of a Yes vote would have been to shut down Massachusetts' two operating nuclear plants and to forbid the startup of any new ones.

If you lived in Massachusetts, how would you have voted? On what would you have based your vote? What *is* nuclear waste, anyway? How much waste do nuclear power plants produce? How dangerous is the waste? What should we do with it? Are nuclear power plants safe? Safe relative to what? Do we really need nuclear power? What are some alternatives? How is nuclear power related to nuclear weapons? Could a Chernobyl-type accident occur in Massachusetts? What is radiation, and what can it do to me? What's so special about things nuclear?

Those questions and many more might have come up as you considered how to vote in the Massachusetts referendum. And Massachusetts voters are not the only ones who are called to make nuclear choices. A year earlier, Maine voters faced a similar referendum. In 1989 it was California's turn, as voters focused on the fate of a single nuclear power plant. At least twelve similar referenda have been held in the United States, and more than 150 U.S. communities have voted to ban nuclear weapons or other nuclear

Figure 1.1 The Pilgrim nuclear power plant in Plymouth, Massachusetts. (Boston Edison)

QUESTION 4

LAW PROPOSED BY INITIATIVE PETITION

Do you approve of a law summarized below, *upon which no vote was taken by the House of Representatives or the Senate before May 4, 1988?*

SUMMARY

The proposed law would provide that, after July 4, 1989, there shall be no further generation of electric power by commercial nuclear power plants in the Commonwealth by means which result in the production of nuclear waste.

| YES | ▶ |
| NO | ▶ |

Figure 1.2 The 1988 ballot question that would have forced Massachusetts' nuclear power plants to close. Voters rejected the proposition by a 2-to-1 margin. (Massachusetts State Elections Division)

materials. (With the exception of the 1989 California vote on the Rancho Seco nuclear plant, voters rejected propositions that would have shut down existing power plants.)

Whether or not you encounter nuclear questions in the voting booth, you can't escape nuclear issues in the news. Often that news calls for nuclear choices, either by you or by your elected representatives. Some recent examples follow.

> New York Times, *November 29, 1989: U.S. WILL START OVER ON PLANNING FOR NEVADA NUCLEAR WASTE DUMP . . . The Government has abandoned a two-year effort . . . to plan the nation's only dump for highly radioactive nuclear waste because it lacks confidence in its work. . . . The decision delays until at least 2010 . . . the opening of a repository . . . for the spent fuel of more than 110 civilian nuclear reactors. . . .*

Will they *ever* find a way to dispose of nuclear waste? Is this a technical problem, or is it really political? Would I want a nuclear waste dump in my state? Should I have voted Yes on the referendum to close down nuclear plants because of this waste problem?

> New York Times, *March 29, 1990: 30-YEAR PLUTONIUM LOSS AT PLANT EQUALS 7 BOMBS . . . Seven nuclear bombs' worth of plutonium escaped into air ducts at the Rocky Flats weapons plant near Denver. . . . "a very substantial quantity of plutonium that was simply not accounted for". . . . as filters became clogged, workers had been punching holes in them so the air, although contaminated, could pass through.*

Seven bombs' worth unaccounted for? How do I know the plutonium hasn't gone to a terrorist group? And what is this about bypassing filters to let plutonium-contaminated air through? Why put a bomb plant near a big city like Denver? Where else are there nuclear weapons plants? What kind of operation is my government running, anyway?

> New York Times, *September 13, 1988: MAJOR RADON PERIL IS DECLARED BY U.S. IN CALL FOR TESTS . . . Cancer threat is called wider than had been believed. . . . the Government today issued a national public health advisory urging that most homes be tested. . . .*

What is radon? Should I get my house tested? Will I need to move? Is my family safe? Does this have anything to do with nuclear power? With nuclear weapons? Why wasn't I told about this earlier?

> New York Times, *September 28, 1988: SCIENTIST SAYS LOW RADON LEVELS MAY BE HARMLESS . . . Government warnings may be exaggerated. . . .*

But the government just recommended that I get my house tested. Who am I supposed to believe?

> New York Times, *November 27, 1988: ATOMS FOR PEACE AND WAR: IS THERE A CLEAR DISTINCTION? . . . Under an executive order issued by President Reagan . . . the Nuclear Regulatory Commission must draw up detailed plans for dealing with a national security emergency, including a plan to seize civilian nuclear power plants to obtain material for weapons.*

Does this mean my local nuclear power plant is a bomb factory? Am I making plutonium every time I turn on an electric light? How much? Could a terrorist get hold of this stuff? Should this presidential order change my vote on nuclear power?

> New York Times, *September 12, 1989: NEW MACHINES CAN DETECT TERRORISTS' BOMBS, USUALLY . . . The new . . . analyzer at Kennedy Airport passes suitcases through a cloud of subatomic particles and analyzes the radiation produced for signs of nitrogen, which is a component of virtually every . . . explosive.*

Here is a beneficial use of nuclear radiation. But will my luggage become radioactive? And can I be sure that my wool sweater won't set off the alarms? I've heard that wool contains a lot of nitrogen.

> Boston Globe, *September 19, 1989: SCREENING SOUGHT IN CANCER LINK TO PILGRIM . . . Two prominent health specialists said . . . that evidence linking the Pilgrim nuclear plant to elevated rates of cancer is so strong that nearby residents should be screened for cancer. . . .*

Should I get tested? I thought nuclear plants were safe. Should I have voted to shut down Pilgrim? Is this one of those media scare stories, or is there something to it? Who can I believe?

New York Times, *March 25, 1990: NEW EXPLOSION THREAT SEEN AT NUCLEAR PLANT . . . tanks that store atomic waste at the Hanford [Washington state] nuclear reservation could explode and spew radiation into the air. . . .*

New York Times, *September 8, 1989: U.S. TO REOPEN REACTOR IN 1990; SAFETY IN CAROLINA IS STILL AT ISSUE . . . The Department of Energy . . . will reopen one of its crippled nuclear reactors, resuming production of tritium . . . vital to nuclear weapons. . . . But the department refused to commit itself to completing safety testing . . . before reopening the plant, in South Carolina.*

So there are nuclear weapons plants in Washington and South Carolina, too! Possible explosions? Doesn't sound too safe. And what is tritium? Why do nuclear weapons need it? Why do we need to make more? Do we need it so badly that we should operate an unsafe reactor?

New York Times, *February 19, 1990: NEW ESTIMATES INCREASE RADIATION RISK IN FLIGHT . . . radiation exposure for flight crews . . . greater than for average nuclear power plant workers. . . .*

New York Times, *March 1, 1990: TESTS OF COCKPIT RADIATION SHOW LEVELS ABOVE A FEDERAL STANDARD . . . more radiation than the limit set . . . for pregnant women. . . . radiation . . . of such high energy that air crews cannot be shielded . . . could lead to a significant increase in the cancer rate among crew members and adult passengers. . . .*

Do I have to worry about radiation on airplanes? Where does it come from? Is it really like working in a nuclear power plant? Does that mean that nuclear power plants are really pretty safe, or that airplanes are dangerous? Should I avoid long airplane trips? How does this new radiation danger compare with other risks I face?

New York Times, *March 29, 1990: 6 HELD IN BRITAIN IN SCHEME TO SEND ATOM GEAR TO IRAQ . . . Devices That Trigger Nuclear Arms and Equipment for Missiles are Seized . . . Government officials have described the Iraqi [nuclear] program in . . . alarming terms. . . . Iraqi program . . ."well advanced". . . . "It demonstrates a clear intention to acquire nuclear weapons."*

Nuclear weapons in the hands of the Mideast's most aggressive nation? Who else is seeking them? Does this have anything to do with nuclear power? What is my government doing to curb the spread of nuclear weapons?

New York Times, *September 25, 1989: ENERGY EXPERTS SEE NUCLEAR POWER AS A CURE FOR ONE ILL ... The cure for the greenhouse effect . . . is a huge shift to nuclear-generated power, speaker after speaker told delegates at a World Energy Conference. . . .*

Maybe nuclear energy isn't so bad—in fact, maybe it's our salvation in a time of climatic crisis. But what about nuclear waste? How do we weigh the long-term burden of waste disposal against the immediate gain of a shift from fossil fuels to nuclear power? Should I encourage nuclear power or oppose it?

New York Times, *August 14, 1988: PRETORIA SAYS IT CAN BUILD A-ARMS ... Asked by reporters whether South Africa already had nuclear weapons, [foreign minister R. F. Botha] said "I'm not going to enlarge on that statement."*

New York Times, *January 29, 1989: GERMAN CONCERN SAID TO AID PAKISTAN A-WEAPONS ... officials have told Congress that Pakistan is . . . close to developing a nuclear weapon. . . .*

New York Times, *October 25, 1989: U.S. CONCERN RISES OVER NORTH KOREA ATOM PLANT ... administration officials were increasingly worried that North Korea may be trying to develop nuclear weapons.*

So it's not just Iraq! How many countries already have nuclear weapons? Is a nuclear conflict among Third World nations likely? Will newcomers to the nuclear weapons club pose a threat to the United States? What can we do to halt the spread of nuclear weapons?

New York Times, *November 15, 1988: EXPERTS CALL RE-ACTOR DESIGN 'IMMUNE' TO DISASTER ... the new reactor . . . is intrinsically safe because its physical characteristics make it immune to meltdown. . . .*

A safe reactor! Really? Is it economical? Does it emit radiation? What about nuclear waste? Is this what we need, or should we be looking at other energy alternatives?

New York Times, *October 18, 1988: NEW PLANS FOR SPACE REACTORS RAISE FEARS OF NUCLEAR DEBRIS . . . The United States and the Soviet Union are drawing up elaborate plans for the use of nuclear power in space, reigniting fears that radioactive debris from an accident will sooner or later come raining down on Earth.*

Are we putting nuclear reactors in space? Why? I thought spacecraft used solar power. Could one of these reactors come down? What about a launch accident? Do we really want nuclear power in space?

New York Times, *September 1, 1989: SEARCH INTENSI-FIES FOR TRITIUM, STILL MISSING AFTER 2 MONTHS . . . Some fear the gas could be used to make A-bombs.*

Tritium again! What is this stuff? How much is missing? How much does it take to make a bomb? Are we careful enough in guarding nuclear materials? Has anyone made a bomb with stolen material? Could they?

New York Times, *December 25, 1988: POTTERY SHARD BOMBARDED IN REACTOR MAY HOLD CLUES TO ANCIENT TRADING . . . a post-doctoral fellow in archaeology . . . took a tiny bit of the bowl and plunged it into the core of a nuclear reactor. When it was removed . . . it emitted gamma rays, and they are being studied for clues to ancient trade patterns in northern Mesopotamia.*

So nuclear reactors do more than generate power or materials for weapons. What else is nuclear technology used for? Does it benefit me in any way? Is it safe for archaeologists to be messing around with nuclear reactors?

New York Times, *November 30, 1989: 50 MX MISSILES TO BE SHIFTED TO TRAINS IN 7 STATES . . . the missiles would be placed on 25 trains . . . stationed . . . in Wyoming, Texas, Louisiana, North Dakota, Washington, Arkansas, and Michigan.*

I've heard that a single MX missile has as much explosive power as all the weapons used by the Allies in World War II. I don't want them in my state—we'd be sitting ducks in a nuclear war. On the other hand, maybe these missiles will help deter war. But can't we find a better use for the billions of dollars this will cost?

Getting Informed

As these news items suggest, nuclear issues are complex. They raise technical, political, moral, and practical questions. Those questions are far from academic; they demand answers and action from citizens, legislators, political activists, scientists, businesspeople, and national leaders. The answers we give and the actions we take have potentially major roles in shaping the future of civilization and of our planet itself.

We are called upon to answer nuclear questions and to make nuclear choices, often without a clear sense of the relevant technical and political realities. How many voters really know what plutonium is, where it comes from, and why it is a crucial material? How many people, flipping on a light switch, really understand what is going on at the power plant that provides their electricity? How many senators, voting to deploy MX missiles, know that each missile contains the explosive equivalent of 250 Hiroshimas, and understand the violence that can do? How many people, alive today because nuclear medical techniques detected their cancers, know that they owe their lives to radioactivity? We dread the prospect of a nuclear-armed Iraq or Libya, but how many of us understand how our own decisions could aid or hinder such countries' efforts to acquire nuclear weapons? Most of us harbor a deep fear of nuclear radiation, but do we know how its dangers compare with risks we willingly take, such as smoking or refusing to use seat belts? Many of us yearn for a world free of nuclear weapons, but what about the political, strategic, and technical challenges on the path to that goal? And, ultimately, do we understand what it is that makes our nuclear technologies so fundamentally different from anything humanity has known before?

This book is designed to provide citizens with a basic understanding of nuclear technology and of the controversies surrounding its use. The book is divided into three parts. Part I deals with the nature of the atom and its nucleus, with nuclear radiation, and with the fundamentals of nuclear energy. Part II examines nuclear power, including our use of energy, the operation of nuclear power plants, nuclear accidents, nuclear waste, and alternatives to nuclear power. Part III describes nuclear weapons, including their operation, their destructive effects, delivery systems for getting them to their targets, strategies for their use or non-use, the feasibility of defense against them, and the prospects for controlling these weapons and preventing nuclear war. But this division into three parts is in some respects only a convenience. Nuclear power and nuclear weapons share the same fundamental physics, and many of their technologies overlap. Some of the thorniest nuclear issues center on connections between nuclear power and nuclear weapons, and these issues will arise repeatedly throughout the book.

This book is for citizens, not scientists or nuclear specialists. It assumes no particular background in science or in nuclear issues. It provides a simplified introduction to nuclear science and technology and the controversies that surround them. The book's goals are to instill a level of nuclear literacy that gives you an understanding of the nuclear issues you continually encounter in the news media and to help you make intelligent choices based on that understanding.

This is not an anti-nuclear book, nor is it a pro-nuclear book. It aims to provide you with an understanding of nuclear technology and the issues that surround it. That's easy where scientific and technological fact are concerned, but harder when things get controversial. Engaging those controversies is as important as understanding the underlying technology, and therefore every attempt will be made to present arguments on all sides. Your author does have strong feelings on some of the issues, and is quite undecided on others. Where a display of personal opinion is unavoidable, it will be clearly indicated as such; otherwise, any arguments presented or questions asked do not necessarily reflect the author's own views.

Reading this book is not an academic exercise. Nuclear technology is an unavoidable part of our world, with the potential to bring us substantial benefit or unimaginable disaster. You will be called to make choices about nuclear issues, and this book should help you make them informed choices.

As a reminder that nuclear issues are regularly before you, the book presents occasional news items of nuclear significance. Most are quite recent, but some older ones are used to illustrate important steps in the evolution of nuclear technology. Many of these items are from the *New York Times,* which (as figure 1.3 shows) gives nuclear matters considerable attention.

Will this book give you all the answers? Will it tell you how to vote on nuclear questions? No! You may find it frustrating to come away from your reading less certain of your opinion on many complex nuclear issues. That shouldn't be surprising—the news items presented at the beginning of this chapter suggest that even nuclear experts often disagree. In fact, one thing you should take from this book is a healthy, critical skepticism about experts' opinions. If the experts agreed, resolving nuclear questions would be easy. But they don't agree. Yet the nuclear issues need resolution— and you are the one who has to resolve them. This book is written

Figure 1.3 The numbers of nuclear-related news articles in selected categories that appeared in the *New York Times* during 1989, obtained by searching *Newspaper Abstracts Ondisc* (University Microfilms, Inc.).

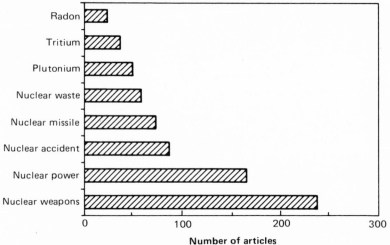

on the premise that those issues are best resolved by citizens who understand the basis of nuclear technology.

Editor's Note

Words that appear in boldface type in the text of a chapter are defined in the glossary at the end of that chapter.

Atoms and Nuclei

2

After the nuclear bombings of Japan in 1945, Albert Einstein re-marked that "the unleashed power of the atom has changed every-thing save our modes of thinking."[1] What is it that makes nuclear technology—"the unleashed power of the atom"—extraordinary? Why is the nuclear age unlike any previous age? The answer, in a nutshell, is that nuclear processes release over a million times as much energy as the more familiar happenings of our everyday world. Whereas a coal-burning power plant consumes many 110-car trainloads of coal each week, a comparable nuclear plant requires only a few truckloads of uranium fuel a year. A single nuclear bomb can destroy a city, a job that thousands of conventional bombs cannot accomplish. But why is there such a difference? Where does the millionfold energy increase come from? And why, with nuclear processes, do we encounter the new phenomenon of radiation? The answers to these questions lie in the atom and its nucleus.

What the World is Made Of . . .

The matter of our world exhibits tremendous variety, from the tenuousness of air to the solidity of steel, from the rugged density of rock to the delicacy of a snowflake, from the green slipperiness of a frog to the savory crunch of a fresh apple. It is remarkable that all these things—and all other things on Earth and throughout most of the known universe—are made from combinations of just three

Figure 2.1 A 110-car trainload of coal at a power plant in Kansas. Fourteen such trainloads arrive at the plant each week. (Earl Richardson, *Topeka Capital-Journal*)

Figure 2.2 A truckload of uranium fuel being unloaded at the Vermont Yankee nuclear power plant. Four such truckloads supply all the fuel needed for the once-in-18-months refueling. (Vermont Yankee Nuclear Power Corporation)

simple building blocks: the **neutron**, the **proton**, and the **electron**. In this chapter we will see how neutrons, protons, and electrons join to constitute the atomic nuclei and then the atoms from which the matter of our world is made.

Physicists have identified scores of so-called subatomic particles, and experiments with ever larger, more energetic, and more expensive machines continue in an effort to understand how these fundamental bits of matter are related. Yet most subatomic particles appear to be of little importance in the day-to-day interactions of matter. They do arise in the physicists' giant accelerators, and in interactions of cosmic rays with Earth's atmosphere, and many played important roles in the early seconds of the universe. But it is only a slight simplification to say that the composition and the behavior of ordinary matter—from a human heart to a nuclear bomb—involve only the interactions of neutrons, protons, and electrons.

Neutrons and protons are so tiny that it would take 13 trillion of them, lined up, to span an inch. Neutrons and protons have very nearly the same mass (for our purposes, the same thing as weight), and that mass is so small that a pound of either would contain 270 trillion trillion particles. You can envision these particles as small spheres, although a physicist might caution that in so doing you are imposing concepts from your everyday world on a subatomic world in which those concepts are not entirely appropriate.

Neutrons and protons differ in an important respect: The neutron carries no **electric charge**, whereas the proton carries one unit of positive electric charge. What *is* electric charge? No one knows, but we do understand how charge behaves, and will see shortly how the behavior of charge helps explain how matter is put together.

Charge is simply a fundamental property of matter, and a subatomic particle either has it or doesn't. A particle with charge has either one unit of positive charge or one unit of negative charge; no other amount seems possible.[2] There is nothing missing or deficient about negative charge; *positive* and *negative* are just names we use to distinguish the two different kinds of charge.

So we have the neutron and the proton: particles of essentially the same size and mass, differing in that the proton carries one unit of positive electric charge, whereas the neutron, as its name implies, is electrically neutral. Together, neutrons and protons are called **nucleons**.

The third particle, the electron, is much less massive than the others—it would take about 2,000 electrons to equal the mass of a neutron or proton. The electron carries one unit of negative electric charge. Even though the electron is much less massive than the proton, its charge is exactly equal but opposite to that of the proton.

. . . and How It Is Stuck Together

To see how our complex world can be made from just neutrons, protons, and electrons, we need to understand how these particles stick together. Nature provides what appear as three fundamental ways—called **forces**—by which pieces of matter can interact and stick together: the **gravitational force**, the **electric force**,[3] and the **nuclear force**. (We used to think there were additional fundamental forces—for example, the magnetic force—but gradually physicists came to realize that seemingly distinct forces were aspects of the same underlying phenomenon. As recently as the 1970s, physicists still counted four fundamental forces; now they are striving to learn if the remaining three are really aspects of a single fundamental interaction governing all that happens in the universe.)

The gravitational force is familiar: It keeps you rooted to the Earth, makes you fall, and holds the Moon in its orbit around the Earth and the Earth in its orbit around the Sun. But gravity is the weakest of the forces, and is significant only for larger objects—things the size of people, missiles, mountains, planets, and stars. Gravity plays essentially no role in the subatomic world of nuclear interactions, and we will neglect gravity as we explore basic nuclear phenomena. Gravity will become important again when we consider the trajectory of a missile or the meltdown of a reactor.

With gravity out of the picture, we have only the electric force and the nuclear force. For our purposes, the electric force manifests itself as a simple interaction between electrically charged particles. Particles with no charge—neutrons—do not experience the electric force. Two particles with the same charge, either positive or negative, are repelled from each other by the electric force. Two particles with opposite charges are attracted. The strength of the attractive or repulsive force depends on the distance between the particles; move them farther apart and the force weakens. However, it does not weaken rapidly with increasing distance, so the electric force is a **long-range force**.

The nuclear force acts only between nucleons—between protons and protons, between neutrons and neutrons, or between protons and neutrons. It is always attractive. When the particles are very close—roughly their own diameter apart—the nuclear force is extremely strong, but it falls off rapidly with increasing distance, quickly becoming insignificant. The nuclear force is thus a strong but **short-range force**. Figure 2.3 compares the ranges and strengths of the electric and nuclear forces.

Figure 2.4 summarizes the three particles and the two forces by which they interact. The forces are characterized by their strength and range, and by the particles between which they act. Here we have everything we need to build the nuclei, atoms, and

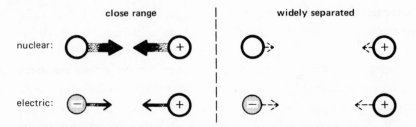

close range | widely separated

nuclear:

electric:

Figure 2.3 Strength and range of the nuclear and the electric force. At top left, a neutron and a proton at close range are strongly attracted by the nuclear force, indicated by the thick arrows. At bottom left, an electron and a proton at the same range experience a much weaker attraction due to the electric force. At top right, the nuclear force between widely separated nucleons has diminished to almost nothing, while the electric force between widely separated electron and proton remains significant.

What the world is made of . . .

◯	neutrons (n)	No electric charge. Mass = 1. Feels nuclear force only.
⊕	protons (p)	Electric charge: +1. Mass = 1.
⊖	electrons (e)	Electric charge: −1. Mass = 1/2,000. Feels electric force only.

. . . and how it is stuck together

nuclear force	Acts between nucleons (n, p), (n, n), (p, p). Always attractive. Strong, but short range.
electric force	Acts between charged particles (e, p), (e, e), (p, p). Opposites attract; likes repel. Weak, but long range.

Figure 2.4

molecules from which all substances are made. We will start with the nuclei.

Building Nuclei

An **atomic nucleus** is simply a group of nucleons—neutrons and protons—bound together by the nuclear force. The simplest nucleus is a single proton; every other nucleus contains a mixture of protons and neutrons. A single proton is known as a *hydrogen nucleus*. The next simplest nucleus is a combination of a proton and a neutron, and is called a *deuterium nucleus*. A combination of one proton and two neutrons is *tritium,* a nucleus that can't last very long. Two other simple nuclei are helium-3, formed of two protons and one neutron, and helium-4, containing two protons and two neutrons. Figure 2.5 shows the five nuclei that have been introduced so far, using the pictorial symbolism of figure 2.4 to represent the individual protons and neutrons. We will explore other nuclei shortly; first, we need to distinguish nuclei from atoms.

Building Atoms

A nucleus contains only protons, which carry positive electric charge, and neutrons, which carry no electric charge. Therefore all nuclei are positively charged, and that means they attract negatively charged electrons. Typically, a nucleus surrounds itself with a number of electrons equal to the number of protons it contains. The resulting object is called an **atom**. An atom may be visualized as something of a miniature solar system, with the nucleus surrounded by orbiting electrons, as the Sun is by the planets. Whereas the planets are held in the solar system by the Sun's gravity, the electrons are held in the atom by the electrical attraction of the nucleus. Although a gross oversimplification, this picture contains the essence of what we need in order to understand the difference between nuclear and more conventional sources of energy. Figure 2.6 depicts atoms of hydrogen and helium, each consisting of a nucleus surrounded by the appropriate number of electrons.

Although figure 2.6 shows the essential configurations of the

Nuclear News: Discovery of the Nucleus

The classic picture of the atom, with its electrons whirling about a tiny nucleus, is commonplace. But science has not always had such a picture. After J. J. Thomson's discovery of the electron in 1897, the atom was viewed as a sort of "plum pudding," with negative electrons embedded like plums in a positively charged "pudding." Then, in 1909–1911, a series of experiments revealed the atom as mostly empty space, with virtually all its mass concentrated in a tiny nucleus at the center. Suggested by Ernest Rutherford of the University of Manchester in England, and performed by his assistants Hans Geiger and Ernest Marsden, these experiments involved bombarding thin sheets of gold with high-energy projectiles called *alpha particles*. Most of the alpha particles passed right through the gold, but, much to the experimenters' surprise, a few bounced back. It was as though one fired bullets into a thin target of hay. They should all go right through; if a few bounced back, one might infer that hard, massive objects such as rocks were embedded in the hay. Similarly, Rutherford inferred the presence of tiny but relatively massive objects—nuclei—within the atoms of his gold foil. And because most of the alpha-particle "bullets" passed undeflected through the gold, most of the gold atom must be empty space. Rutherford's nuclear atom competed for several years with Thomson's "plum pudding" model before it became widely accepted.

News source: "British Association. Work of the Sections. Section A.—Mathematical and Physical Science," *The Times* (London), September 12, 1913.

---

atoms it depicts, it is misleading in an important respect: its scale. In a real atom, the distance between the nucleus and the surrounding electrons is much larger than our figure suggests. That distance is, in fact, over 10,000 times the diameter of the nucleus. If figure 2.6 were drawn to scale, the nucleus would be an invisible dot. If the nucleus of an atom were the size of a basketball, the electrons would be a mile away. Between the nucleus and its electrons would be a mile of emptiness. Atoms—and therefore everything that is made from atoms—are mostly empty space. (Even though atoms are over 10,000 times larger than nuclei, real atoms are still tiny; it would take more than 200 million of them to span an inch.)

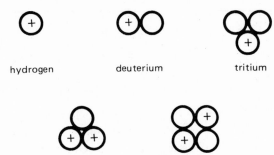

hydrogen   deuterium   tritium

helium-3   helium-4

Figure 2.5 The hydrogen nucleus is a single proton. Other simple nuclei are deuterium (one proton, one neutron), tritium (two neutrons and one proton), helium-3 (two protons and one neutron), and helium-4 (two protons and two neutrons).

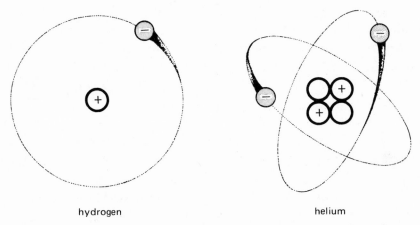

hydrogen   helium

Figure 2.6 Atoms of hydrogen and helium. An atom consists of a nucleus surrounded by as many electrons as there are protons in the nucleus.

Building Molecules

Two or more atoms can join together to form a **molecule**. The water molecule, for example, consists of two hydrogen atoms and an oxygen atom; its composition is reflected in its chemical formula, H_2O. This bonding of atoms to form molecules is what chemistry is all about, and the rearrangement of atoms into a new molecular configuration is called a **chemical reaction**. Many conventional sources of energy involve chemical reactions. Burning coal, for example, combines carbon atoms in the coal with oxygen from the atmosphere, giving rise to carbon dioxide gas and in the process releasing energy (figure 2.7). Burning gasoline in your car's engine breaks up complicated molecules containing mostly hydrogen and carbon, and produces mostly carbon dioxide and water, along with energy. The high explosives used in conventional weapons are substances that undergo rapid chemical reactions, releasing their energy in a short time. The atoms in the food you eat are rearranged through chemical reactions in your body, supplying the energy that keeps you alive. Chemical reactions are important and common-place in our lives.

For our purpose—understanding the nuclear difference—the important thing about chemical reactions is that they involve the electrons at the outer fringes of the atoms. Those electrons interact through the relatively weak electric force, and therefore the energy involved in taking apart or putting together molecules is relatively small—chemical reactions are not highly energetic. To picture the

Figure 2.7 A chemical reaction. Here one atom of carbon joins with two atoms of oxygen to form carbon dioxide; this is the basic reaction involved in burning coal. Energy is also released; the number 4.1 gives the amount of energy in units used by physicists working with atomic and nuclear processes. Black dots represent atomic nuclei, which are not affected by the rearrangement of the atoms.

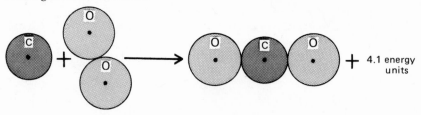

energy of a chemical reaction, you might think of a simple molecule as consisting of a couple of tennis balls connected by a weak rubber band. Because the rubber band—like the electric force—is weak, it doesn't take much effort to separate the balls and rearrange them into some other configuration. And because the electrons of an atom are so far from its nucleus, the nuclei of two interacting atoms remain widely separated. The nuclear force, which acts only at short distances, therefore plays no role in chemical reactions. In a chemical reaction, the nuclei of the interacting atoms remain essentially unchanged.

The Nuclear Difference

We have seen how chemical reactions involve relatively little energy, and how they result in a rearrangement of atoms that does not affect their underlying nuclei. **Nuclear reactions**, in contrast, occur when the nucleus itself changes. This can happen if two nuclei join, or if a nucleus ejects some of its nucleons, or if a nucleus is struck by another particle. Because the nuclear force is so strong at distances the size of a nucleus, nuclear reactions involve a lot of energy. (Picture now a very tight, thick rubber band holding together some tennis balls, here representing nucleons. You must work hard to pull the balls apart, and if the rubber band snaps, watch out!) Here, then, is the nuclear difference. It is, fundamentally, the difference between the weaker but long-range electric force (which governs the interactions of everyday chemical reactions) and the much stronger but short-range nuclear force (which comes into play only in reactions involving the nucleus). Because of the relative strengths of the forces, that difference means that a typical nuclear reaction releases several million times the energy of a chemical reaction (figure 2.8). That difference—based ultimately on the difference between the electric and nuclear forces—is, in turn, responsible for the dramatic differences between nuclear and conventional weapons, and between the fuel requirements of nuclear and coal-burning power plants (recall figures 2.1 and 2.2).

You might wonder how the energy of a nuclear reaction compares with energies you are familiar with. For example, could a single nuclear reaction boil a cup of water? The answer is decidedly

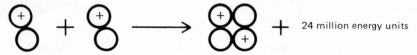

Figure 2.8 A nuclear reaction. Here two deuterium nuclei join to form a helium-4 nucleus. The physical size of the nuclei is less than 1/10,000 that of the atoms in figure 2.7, but the energy released is almost 6 million times that of the chemical reaction shown in the earlier figure.

No. A single nuclear reaction, although several million times more energetic than a single chemical reaction, releases an amount of energy that is still minuscule by our everyday standards. It would take 22 thousand trillion reactions like that of figure 2.8 to boil a cup of water. But nuclei are so small that the mass of all those reacting nuclei would be only 4 billionths of an ounce, showing that nuclear fuel is a concentrated source of energy. You would have to burn 7 million times as much coal to boil the same cup of water.

Whereas chemical reactions—burning coal or gasoline, metabolizing food, synthesizing plastics, and so on—are commonplace, nuclear reactions are rare under the conditions that prevail on Earth today. Our species harnessed fire—a chemical reaction—in prehistoric times, but only in the last half-century have we tended nuclear "fires" and their violent cousins, nuclear explosions. Nuclear reactions are common, though, in some parts of the universe; in particular, the Sun and other stars shine because of nuclear reactions in their interiors.

Incidentally, you will notice that this book never speaks of "atomic energy," "atomic bombs," "atomic power plants," or "atomic warfare;" or of "splitting the atom." The adjective *atomic* is ambiguous; since the interaction of atoms is involved in everyday chemical reactions, the energy they release might as well be called "atomic." Even Einstein's "unleashed power of the atom" suffers the same ambiguity. The reactions, the reactors, the bombs, the wars, the technologies that we are interested in here are distinctly *nuclear,* since their essential factor is the rearrangement of the atomic nucleus. And if we split anything, it will be a nucleus. I use the adjective *nuclear* to make all this absolutely clear.

Elements and Isotopes

An **element** is a substance that behaves chemically in a unique and identifiable way, and whose most basic particle is a single atom. Oxygen is an element; so is hydrogen. You can have only one oxygen atom and still have oxygen, but if you break that atom further it no longer behaves as oxygen. Water, H_2O, is *not* an element; the smallest piece of water you can have is a single molecule, consisting of two hydrogen atoms and one oxygen atom. If you take the molecule apart, you have hydrogen and oxygen but you no longer have water.

What gives the atoms of a particular element their unique chemical behavior? "Chemical behavior" means how they interact with other atoms, forming the multitude of different substances that make up our world. We have already seen that chemical reactions involve only the electrons that swarm in a distant cloud around the nucleus. So what determines the chemical behavior of an atom? Simply this: the number of electrons it contains. And what determines that? Since an atom forms when a nucleus attracts to itself as many electrons as it has protons, it is the number of protons in its nucleus that ultimately determines to what chemical species a given atom belongs.

The number of protons in a nucleus is called the **atomic number**. Hydrogen, as figure 2.9 shows, has atomic number 1, helium has atomic number 2, and carbon 6. Although figure 2.9 does not show all the individual particles, iron has 26 protons, gold 79, and uranium 92. An element's name and its atomic number are synonymous; you can't talk about oxygen without talking about atoms whose nuclei have eight protons each, and if you talk of an element with eight protons in its nucleus then you're talking about oxygen. In addition to its name and its atomic number, each element also has a unique one- or two-letter symbol; hydrogen is H, helium He, oxygen O, iron Fe, and uranium U. Table 2.1 gives the names, atomic numbers, and symbols of selected elements.

If we were chemists, we would be content to know only the atomic number—the number of protons—of a nucleus; that alone determines the species of chemical element. But since we are concerned with nuclear matters, we need to characterize nuclei further.

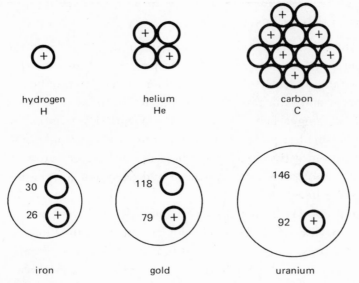

Figure 2.9 Some nuclei with element names and symbols. Figures are only suggestive; particles within a nucleus do not remain locked into a specific pattern.

Table 2.1
Selected elements.

Atomic number	Element name	Symbol
1	hydrogen	H
2	helium	He
6	carbon	C
7	nitrogen	N
8	oxygen	O
13	aluminum	Al
26	iron	Fe
38	strontium	Sr
79	gold	Au
86	radon	Rn
92	uranium	U
94	plutonium	Pu

Figure 2.5 introduced the nuclei of hydrogen, deuterium, tritium, helium-3, and helium-4. Why do the last two have such similar names? Because they are both nuclei of the same chemical element, helium. You can see in figure 2.5, and again in figure 2.10, why this is. Both contain two protons, and therefore both would form atoms with two electrons. Those atoms would behave the same chemically, even though one's nucleus had only one neutron and the other's two neutrons. As far as the chemist is concerned, both are atoms of the same substance: helium. The names helium-3 and helium-4 reflect the total numbers of nucleons: two protons and one neutron in helium-3, two protons and two neutrons in helium-4. The total number of nucleons—protons and neutrons—is called the **mass number** of a nucleus. Since protons and neutrons have nearly the same mass, the mass number gives approximately the total mass of a nucleus.

So helium-3 and helium-4 are both nuclei of helium, since the atoms to which they give rise both have the same chemical behavior. But they *are* different, and that difference manifests itself at the level of nuclear reactions. That is why we need to distinguish nuclei of the same element that have different numbers of neutrons and therefore different mass numbers. Such nuclei are called **isotopes**. Helium-3 and helium-4 are two isotopes of helium. And, as figure 2.10 shows, nuclei of hydrogen, deuterium, and tritium each have only one proton. They are really isotopes of the same element, namely hydrogen. Ordinary hydrogen could be called hydrogen-1, deuterium could be called hydrogen-2, and tritium could be called hydrogen-3. The use of separate isotope names is a confusion that, fortunately, is limited to hydrogen.

To a chemist, He is the symbol for helium. But for nuclear purposes that doesn't tell us enough, so we elaborate on the chemical symbol by adding the atomic and mass numbers. The atomic number goes in front of the element symbol, at the bottom; the mass number goes in front, at the top. Thus the helium isotopes helium-3 and helium-4 are written ^3_2He and ^4_2He, respectively. Ordinary hydrogen is ^1_1H, deuterium is ^2_1H, and tritium is ^3_1H. Soon we will be very much concerned with two important isotopes of uranium, uranium-235 and uranium-238; since uranium has atomic number 92, their symbols are $^{235}_{92}\text{U}$ and $^{238}_{92}\text{U}$. Strictly speaking, the letter(s) and the atomic number in a symbol are redundant; the

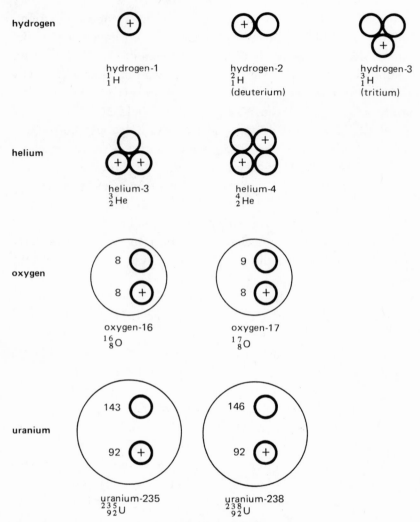

Figure 2.10 Isotopes of a given element have the same number of protons but differ in the number of neutrons. Shown here are three isotopes of hydrogen and two each of helium, oxygen, and uranium. Although isotopes behave the same chemically, their nuclear behavior is different; for example, only the rare isotope uranium-235 can be used directly as fuel in nuclear reactors and bombs.

atomic number 92 and the symbol U mean the same thing, namely uranium. Sometimes you will see a nuclear symbol written with just the mass number, for example ^{235}U or U-235; that's enough to tell the element (uranium) and the particular isotope (the one with a total of 235 nucleons). If you need the number of neutrons, you can get it by subtracting: $^{235}_{92}$U, for example has 235 total nucleons and 92 protons, so there are 235−92 or 143 neutrons. Figure 2.11 shows the meanings of all parts of a nuclear symbol.

We have already encountered a number of elements and some of their isotopes. Each isotope represents a unique combination of protons and neutrons, bound by the nuclear force to form a nucleus. How many different kinds of nuclei can we make? Is any combination of protons and neutrons a viable nucleus? The answer is No. The range of possible nuclei is distinctly limited.

Nuclei consist of protons and neutrons held together by the nuclear force—a strong but short-range force. But the electric force is also present, and acts to repel the protons in a nucleus. A combination of protons alone is not possible; neutrons must always be present in a nucleus to "dilute" the repulsive effect of the electric force. The lighter nuclei generally contain nearly equal numbers of protons and neutrons, and in the most common isotopes of helium ($^4_2$He), carbon ($^{12}_6$C), and oxygen ($^{16}_8$O) the numbers of protons and neutrons are exactly equal. Less common isotopes of these elements

---

Figure 2.11 Anatomy of a nuclear symbol. The number of neutrons is the difference between the mass number and the atomic number.

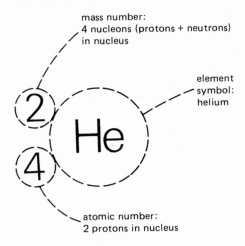

mass number:
4 nucleons (protons + neutrons)
in nucleus

element
symbol:
helium

atomic number:
2 protons in nucleus

include helium-3 ($^3_2$He), carbon-13 ($^{13}_6$C), oxygen-17 ($^{17}_8$O), and oxygen-18 ($^{18}_8$O). In fact, these particular isotopes are the only nuclei of helium, carbon, and oxygen you can make and have stick together forever. (Nuclei that can stick together indefinitely are called **stable nuclei**.) If, for these and other light elements, you deviate too much from equal neutron and proton numbers, the resulting nucleus simply will not stick together; it is **unstable**. Try to make oxygen-19 ($^{19}_8$O), for example. It just won't stick. There are too many neutrons, and the nucleus will soon fly apart; we will see just how in the next chapter. Try to make oxygen-14 ($^{14}_8$O) and again the nucleus comes apart; now there are too many protons.

You can see why a nucleus with too many protons might tend to come apart: the electrical repulsion counters the attractive nuclear force. With larger nuclei, this effect becomes more important. Why? Because protons at opposite sides of a large nucleus are so far apart that the attractive but short-range nuclear force has little effect. But the long-range electric force is still present and tends to repel them (figure 2.12). To counter this electric repulsion, more nuclear "glue," in the form of neutrons that feel only the nuclear attraction, is needed. Therefore larger nuclei tend to have more neutrons than

---

Figure 2.12   Two widely separated protons in a large nucleus experience a mutual repulsion due to the long-range electric force. But because of its short range, the attractive nuclear force between them is insignificant. To keep the nucleus from flying apart, an excess of neutrons over protons is therefore needed.

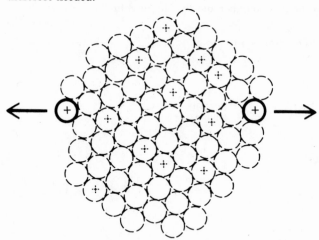

---

protons. Figure 2.13 plots number of neutrons versus number of protons for the known stable nuclei. Each little square represents a stable nucleus, specified by its neutron and proton numbers. For lighter nuclei (those with fewer nucleons, near the lower left of figure 2.13) the stable isotopes lie very close to the line representing equal numbers of protons and neutrons. But heavier nuclei deviate from this line more and more, as they require more neutrons to counter the electric repulsion of their widely separated protons. Thus the stable nuclei lie in a curved band that bends increasingly upward; we will soon see that the shape of this band explains why the waste products of nuclear reactors and weapons are so dangerous. Above atomic number 83 (bismuth, the element with 83 protons in its nucleus) there are no stable isotopes. For these large nuclei, the repulsive electric force ultimately ends. Nuclei with more than 83 protons are all unstable, and ultimately they must come apart in one way or another.

Summary

We have now met the few simple ingredients that make up our world: protons, neutrons, electrons, and the electric and nuclear forces that bind them into the nuclei, atoms, and molecules from which all else is made. We have seen how the relatively weak electric force is responsible for ordinary chemical reactions—interactions that involve only the electrons in distant orbits around their nuclei, and that leave the nuclei unchanged. These chemical reactions are responsible for the energy released in burning coal or gasoline, in exploding TNT, or in metabolizing food inside your body.

Nuclear reactions, in contrast, involve rearrangement of the protons and neutrons that make up the atomic nucleus. Because these particles are so tightly bound by the strong nuclear force, the energy involved in nuclear reactions is millions of times that of chemical reactions. That single fact constitutes the nuclear difference, which explains why a coal-burning power plant consumes trainloads of coal each week whereas a few truckloads of uranium will fuel its nuclear counterpart for a year, and why a nuclear bomb can destroy a city whereas a conventional one may destroy only a few buildings. And it is this nuclear difference that makes warfare

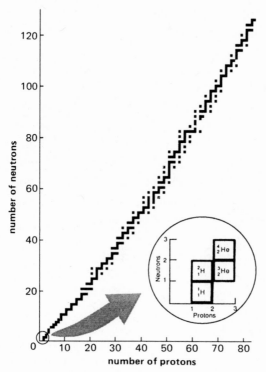

Figure 2.13 Graph of neutron number versus proton number for stable nuclei. Each stable nucleus is indicated by a small black square. The diagonal line corresponds to equal proton and neutron numbers. For small nuclei the numbers are nearly equal, but for large nuclei there is an excess of neutrons (see figure 2.12). A small section of the graph is shown magnified. Note that the squares in a vertical column have the same number of protons and therefore represent isotopes of the same element. (prepared using the MacNuclide Program, courtesy C. A. Stone, Center for Analytical Chemistry, National Institute of Standards and Technology, in collaboration with Edgardo Browne and Richard Firestone)

Nuclear News: The Largest Nuclei

The heaviest naturally occurring element is uranium (atomic number 92), most of which is the isotope U-238. Since the 1930s, still heavier nuclei have been created artificially by bombarding uranium and other elements with lighter particles, including neutrons and light nuclei. The best known of these so-called **transuranic** elements is plutonium (atomic number 94), which is produced in nuclear reactors and used in nuclear weapons. Starting in 1940, elements 93–103 were synthesized and studied at the Lawrence Berkeley Laboratory in California; these include americium (atomic number 95), berkelium (97), californium (98), and einsteinium (99). The last in the series, lawrencium (atomic number 103), was created in 1961.

After 1961, creation of heavier nuclei slowed. The isotope einsteinium-252 ($^{252}_{99}\text{Es}$) is a useful raw material for building still heavier nuclei, but Es-252 is itself an artificially produced nucleus and is in short supply. By the early 1980s, however, researchers at the Institute for Heavy Ion Research in Germany had produced elements through atomic number 109. To avoid competition among rival groups that would like to name the new elements, a universal naming scheme has been adopted that forms element names out of the digits of the atomic number. Element 109, for example, is unnilennium, and the particular isotope synthesized is unnilennium-266 ($^{109}_{266}\text{Une}$). Finding out which heavy nuclei are possible helps physicists develop theories of nuclear structure.

Although the heaviest nuclei are unstable and do not last long (Une-266 lives only about 5/1,000 second before it comes apart), nuclear physicists expect an "island" of more stable isotopes in a range about atomic number 114. Researchers eagerly await the creation of these entirely new elements.

News source: "Despite Setbacks, Scientists Seek Heavy Elements," *New York Times,* June 7, 1988.

in the nuclear age a serious threat to the continuing existence of human civilization.

Although we have not yet explored nuclear reactions in detail, we have seen how protons and neutrons join together to form nuclei. Not all combinations result in stable nuclei, which can stick together indefinitely. Here we have stressed stable nuclei, but we will see in the next chapter that unstable nuclei can also exist, although not forever. Those unstable nuclei are important factors—sometimes desirable, sometimes not—in most of today's nuclear technologies.

Notes

1. Einstein, quoted in *The New York Times Magazine,* August 2, 1964, p. 54.

2. The quarks—more fundamental particles that make up protons and neutrons—seem to have charges of one-third or two-thirds of the fundamental unit. Quarks are not essential to an understanding of nuclear physics at the level needed here, so we will not consider them further.

3. I am using the term *electric force* to include not only the forces of electrical attraction and repulsion, but also the magnetic force and the so-called weak nuclear force, all of which are now understood as aspects of the same fundamental force. That force is usually called the *electroweak force.*

Further Reading

Paul G. Hewitt, *Conceptual Physics* (Scott, Foresman, 1989). A delightfully written, nonmathematical introduction to physics. Part 7 covers atomic and nuclear physics.

Raymond Murray, *Nuclear Energy* (Pergamon, 1988). Part I affords a detailed look at the physics behind nuclear energy, at a more mathematical level than in this book.

Glossary

atom A nucleus surrounded by a number of electrons equal to the number of protons in the nucleus. The electrons are relatively far from the nucleus, making the atom more than 10,000 times the size of the nucleus. An atom is the smallest particle of a chemical element.

atomic nucleus A cluster of protons and neutrons bound together by the nuclear force. Except for the hydrogen nucleus, which consists of a single proton, all nuclei contain both protons and neutrons.

atomic number The number of protons in the nucleus of an atom. The atomic number determines the element; for example, hydrogen has atomic number 1, helium 2, oxygen 8, and uranium 92.

chemical reaction An event in which atoms are rearranged into a new molecular configuration, as in the joining of two hydrogen atoms and one oxygen atom to make a water molecule, H_2O. The nuclei of the interacting atoms are essentially unaffected in a chemical reaction, and an individual chemical reaction involves far less energy than a nuclear reaction.

electric charge A fundamental property of matter possessed by electrons and protons. Electric charge comes in two kinds, called *positive* and *negative,* and manifests itself in an attraction between opposite charges and a repulsion between like charges.

electric force A force that acts between electrically charged particles. The electric force between oppositely charged particles is attractive; between particles of like charge, it is repulsive. At close range—roughly the size of a nucleon—the electric force is much weaker than the nuclear force, but with increasing distance it falls off less rapidly than the nuclear force. The electric force holds atoms together and is responsible for joining atoms to form molecules.

electron A subatomic particle whose mass is about 1/2,000 that of a proton or a neutron. The electron carries one unit of negative electric charge, and feels only the electric force. Electrons surround nuclei to make complete atoms.

element A substance that behaves chemically in a unique and identifiable way, and whose most basic particle is a single atom. The chemical behavior of an element is determined by the number of electrons in its atoms, which in turn is determined by the number of protons in its nucleus. Therefore all atoms of a given element have the same number of protons in their nuclei, although they may differ in the number of neutrons. Hydrogen, helium, oxygen, and uranium are among the 92 naturally occurring elements.

force An interaction between particles of matter that manifests itself as an attraction or a repulsion.

gravitational force An attractive force that exists between any two particles of matter. Gravity is the weakest of the fundamental forces, and is insignificant at the atomic and nuclear scales.

isotope Isotopes of a given element are nuclei that differ in the number of neutrons they contain. Since all isotopes of a given element have the same number of protons, they must differ in mass number. Helium-3, for example, is an isotope of helium containing two protons and one neutron; helium-4 has two protons and two neutrons.

long-range force A force whose strength decreases relatively slowly with increasing distance between two particles. Both the electric force and the gravitational force are long-range forces.

mass number Total number of nucleons—protons and neutrons—in an atomic nucleus.

molecule A group of atoms joined through electrical interactions involving their outermost electrons. A molecule is the smallest particle of a chemical compound.

neutron A subatomic particle that is one constituent of the atomic nucleus. The neutron has nearly the same mass as the proton, but carries no electric charge. It feels only the nuclear force.

nuclear force A force that acts between nucleons. The nuclear force is always attractive. It is very strong at close range—roughly the diameter of a nucleon—but its strength drops rapidly with increasing distance. The nuclear force is what binds nucleons together to form atomic nuclei.

nuclear reaction An event in which nucleons are rearranged, giving rise to one or more new nuclei. The joining of two deuterium nuclei to form a helium-4 nucleus is an example; so is the splitting of a uranium nucleus into smaller nuclei. Because the nuclear force is so strong, nuclear reactions involve a great deal of energy.

nucleon A proton or a neutron, either of the constituent particles of the atomic nucleus.

nucleus See **atomic nucleus**.

proton A subatomic particle that is one constituent of the atomic nucleus. The proton has nearly the same mass as the neutron. It carries one unit of positive electric charge, and feels both the nuclear and the electric force.

short-range force A force whose strength decreases rapidly with increasing distance between two particles. The nuclear force is a short-range force

stable nucleus A nucleus that can exist indefinitely, without spontaneously coming apart.

unstable nucleus A nucleus that cannot exist indefinitely, but eventually will spontaneously come apart.

Radioactivity: When Things Come Apart

3

In the preceding chapter, we saw how atomic nuclei are made up of protons and neutrons. But not all combinations of these nucleons can stick together indefinitely. Those that do are the stable nuclei shown in figure 2.13; those that don't are unstable. Ultimately, anunstable nucleus must come apart. The coming-apart process is called **radioactive decay**, and unstable nuclei or materials containing them are said to be **radioactive**. Some elements may have both stable and unstable isotopes; the latter are called **radioisotopes**. The term *radioactivity* was coined by the physicist Marie Curie, who won Nobel Prizes in both physics and chemistry for her pioneering work on radioactive decay.

Radioactive Decay

How does a radioactive nucleus come apart? Although there are many ways, we will focus on three of the most common: alpha decay, beta decay, and gamma decay.

In **alpha decay**, a nucleus spits out two protons and two neutrons, bundled together as a helium-4 nucleus. This helium-4 nucleus is called an **alpha particle**, a name that dates to a time near the turn of the century when it was not yet known that the emitted particles were in fact helium nuclei. Alpha decay is common among the larger unstable nuclei, which need to rid themselves of excess protons (recall figure 2.12 and related discussion). The emitted alpha

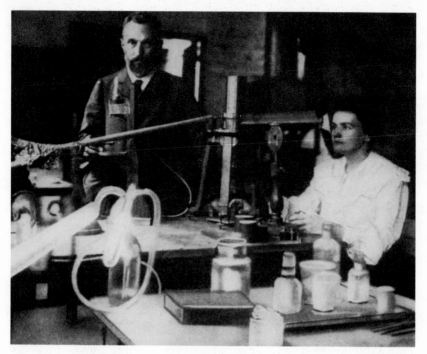

Figure 3.1 Marie and Pierre Curie in their Paris laboratory. (Niels Bohr Library, American Institute of Physics)

particle carries off two protons, dropping the atomic number of the remaining nucleus by 2. Since the alpha particle contains a total of four nucleons, the mass number drops by 4. In a typical alpha decay, uranium-238 ($^{238}_{92}U$) emits an alpha particle ($^{4}_{2}He$), leaving a nucleus of thorium-234 ($^{234}_{90}Th$). Figure 3.2 shows this decay, both pictorially and using nuclear symbols. Note how the sum of the atomic numbers on the right equals the atomic number on the left; the same is true for the mass numbers.

Unstable nuclei with too many neutrons would like either to rid themselves of neutrons or to gain protons. Remarkably, they do both at once. In the process called **beta decay**, a neutron turns itself into a proton and an electron. The electron, also called a **beta particle**, flies out of the nucleus, leaving the nucleus with one more proton and one fewer neutron than it previously had. Since there is one more proton, the atomic number increases by 1. But the total number of nucleons remains the same, so the mass number is unchanged. Figure 3.3 shows a typical beta decay, that of carbon-

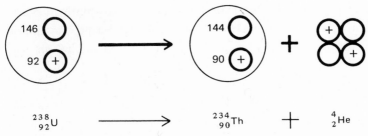

$$^{238}_{92}U \longrightarrow ^{234}_{90}Th + ^{4}_{2}He$$

Figure 3.2 Alpha decay of uranium-238. The uranium nucleus emits an alpha particle (a helium-4 nucleus), leaving a nucleus of thorium-234.

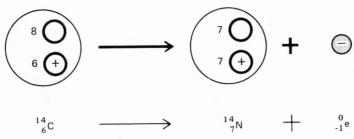

$$^{14}_{6}C \longrightarrow ^{14}_{7}N + ^{0}_{-1}e$$

Figure 3.3 Beta decay of carbon-14. A neutron in the carbon-14 nucleus turns into a proton and an electron; the electron is ejected, leaving a nucleus of nitrogen-7. Note that the mass number—6 + 8 for C-14 and 7 + 7 for N-14—remains unchanged.

14. The end product, nitrogen-7, is the most common stable nucleus of nitrogen. In writing the decay symbolically, I have indicated the electron as $^{0}_{-1}e$; since it carries one unit of *negative* charge, the electron's atomic number is −1, and its mass number is 0 since its mass is far less than that of a nucleon. Using these numbers, the sum of the atomic numbers on the right is equal to that on the left, and similarly for the mass numbers. This equality must always hold in a nuclear reaction of any kind.

You might wonder how the electron got mixed up with beta decay. Aren't nuclei made up only of protons and neutrons? Was the electron hiding in there somehow? Or is a neutron really made up of a proton and an electron? The answer to both questions is No. But a neutron can, through one manifestation of the forces we have subsumed under the term *electric*, spontaneously turn into a proton and an electron.[1] In fact, a free neutron—one that is not part of a nucleus—will do so in less than an hour if left to itself. And when a neutron finds itself inside an unstable nucleus with an excess of neutrons, it will also undergo this change into a proton and an electron. Only when they are constituents of stable nuclei can neutrons last indefinitely.

Sometimes a nucleus is struck by another particle that bounces off it or goes right through it without causing a nuclear reaction. Then the nucleus retains its identity, but it may be "shaken up" in the process. That is, it may acquire some excess energy. The nucleus is like a gong that has been struck by a hammer; originally it was quiet, but now it's vibrating. A nucleus with excess energy is said to be **excited**. Unlike the gong, which starts getting rid of its excess energy immediately in the form of sound, the nucleus may store its excess energy for while. It then emits the energy suddenly, in the form of a little energy bundle called a **gamma ray**. A gamma ray is a high-energy version of ordinary visible light, and is yet another manifestation of the electric force. Once the nucleus has shed its excess energy by this process of **gamma decay**, it returns to its original quiet state. Figure 3.4 depicts a gamma decay.

Nuclear Radiation

Each of the radioactive decays discussed above results in a modified nucleus and a much smaller entity—either an alpha particle (helium-4 nucleus), a beta particle (electron), or a gamma ray. In all cases, those smaller entities are highly energetic. The alpha particles from

Figure 3.4 Gamma decay. An excited nucleus (the arrows suggest its vibration) sheds its excess energy by emitting a gamma ray, depicted as a wavy line. Since no nucleons or electric charge leave the nucleus, it retains its identity.

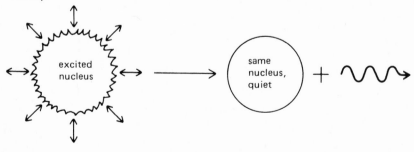

uranium-238 decay are traveling at nearly 10,000 miles per second, and the beta particles from carbon-14 decay are going almost 150,000 miles per second. Gamma rays go at the speed of light— 186,000 miles per second—and they pack about a million times the energy of visible light. Where does all this energy come from? Again, the answer lies in the nuclear difference. The forces binding the nucleus are so strong that comparatively large energies are involved any time a nucleus is disrupted.

The energetic particles emitted in radioactive decay are collectively termed **nuclear radiation**. Because of its high energy, nuclear radiation can cause serious damage to atoms and molecules in its path. That is the reason for concern about exposing humans to radiation. Other living things are also harmed, and even nonliving materials suffer radiation damage. Radiation effects will be discussed in the next chapter; here, we will explore further the physical aspects of radiation and radioactivity.

Although all nuclear radiation is highly energetic, the various forms of radiation differ in their ability to penetrate matter. Alpha particles, the slowest-moving of the three, have relatively little penetrating power; typically, alpha particles can be stopped by a sheet of paper, a layer of clothing, or an inch of air. Thus it is relatively easy to shield against alpha radiation—unless alpha-emitting material ends up on or inside the body. Radiation-induced lung cancers, for example, can result when alpha emitters lodge in the lungs.

Beta particles—electrons—are much lighter than alpha particles, and travel much faster. They can penetrate a fraction of an inch in solids and liquids (including the human body), and several feet in air.

Both alpha and beta particles are ultimately slowed because they are electrically charged particles that interact strongly with the electrons of atoms through which they pass. Gamma rays, in contrast, are electrically neutral. They do not interact as strongly as do alpha and beta particles, and therefore they are highly penetrating. The penetrating power of gamma rays depends on their exact energy; the highest-energy gamma rays encountered in nuclear technology may require several feet of material for adequate shielding (figure 3.5). Figure 3.6 summarizes the penetrating power of nuclear radiation.

Figure 3.5 An operator at the U.S. Department of Energy's nuclear weapons facility at the Hanford Reservation, in the state of Washington, controls remote arms handling highly radioactive material. Nearly 3 feet of leaded glass shield the operator from the intense radiation. (Westinghouse Hanford Company)

Measuring Radiation

A technician refueling a nuclear reactor is exposed briefly to highly radioactive spent fuel. A biologist working with radioactive material spills a carbon–14 solution in the lab. A German citizen steps outdoors as a rainstorm brings down radioactive fallout from the 1986 Chernobyl nuclear accident. An airline pilot is regularly bombarded by cosmic rays. How serious are their radiation exposures? To answer that question, we need ways to describe amounts of radiation and radioactivity.

A chunk of radioactive material contains a great many nuclei, some of which are undergoing radioactive decay. A simple way to characterize the level of radioactivity is to state how many decays occur in a given time. That number is called the **activity** of the given chunk of material. Traditionally, activity is measured in **curies**, a unit named in honor of Marie Curie. One curie is defined as

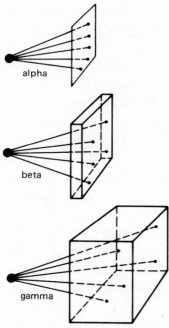

Figure 3.6 Alpha radiation can be stopped by a sheet of paper and beta radiation by a book, but gamma radiation may require a substantial thickness of concrete or even lead.

37 billion disintegrations per second, and is approximately the activity of one gram of pure radium-226, the isotope that Marie Curie discovered. One curie is a substantial level of radioactivity. Consumer products that use radioactive materials—home smoke detectors, for example—have activity levels typically measured in millionths of a curie. The Three Mile Island nuclear accident, the worst in the United States to date, released 10–25 curies of iodine-131, an isotope with significant health effects. The Chernobyl accident, in contrast, released about 7 million curies of I-131. The detonation of a medium-size nuclear weapon produces approximately 100 million curies of I-131. In modern scientific usage, the curie has been superseded by another unit: the becquerel, equal to one decay per second.

Activity, as measured in curies or becquerels, is a property of a source of radiation; it says nothing about objects that might be exposed to the radiation. The effect of radiation on an object depends on the amount of energy the radiation deposits in that object;

the **rad** is a unit describing that energy. Radiation's effects on humans depend not only on the energy absorbed, but also on the type of radiation. Another unit, the **rem**, measures radiation dose in terms of its effect on humans. The rad and the rem have been replaced recently by new units, the gray and the sievert; however, we will stick with the older units, which are more commonly found in the popular press. And, although this is not strictly correct, we will consider the rad and the rem essentially equivalent.

The next chapter will describe the effects of various radiation doses; for now, note that the average citizen receives about 0.3 rem from natural radiation sources in the course of a year. A dose of 400 rems kills about half the people receiving it. Typical low-level radiation exposures are far less than 1 rem, and are often measured in millirems (thousandths of a rem). Yearly exposure to natural background radiation, for example, is about 300 millirems. Table 3.1 summarizes the units of radioactivity and radiation.

Detecting Radiation

Radiation is invisible, odorless, tasteless, and generally quite un-detectable by the human senses. That is one reason for the wide-spread public fear of radiation. In a few unfortunate accidents, people have received fatal radiation doses without knowing that anything was happening.

Technologically, though, radiation is easily detected, and its

Table 3.1
Units of radioactivity and radiation.

Units	What they measure
curie (Ci), becquerel (Bq)	Activity; rate of radioactive decay in a given sample of radioactive material. (1 Ci = 37 billion decays per second; 1 Bq = 1 decay per second.)
rad, gray (Gy)	Energy absorbed in material exposed to radiation. (1 Gy = 100 rad.)
rem, sievert (Sv)	Radiation dose in terms of its biological effect; like the rad and the gray but adjusted for effects of different types of radiation on human tissue. (1 Sv = 100 rem.)

characteristics can be determined with great accuracy. A wide variety of radiation detectors are in routine use. They range from simple, inexpensive devices that can tell if radiation is present to sophisticated laboratory instruments that measure the precise energies of the particles making up the radiation. Here, we will examine just a few of the devices used for radiation detection.

The **geiger counter**, one of the simplest and oldest detectors, is widely used. It consists of a hollow metal tube filled with gas, with a window for radiation to enter (figure 3.7). Down the center of the tube is a wire. The wire is connected to the positive terminal of a battery, and the metal tube to the negative terminal. A loudspeaker (actually, an amplifier that drives a loudspeaker) is also connected in the circuit. Normally the gas doesn't conduct electricity, so nothing happens. But when radiation enters the tube, it tears electrons off atoms in the gas. These electrons are attracted to the positive wire, and make a burst of electric current in the circuit. Each burst causes the loudspeaker to emit a clicking sound, so the rate of clicking is a measure of the amount of radiation entering the tube. A meter is usually provided to give a visual indication of the radiation level. Figure 3.8 shows a geiger counter in use.

In many materials, the passage of radiation strips electrons from their atoms. When the electrons and atoms recombine, they

Figure 3.7 Simplified diagram of a geiger counter. Radiation entering the tube tears electrons off their atoms. The electrons are attracted to the positive wire at the center, giving rise to a burst of electric current that makes a click in the loudspeaker.

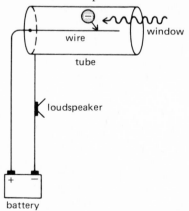

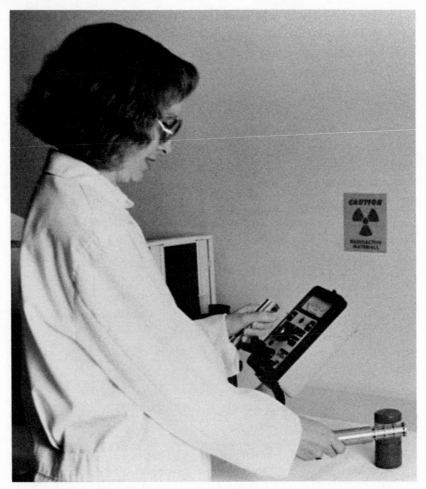

Figure 3.8 A geiger counter in use. The tube itself is on the end of a cable connecting it to a box housing a battery, a loudspeaker, and electronic circuits. (Victoreen, Inc.)

emit a flash of light. Light-sensitive devices record the flash and, by measuring its intensity, can determine the radiation energy. Detectors working on this principle are often made from crystals, and are used in laboratory instruments for radiation analysis.

Radiation leaves permanent records in some materials. These materials are used in radiation detectors that record long-term exposure; such detectors are often worn by workers subject to on-the-job radiation exposure (figure 3.9). The presence of low-level radioactive contaminants such as radon gas in the home can be detected by similar means.

Exposure and Contamination

If you stand near radioactive material, you will be exposed to radiation. The longer you stay, the greater your radiation dose. Move away and the exposure stops. If, on the other hand, radioactive material gets on you—or worse, inside you—then you are continuously exposed to its radiation; you are **contaminated** with the radioactive material. All unnecessary exposure to radiation should be avoided. But contamination is especially dangerous, since the exposure will continue until the contaminant is removed. If radioactive material lands on your skin or clothing, simple washing may be enough for decontamination. But if you eat food containing biologically active radioisotopes—such as strontium-90, which is incorporated into bone, or iodine-131, which is absorbed by the thyroid gland—decontamination may be particularly difficult.

Half-Life

Unstable nuclei must eventually undergo radioactive decay. But how soon? For a single nucleus, there is no way to tell; radioactive decay is a truly random event. But large numbers of nuclei show a simple pattern to their decays. If you start with 1,000 radioactive nuclei, you will find that after a certain time about 500 will have decayed. That time is called the **half-life**. You can't predict which 500 will decay, and the number may vary slightly from one experiment to another, but on average the nuclei decay with remarkable regularity. If you wait another half-life, half the remaining nuclei

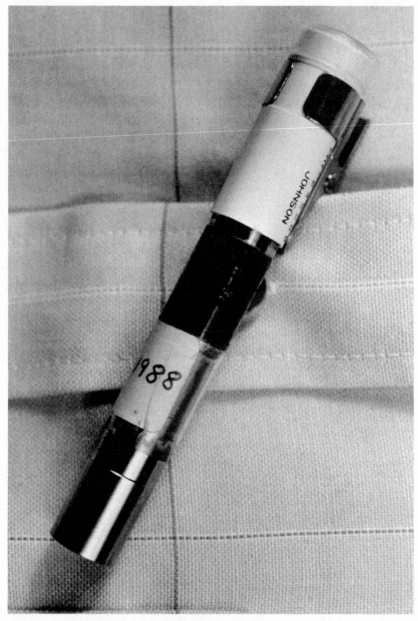

Figure 3.9 This simple radiation detector clips onto the clothing of a nuclear power plant worker. Called a thermoluminescent dosimeter (TLD), the device measures cumulative radiation exposure through radiation-induced changes in a crystal. After the exposure is read, the device can be reset for another long-term measurement. (Vermont Yankee Nuclear Power Corporation)

will decay, leaving only 250 from the original sample. In another half-life, 125 of those will decay, leaving only 125. The process continues until all nuclei have decayed.

Figure 3.10 shows that a given radioactive sample is substantially decayed after a few half-lives have passed; after seven half-lives, only 1/128—less than 1 percent—of the original sample remains. Eight half-lives, and it's down to 1/256; nine, and it's only 1/512; ten half-lives, and only 1/1,024 of the original nuclei remain undecayed. A good rule of thumb is that after ten half-lives only about 1/1000 of the original radioactive nuclei remain. What about after twenty half-lives? After ten half-lives, only 1/1,000 of the original nuclei remained; after another ten half-lives, only 1/1,000 of those 1/1,000 are still undecayed—or a millionth of the original nuclei. Wait another ten half-lives and only 1/1,000 of that millionth remains; your sample now—after thirty half-lives—has only a billionth of its original nuclei. So it doesn't take very many half-lives for essentially all of a radioactive sample to decay.

But how much *time* does it take? That depends on the duration of the half-life. Half-lives vary dramatically from one radioactive isotope to another. Excited states of some nuclei decay with half-lives around a thousandth of a trillionth of a second, whereas ura-

---

Figure 3.10 Decay of a sample of radioactive material. After one half-life, only half of the original 1,000 nuclei remain undecayed; after two half-lives, only half of that number, or 250, remain. The process continues, with half the remaining nuclei decaying in each subsequent half-life. By the time seven half-lives have passed, less than 1 percent of the original nuclei remain.

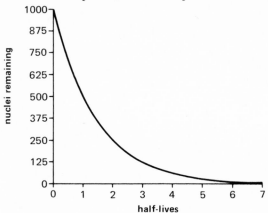

nium–238 has a half-life of 4.5 billion years. Other nuclei lie between these extremes; half-lives of minutes to years are common for the most widely studied nuclei. Table 3.2 lists the half-lives of some typical radioactive isotopes, many of which will concern us as we explore nuclear technologies and their consequences.

Looking at table 3.2, you can see that tritium decays with a half-life of about 12 years. You would have to wait 10 times this long, or about 120 years, for a given sample of tritium to decay to 1/1,000 of its original level. On the other hand, a few years is enough for a small but significant change in the amount of tritium present. Thus the tritium used to boost the explosive yield of nuclear weapons must be replenished every few years.

It would take 45 billion years for a sample of U–238 to drop to 1/1,000 of its original level. That's about three times the age of

Table 3.2
Half-lives of selected isotopes.

Isotope	Half-life	Significance
carbon–14 ($^{14}_{6}C$)	5,730 years	Radioactive isotope of carbon, formed by cosmic rays in Earth's atmosphere. Used in archaeological dating.
iodine–131 ($^{131}_{53}I$)	8.04 days	Formed during nuclear fission, and released in nuclear weapons explosions and reactor accidents. Concentrates in milk and is absorbed by the thyroid gland.
plutonium–239 ($^{239}_{94}Pu$)	24,110 years	Produced in nuclear reactors and used as a principal ingredient in nuclear weapons. Sustains a vigorous chain reaction.
radium–226 ($^{226}_{88}Ra$)	1,600 years	Highly radioactive isotope discovered by Marie and Pierre Curie. Forms in the decay chain of uranium–238.
radon–222 ($^{222}_{86}Rn$)	3.82 days	Radioactive gas formed from the decay of radium–226 in the decay chain of uranium–238. Seeps into houses, where it may give occupants substantial radiation exposure.

the universe! Plutonium-239 decays much more rapidly, with a 24,000-year half-life. But you would still have to wait 240,000 years for a plutonium sample to drop to 1/1,000 of its original level. Strontium-90, a significant component of nuclear waste, has a half-life of 28 years, so the time it takes to decay substantially is somewhat longer than a human lifespan. But oxygen-15, an isotope used in medical studies, has a 2-minute half-life. A patient may be injected with oxygen-15 and after 20 minutes only 1/1,000 of the radioactive material will remain. After an hour—thirty half-lives—only a billionth remains. For that reason, oxygen-15 and other short-lived isotopes are particularly safe for use in medical studies.

Suppose we have a chunk of uranium-238 and another chunk of strontium-90, and suppose further that the two chunks have equal numbers of nuclei. How do their activities compare? We

Isotope	Half-life	Significance
strontium-90 ($^{90}_{38}$Sr)	29 years	Fission product that mimics calcium, concentrating in bones. A particularly dangerous constituent of fallout from explosions of nuclear weapons.
tritium ($^{3}_{1}$H)	12.26 years	Used in biological studies and to enhance explosive yield of nuclear weapons. Short half-life means weapons must be refurbished every few years.
uranium-235 ($^{235}_{92}$U)	704 million years	Used as fuel in nuclear reactors and in early nuclear weapons.
uranium-238 ($^{238}_{92}$U)	4.46 billion years	Predominant uranium isotope, comprising 99.3 percent of natural uranium. Cannot sustain a chain reaction.
unnilennium-266 ($^{266}_{109}$Une)	0.005 second	Heaviest nucleus produced as of 1990; first made with particle accelerators in early 1980s.

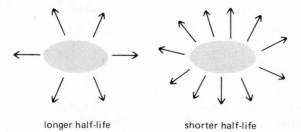

longer half-life   shorter half-life

Figure 3.11   These samples of two different radioactive materials contain the
same number of nuclei. Since it decays at a faster rate, the material with the
shorter half-life has the higher activity level.

know that in 4.5 billion years half the U-238 nuclei will have
decayed. But it will take only 28 years for the Sr-90 nuclei to decay.
So the strontium must be decaying at a much greater rate—greater
by something like the ratio of 4.5 billion years to 28 years. Given
equal quantities of different radioactive materials, those with the
shorter half-lives will therefore be more highly radioactive (figure
3.11). That is one reason why the relatively short-lived waste products from nuclear power plants are much more dangerously radioactive than the long-lived nuclear fuels that go into the plants.

Where Do Radioactive Materials Come From?

Nearly all the nuclei that make up our world were originally created
through nuclear reactions in stars that existed long before the Sun
and the Earth were formed. The violently explosive deaths of those
stars were the events we call *supernovas;* those explosions spewed
into space the materials that would later become our solar system.
The lighter elements were created while the stars shone steadily,
through a process we will look at in later chapters. But elements
heavier than iron formed only in the intensely violent supernova
explosions themselves. Among the products of these explosions
were radioactive isotopes of uranium and other elements.

The Earth and the Sun were formed about 5 billion years ago,
and the nuclei that constitute them even earlier. So radioactive nuclei
that were incorporated into our planet have had plenty of time to

Nuclear News: Fallout from Chernobyl

The 1986 accident at the Soviet Union's Chernobyl nuclear power plant spread radioactive fallout across much of Europe. One of the more dangerous isotopes released in the accident was radioactive iodine-131, whose half-life is about 8 days. I-131, like non-radioactive iodine, is absorbed by the thyroid gland; there the radioactive iodine can cause thyroid cancer. I-131 enters the human body through ingestion of milk contaminated when cows graze in pastures exposed to fallout.

Knowing the half-life of iodine-131 allowed health officials to estimate how long it would take for contamination to drop to what they considered safe levels. In Romania, for example, I-131 levels rose to the point where 2,900 decays were occurring each second in each liter of milk (since 1 becquerel is defined as 1 decay per second, the activity is 2,900 Bq/liter). This level greatly exceeded the Romanian safety guideline of 185 Bq/liter. You have to divide 2,900 in half four times—getting 1,450, then 725, then 363, then 182—before the level drops below the safety guideline. That means Romanians had to wait four half-lives, or just about a month, until their milk was considered safe to drink.

Other nations fared differently, depending on how much fallout they received and on what levels they deemed safe. Sweden also reported I-131 levels of 2,900 Bq/liter in milk. But with a safety guideline set at 2,000 Bq/liter, Swedes needed to wait less than one half-life—under a week—before their milk was considered safe. And Austria, which registered I-131 levels of 1,500 Bq/liter, had a safety guideline of 370 Bq/liter. As a result, Austrians had to wait just about two half-lives, or 16 days. Figure 3.12 summarizes the situation for citizens of Sweden and Romania faced with I-131 contamination after the Chernobyl accident.

When Chernobyl-contaminated milk became safe to drink depended only in part on physical circumstances—the half-life of iodine-131 and the amount of that isotope in milk. Equally important were the guidelines set by the European governments. With safety guidelines varying from 185 to 2,000 becquerels per liter, you can see that the question of "safe" levels of radioactivity is a controversial one.

News sources: "U.S. Gives Warning on Moscow Milk," *New York Times*, May 25, 1986; Christopher Flavin, *Reassessing Nuclear Power: The Fallout from Chernobyl*, Worldwatch Institute Paper 75, 1987.

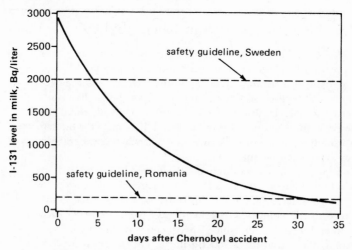

Figure 3.12 Romania and Sweden both received the same level of iodine-131 contamination in milk after the Chernobyl accident. But their different safety guidelines meant that Swedes had to wait less than 5 days before I-131 had decayed to what their government considered safe levels, whereas Romanians had to wait a month.

decay. Even uranium-238, with its 4.5-billion-year half-life, is less than half as abundant as it was when the Earth was new. For isotopes with half-lives substantially less than the age of the Earth, so many half-lives have passed that essentially all the nuclei originally present have long since decayed. For example, Earth's age is equal to nearly 200,000 half-lives of plutonium-239. Even if the Earth had been pure Pu-239 (an impossibility for many reasons), dividing in half 200,000 times would have left virtually none of the original Pu-239. (This example is only suggestive; any Pu-239 formed in ancient stars would actually have been long gone *before* the Earth formed.)

Why, then, should we ever find radioactive isotopes with half-lives much less than the age of our planet? There are several answers to that question.

First, we do expect to find long-lived isotopes, such as uranium-238. What happens when these decay? They form other nuclei, which themselves may be radioactive—and whose half-lives need not be long. Uranium-238, for example, decays by emitting an alpha particle, giving rise to thorium-234 (recall figure 3.2). It turns out that Th-234 is also radioactive. It decays by beta emission, with a half-life of about 24 days. So although its half-life is short, thorium-234 is present wherever uranium is found. Beta decay turns a neutron into a proton and an electron, raising the atomic number by 1 while leaving the mass number unchanged (see figure 3.3). Thus the thorium becomes protactinium-234 ($^{234}_{91}$Pa), which also decays by beta emission. The resulting nucleus, uranium-234, marks the start of a chain of alpha decays that produces, among other things, radium-226 and radon-222. Figure 3.13 illustrates the **decay chain** that leads to radon. The chain continues beyond radon, eventually reaching the stable isotope lead-206. A similar decay sequence starts with thorium-232, a naturally occurring isotope whose 14-billion-year half-life is even longer than that of uranium-238.

Another way nature forms radioactive materials is with **cosmic rays**. These rays are high-energy particles emitted in astrophysical processes. They travel through space, and some reach Earth. Cosmic rays are energetic enough to cause nuclear

Figure 3.13 The decay of uranium-238 results in a number of shorter-lived nuclei that are found wherever uranium is present. Here the decay chain is shown on a table of mass number versus atomic number, and is carried as far as radon-222. The sequence continues on to the stable isotope lead-206.

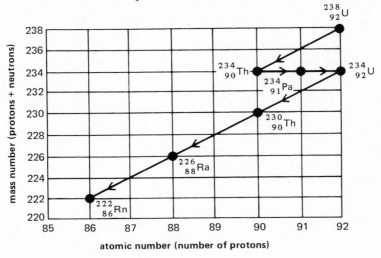

Nuclear News: Radon

Among the decay products of uranium-238 is the radioactive gas radon-222. Radon is chemically inert, meaning its atoms do not combine with others. Its gaseous nature and its chemical inactivity allow radon to move readily through soil to the atmosphere, where it is normally diluted to harmless levels. But when radon in the soil encounters the basement of a house, it can enter through cracks, drains, or openings for pipes and wiring, or by diffusing through porous foundation walls. Radon decays, with a half-life of just under 4 days, to a sequence of radioisotopes that are chemically active and are readily absorbed by the linings of human lungs.

How serious is radon contamination in our homes? It was not until the mid-1980s that scientists recognized the extent of indoor radon exposure. In fact, indoor radon is now known to be the dominant source of radiation for most Americans, greatly exceeding what we get from other natural sources, from medical procedures, or from nuclear power plants and the testing of nuclear weapons. For the average American home, where radon activity measures about 50 decays per second per cubic yard, the radon increases one's chance of fatal lung cancer by about 0.5 percent. This risk is far less than the nearly 30 percent increase in the chance of death due to cigarette smoking, and is only one-fourth as great as the one-in-fifty chance the average American has of dying in a car accident. But it is equal to the risk of dying in a fall or a fire at home, and it greatly exceeds the risks associated with other environmental pollutants, many of which are regulated to prevent cancer risks from exceeding one in a million. Put another way, the radiation dose the average American receives from indoor radon *each year* exceeds the average *lifetime* dose to Europeans resulting from the Chernobyl accident. Should we be alarmed? Should we do something? Should we stop worrying about nuclear accidents or other radiation sources that are less significant than indoor radon? These are nuclear choices, and they are not easy choices.

In some American homes—a small percentage, but still numbering perhaps 100,000—the radon level is more than 10 times the average. In these homes, the risk from indoor radon is comparable to the risks from car accidents. Factors that increase a home's radon concentration include location (since uranium content of soils varies with geological factors), type of soil (since clay soil inhibits the flow of radon whereas sand and gravel offer little resistance), the material and condition of the foundation (since cinder blocks or cracks in concrete offer easy passage to radon), and the rate at which air infiltrates the house through poorly fitting windows or other loose construction. Ironically, radon problems may be exacerbated in tight homes designed for energy conservation. Fortunately, high radon levels are relatively easy to cure. By venting the soil below the foundation, radon is readily diverted to the outside atmosphere. The cost of doing this to an existing house is typically \$1,000–\$2,000. And by installing a simple under-foundation vent pipe at the time of construction, radon contamination in new homes can be effectively prevented at a cost of only about \$200.

News sources: "Report Doubles the Estimate of U.S. Radiation Exposure," *New York Times,* November 20, 1987; "Major Radon Peril is Declared by U.S. in Call for Tests," *New York Times,* September 13, 1988.

reactions when they strike nuclei in Earth's atmosphere, and some of these reactions result in radioactive nuclei. Carbon-14, for example, is formed high in the atmosphere by the interaction of cosmic-ray neutrons with the ordinary nitrogen-14 that makes up most of our air:

$$^{14}_{7}N + ^{1}_{0}n \rightarrow ^{14}_{6}C + ^{1}_{1}H.$$

A proton ($^{1}_{1}H$) is also emitted in the process. Although carbon-14 decays with a half-life of 5,770 years, it is constantly being replenished through cosmic-ray interactions. Thus there is a nearly constant level of carbon-14 in Earth's atmosphere, and some of this radioactive carbon finds its way into living things. We will see in the next chapter how this carbon-14 can be used to determine the ages of ancient objects.

Finally, there are many radioactive substances that we humans produce with our twentieth-century nuclear technology. Most of these are by-products of the nuclear-power and nuclear-weapons industries. Of particular concern are the products of nuclear fission in power plants and weapons; many of these fission products are intensely radioactive, yet have half-lives long enough that they will be dangerous for centuries.

Background Radiation

The existence of naturally and artificially produced radioactive materials means that we are all unavoidably exposed to radiation. How does this exposure come about? How much of it is due to natural and how much to artificial sources of radiation?

For the average American, the single most important source of radiation is radon-222. Other decay products found in rocks and soils account for an additional 8 percent of Americans' radiation dose. We are exposed not only because we live on Earth's soil-covered rocky crust, but also because we derive building materials from soil and rock. Residents of brick and cinder-block houses, for example, receive more radiation than their neighbors in wooden houses. In areas where uranium is prevalent, exposure to radon and other uranium decay products may increase substantially.

Cosmic rays are a second natural source of radiation, contributing another 8 percent to the average American and much more for those who spend a lot of time in airplanes.

A third source is more surprising: our own bodies. We are all slightly radioactive. In fact, about 15,000 decays of potassium-40 occur every second in the average adult's body. Potassium-40 and other radioactive substances found in air, water, and soil are incorporated into the food we eat, then into our own tissues. The radiation we give ourselves is significant—about 11 percent of our total dose. In all, about 82 percent of our average radiation exposure comes from natural sources, for an annual natural radiation dose of about 300 millirems (0.3 rem).

Of the remaining 18 percent of our average radiation exposure, nearly all comes from medical applications of radiation. Less than 4 percent is attributable to nonmedical causes, including 3 percent from such consumer products as television sets and smoke detectors and less than 1 percent from nuclear power and weapons activities. The average exposure from all sources amounts to about 360 millirems (0.36 rem) per year.

The radiation doses suggested in figure 3.15 are, of course, averages. Actual doses vary significantly among individuals, often as a result of choices they make. If you choose to live in Denver or to work on an airplane, your higher altitude results in greater

Figure 3.14 Scot Chapin of Chapin Environmental Services checks an exhaust fan that vents radon-contaminated air from beneath a house's basement. (Chapin Environmental Services)

Nuclear News: Flying and Radiation

Cosmic radiation, originating in the Sun and other astronomical objects, provides a relatively small part of the average human being's normal radiation exposure. But airline crews and even passengers who spend a lot of time at high altitudes may experience much higher levels of cosmic radiation. In 1990, the U.S. Department of Transportation released a study showing that radiation doses to some flight crews could exceed those experienced by workers in nuclear power plants. On rare occasions, associated with bursts of radiation from solar flares, radiation levels in commercial aircraft exceed levels that would require high-radiation warnings in a nuclear power plant. The study showed that 100,000 airline workers flying for 20 years could develop 1,000 excess cancers as a result of exposure to cosmic radiation. An author of the Transportation Department's report urged that passengers in the crucial eighth to fifteenth weeks of pregnancy avoid flying over high-radiation routes.

The Federal Aviation Administration now finds itself wrestling with nuclear choices: Should airline crews be classified as radiation workers because of their exposure to cosmic radiation? Are the expected cancer deaths enough to warrant remedial action? How does this newfound risk weigh against the benefits of modern air transportation?

News source: "Radiation Exposure Is Termed a Big Risk for Airplane Crews," *New York Times,* February 14, 1990; "New Estimates Increase Radiation Risk in Flight, *New York Times,* February 19, 1990.

exposure to cosmic rays. Should you work in a nuclear power plant, you may legally receive up to 5 rems per year—over 10 times the average background dose. The city of Reading, Pennsylvania, sits on top of a geological feature rich in uranium and thorium; if you choose to live near Reading, your exposure to radon and other uranium decay products could be significantly enhanced. And you can lower your medical exposure by avoiding x-ray examinations and other procedures involving radiation.

But should you avoid medical x rays? Should you move out of the mountains to reduce your exposure to cosmic rays? Should you give up your job as a pilot, or sell your house in Reading? Wherever you live, should you install a basement ventilation system to lower the radon level in your home? And what about the smoke

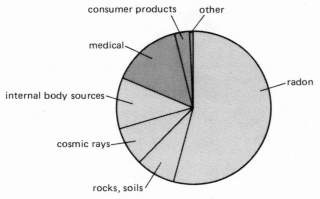

consumer products other

medical

internal body sources

cosmic rays

rocks, soils

radon

Figure 3.15 Sources of radiation exposure for the average American. Lighter areas show natural sources of radiation; darker areas represent radiation that is artificially produced. The category "other" includes the effects of nuclear power plants. These are average values; individual exposures vary considerably with geographic location, occupation, and other factors. (data source: National Council on Radiation Protection)

detector that warns you of a house fire? Chances are it contains about a microcurie of the radioactive isotope americium-241. Should you get rid of it to avoid radiation exposure? These and other questions force us to make nuclear choices. In the next chapter we will look at some effects and uses of radiation that bear on these choices.

Summary

Unstable nuclei inevitably decay, emitting alpha, beta, or gamma radiation. This nuclear radiation is invisible but highly energetic, and is readily detected by a variety of technological means. The rate of decay of a radioactive sample is its activity, measured in curies or becquerels. Each radioisotope has a characteristic half-life, the time it takes half the nuclei in a sample of that isotope to decay; half-lives range from fractions of a second to billions of years. The effect of radiation on materials or living things is measured in rads or rems. We are all exposed to background radiation from natural and artificial sources; for the average U.S. citizen, the total background exposure is about 360 millirems (0.36 rem) per year. Radiation exposure varies greatly with occupation, geographical location, house construction, medical treatments, and other factors.

Notes

1. In fact, a third particle is involved: an electrically neutral particle of zero or negligible mass, called a *neutrino*. The neutrino has essentially no interaction with ordinary matter, and we will neglect its existence.

Further Reading

Paul G. Hewitt, *Conceptual Physics* (Scott, Foresman, 1989). Part 7, which covers atomic and nuclear physics, includes a section on radioactive decay.

David W. Lillie, *Our Radiant World* (Iowa State University Press, 1986). A well-written primer on radiation from both natural and artificial sources, written at about the level of this book and covering in much more depth the material of chapters 3 and 4. Mostly objective, but the author's industrial background shows. Excellent bibliography.

Glossary

activity The rate at which a sample of radioactive material decays, measured in curies or becquerels.

alpha decay Radioactive decay by emission of a helium-4 nucleus, also called an alpha particle. The remaining nucleus has its atomic number reduced by 2 and its mass number by 4.

alpha particle A helium-4 nucleus ($^4_2$He), consisting of two protons and two neutrons, that is emitted in radioactive decay.

background radiation Radiation from natural or artificial sources in the everyday environment.

beta decay A radioactive decay process in which a neutron turns into a proton and an electron. The electron is ejected from the nucleus, resulting in a new nucleus with atomic number increased by 1 and mass number unchanged.

contamination Radioactive material in an undesired location.

cosmic rays High-energy particles from space that provide an important component of background radiation.

curie A unit of radioactivity, equal to 37 billion decays per second.

decay chain A series of isotopes formed as a result of successive radioactive decays.

excited nucleus A nucleus containing excess energy. Often such a nucleus gives up the excess energy by emitting a gamma ray.

gamma ray A bundle of energy emitted by an excited nucleus.

geiger counter A radiation detector in which radiation strips electrons from atoms in a gas-filled tube, resulting in a burst of electric current.

half-life The time it takes for half the nuclei in a given radioactive material to decay.

nuclear radiation High-energy particles—alpha, beta, or gamma—emitted by radioactive nuclei.

rad A unit that measures the energy an object absorbs when exposed to radiation.

radiation See **nuclear radiation**.

radioactive decay The process in which an unstable nucleus comes apart, usually by emitting a particle.

radon-222 A radioactive gas formed in the decay sequence of uranium-238 and responsible for over half the average radiation exposure of citizens of the United States.

rem A unit of radiation dose that describes the effect of radiation on the human body. In this book the rem and the rad are treated interchangeably, although this is not strictly accurate.

Effects and Uses of Radiation

4

Why are people afraid of radiation? For good reasons. First, radiation can do grievous harm to biological systems, including human beings. Second, radiation is invisible and undetectable without special equipment. Third, nuclear technology has created new sources of potentially hazardous radiation.

But the dangers of radiation do not tell the whole story. One's risk from radiation exposure depends on the dose of radiation one receives. "How much?" is an important question to ask when interpreting news reports about radiation exposure and nuclear accidents. Furthermore, some uses of radiation are distinctly beneficial. There are many people alive today who would not be had radiation not been used in the diagnosis or treatment of otherwise fatal diseases. Many others are helped by radiation in less obvious ways—for example, through advances in medical, environmental, or nutritional science. Airline passengers fly more securely, their luggage probed with radiation to search for terrorist bombs. And we are all enriched by the knowledge of our own past that archaeologists obtain through radiation dating. This chapter explores the harmful effects of radiation and also samples some of its beneficial uses.

Biological Effects of Radiation

Radiation consists of high-energy particles, mostly alpha, beta, and gamma rays. With their high energy, these particles can knock electrons out of atoms. This electron removal, called **ionization**, is fundamentally the way radiation works its biological harm. (Radiation capable of ionizing atoms is called **ionizing radiation**, and includes not only nuclear alpha, beta, and gamma radiation but also x rays.) Molecules containing ionized atoms are chemically very active. When those molecules are part of living tissue, they may undergo chemical reactions whose products are detrimental to life. Water, for example, is the major constituent of living things. Ionization in the water molecule leads to formation of substances that act as cell poisons. With high enough radiation doses, cell function may be disrupted by such radiation-induced poisons. Radiation striking more complex biological molecules, such as proteins or nucleic acids, may break the molecules and thereby prevent their proper functioning. Loss of cell vitality, decreased enzyme activity, initiation of cancer, and genetic changes are among the possible outcomes.

Genetic Effects

If radiation strikes the germ cells in the reproductive organs of a living organism, it may damage the genes and chromosomes that determine hereditary characteristics of the organism's offspring. DNA—the double-helix molecule that carries the genetic message—is especially sensitive to radiation. When DNA in reproductive cells is damaged, offspring inherit changed genetic characteristics called **mutations**. A few mutations are obvious and grotesque, but most are more subtle. Yet over 99 percent of mutations are in some way detrimental to the organism. In fact, the mutations we see in living organisms represent only a small fraction of the mutations that have actually occurred. Most mutations result in death before a developing organism has sprouted, hatched, or been born.

How much radiation is required to cause mutations? Since the mutagenic (mutation-causing) properties of radiation were discovered, in the 1920s, generations of laboratory animals have been bred under controlled conditions where genetic effects of radiation can be studied. For animals exposed to relatively high but nonfatal doses of radiation, we know that the rate of mutation increases in proportion to the dose. But even with no radiation exposure, some

Figure 4.1 A radiation-induced mutation. The mouse at left has deformed feet, resulting from its mother's exposure to gamma radiation. (Oak Ridge National Laboratory)

mutations still occur. These spontaneous mutations arise naturally, although some may be induced by chemical agents or by natural background radiation. In humans, the spontaneous mutation rate is about twelve mutations per 1,000 live births. The important question to ask about radiation-induced mutations is how their rate compares with the spontaneous mutation rate. An answer to that question can be given for mice. "Megamouse" experiments—involving literally a million mice—reveal that the dose needed to double the spontaneous mutation rate in mice is in the range of 10–200 rads.

But what about human beings? The largest sample population we have for study are the roughly 100,000 survivors of the nuclear bombings of Hiroshima and Nagasaki. Even this large number is too small for statistically reliable studies, and furthermore we have no accurate measure of the Japanese survivors' radiation exposure. Analysis of the Japanese survivors does show slight increases in genetic mutations over what would be expected for a normal population, but the increases are so small that they could be accounted for by chance.[1] The data suggest that the radiation dose needed to double the mutation rate is at least 100 rems, which is consistent with the results of animal experiments.

Now, 100 rems is a large dose of radiation—about 300 times

the yearly natural background dose. What about the genetic effects of very low levels of radiation, such as the 1 millirem received by residents within 50 miles of the Three Mile Island accident? We might try to estimate those effects by extrapolating downward. A dose of 100 rems supposedly doubles the mutation rate, giving twelve radiation-induced mutations per 1,000 births in addition to the twelve spontaneous mutations. Then 50 rems should cause half this number, or six mutations; 1 rem should cause 1/100 of 12, or 0.12, mutations, and 1 millirem should cause only 0.00012 mutations per 1,000 live births. With about 30,000 live births per year in the population living within 50 miles of TMI, we should expect only 0.0036 radiation-induced mutations (0.00012 mutations/1,000 births times 30) each year as a result of the TMI accident. That this number is far below 1 indicates the improbability of *any* radiation-induced genetic effects, even in the 50 years following the accident.

Actually, our example may overestimate the mutation rate from Three Mile Island. (The Presidential Commission investigating TMI estimated between 0.0001 and 0.002 cases of radiation-induced genetic damage per year.) The reason for the overestimate is that many living cells contain **repair enzymes**, which repair damage to genetic material and thereby prevent mutations. At low radiation doses, the repair enzymes probably reduce the mutation rate below what would be expected from high-dose studies. At high doses, the repair enzymes themselves may be destroyed, leading to a mutation rate that does depend directly on dose.

Most scientists believe that genetic effects of public exposure to radiation from nuclear technologies are insignificant. Some, however, argue that long-term effects of even minuscule exposure may propagate far into future generations, producing a genetically weakened human race. Do we accept the benefits of nuclear technology today in return for such a remote but potentially devastating possibility?

Somatic Effects

Somatic cells are the ordinary body cells that are not involved in reproduction. Somatic effects of radiation are therefore effects on an individual, effects that are not passed on to future generations.

Although very high radiation exposures in humans are rare, there have been enough cases—associated with the bombings of Japan, with fallout from nuclear tests, and with accidents at nuclear facilities—that the effects are well known. Doses of many hundreds of rems destroy cells that normally divide rapidly, diminishing the body's ability to replenish its red and white blood cells and the cells that line the intestinal tract. Nausea and vomiting are the first symptoms of acute **radiation sickness**, and they typically appear a few hours after exposure. There follows a lull of several days to a week or more during which the victim may feel fine. But then, as red blood cells die without being replaced, anemia sets in. The anemia is compounded by intestinal bleeding, complicated further by the loss of blood-clotting factors. As white blood cells go unreplaced, the body's immunity to disease declines. Death may follow in a matter of weeks or months. At a radiation exposure of 400 rems, about half the victims die of radiation sickness. Survival rates may be enhanced by blood transfusions and by bone-marrow transplants, which allow the body to make new red blood cells.

People receiving somewhat lower doses than 400 rems exhibit similar symptoms but generally recover in several months. The same is true for those lucky enough to survive higher doses. These victims seem to recover completely and, except for an increased cancer risk, go on to lead normal lives.

Whereas doses above 100 rems cause radiation sickness, exposures to tens of rems may produce no obvious effects. How are we to determine whether such radiation exposures are harmful? And how can we possibly establish harmful effects at the much lower doses received by the general public in typical nuclear incidents or through medical procedures? The answers lie in statistical analysis of large populations exposed to low-level radiation.

The most significant effect of low-level radiation is an increased incidence of cancer. Numerous studies have confirmed this effect. For example, in the 1950s x-ray technicians developed leukemia at many times the rate of the general public. In the 1920s, watch hands were painted with a radium-containing paint to make them visible in the dark. Factory workers painting watch hands often licked their brushes, ingesting radium in the process. Nearly all of them eventually died of bone cancer or radiation-induced anemia. Uranium miners, exposed to radioactive radon gas trapped in the mines, have developed lung cancer at a rate much higher than the average. Many early nuclear scientists—including Marie Curie and her daughter Irene—died from leukemia that was undoubtedly caused by radiation exposure. (Pierre Curie was spared this fate; he was killed in the street when a horse bolted from its carriage.)

Today, we know enough to avoid painting our watches with radium. Strict controls have been imposed on uranium mines, dropping radon exposure to the point where perhaps a million American homes have higher radon levels than uranium mines. Scientists and medical personnel are now conscious of radiation's dangers, and take precautions to minimize exposure. Are we then safe from radiation? Or do much lower levels still pose a risk? Unfortunately it is nearly impossible to carry out a truly controlled study on a large human population, so controversy about the effects of low-level radiation continues.

A look at survivors of Hiroshima and Nagasaki shows why the issue of low-level radiation is still clouded. The radiation dose can be estimated for those individuals whose locations at the times of the bombings are known; in general, the farther from the explosion, the lower the dose. Figure 4.2 shows the number of leukemia

Figure 4.2 Incidence of leukemia in Hiroshima survivors. Shown are both the actual number of cases and the number expected in an unradiated population. At high doses there are clearly excess leukemia cases attributable to radiation, but for doses less than 50 rads the Hiroshima survivors actually show less leukemia than expected. (data source: David Lillie, *Our Radiant World* [Iowa State University Press, 1986], p. 100)

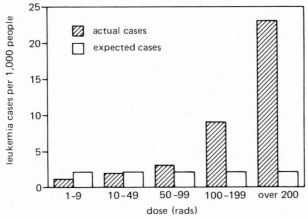

cases in Hiroshima residents exposed to various radiation doses from the nuclear bomb. It also shows the number *expected* in a population of the same size. Where the actual number of cases greatly exceeds the expected number, we can assume that the additional cases are due to radiation exposure. And sure enough, large doses of radiation result in greatly increased leukemia incidence—over 10 times the expected rate for exposures over 200 rads.[2] But look again at the figure. From 10 to 49 rads the expected and actual rates are nearly the same, and from 1 to 9 rads the Hiroshima survivors actually show lower rates of leukemia than an unexposed population. What is going on here? Does low-level radiation actually prevent cancer? That is hardly likely, although a handful of scientists affiliated with the nuclear industry have tried to argue as much. More likely, what the data show is that, even at relatively high exposures (1–50 rads), effects other than radiation become important in determining the incidence of cancer. On the other hand, anti-nuclear scientists would argue that Hiroshima survivors should hardly be compared with the general population; after all, they survived not only the nuclear blast but also the disease and social disruption that followed. Perhaps they survived because they were generally heartier individuals, less prone to leukemia and a host of other ills. Perhaps their leukemia rates, even at the lower doses, are still above normal for such a robust group.

A simple way to summarize radiation's carcinogenic effects is through the **cancer dose**, defined as the total radiation dose that, if spread throughout a population, would cause one additional cancer death. Until 1990, most agencies dealing with radiation protection assumed cancer doses between 5,000 and 10,000 rems. In 1990, as a result of continuing analysis of the Japanese survivors and refinement of mathematical models, the estimated cancer dose was lowered to about 2,000 rems.[3] This means that if a population of, say, 1,000 people were exposed to a total dose of 2,000 rems (2 rems per person), then one of those people could be expected to die of radiation-induced cancer. If, on the other hand, a million people were exposed to 1 millirem each (a typical exposure in the Three Mile Island accident), then the total dose to the population would be only 1,000 rems, and even one cancer death would be unlikely. Or, if a single individual were exposed to 1 rem, then that individual would have one chance in 2,000 of developing cancer as a result of the radiation exposure. However, not all scientists agree with the "official" studies. John Gofman, a former associate director of the Lawrence Livermore Laboratory (a nuclear weapons lab) and a leading critic of the U.S. government's radiation safety standards, suggests that the cancer dose may be as low as 150 rems. In a particularly controversial study, Dr. Thomas Mancuso and colleagues studied health records of workers at the Hanford Reservation (a military nuclear facility in Washington state) and suggested a cancer dose as low as 120 rems. Pro-nuclear scientists discredit Mancuso's methodology and claim that the work is inconsistent with studies of larger populations. Anti-nuclear scientists point to the sudden removal of Mancuso's government funding just a year before the results were published. Was the government trying to squelch results unflattering to its nuclear programs? Or was the work truly defective? Who is to decide, and how is the public to know? (The government's 1989 decision to release health records of some 600,000 people who have been involved in the nuclear weapons program may help lay these controversies to rest.)

Despite the ongoing controversy, it appears that low-level radiation exposure is not among the most serious threats to our lives. Table 4.1 shows that many dangers we willingly accept in everyday life carry far greater risks than low-level radiation from, say, nuclear power plants. The table gives the effects of each risk as an average decrease in lifespan associated with that risk. These figures are averages. Some people get cancer from radiation exposure, and some don't. Those who do lose a lot of life, the others none. The average figures come from weighing the fraction of people who do get cancer and the amount of life they lose. Similarly, not everyone gets killed in car accidents, but enough do that the risk of driving for one year averages to a lifespan reduction of 2.8 days. You would be horrified at the thought of being among the people most seriously irradiated at Chernobyl, yet, as table 4.1 shows, you complacently face a greater risk from drunk drivers. You might consider moving if a nuclear power plant were built in your neighborhood. On the other hand, suppose your new location requires that you drive 10 miles further in your round-trip commute to work each day. In 10 years of 250 working days each, the risk associated with this extra driving will give an average lifespan reduction of nearly

Table 4.1
Average decreases in lifespan associated with various risks and activities, for residents of the United States.

Risk or activity	Average decrease in lifespan
Cross a street	24 seconds
Drive 10 miles	4 minutes
Receive 1 mrem of radiation (average exposure to residents within 50 miles of Three Mile Island)	6 minutes ("official" estimate*); 72 minutes ("anti-nuclear" estimate*)
Smoke 1 cigarette	10 minutes
Fly coast to coast	1.7 hours
Earthquake (lifetime risk)	2.4 hours
Average public exposure to radiation from nuclear power plants (lifetime risk)†	23 minutes ("official" estimate); 4.6 hours ("anti-nuclear" estimate)
Tornado (lifetime risk)	12 hours
Drive coast to coast	20 hours
Spend your life living near nuclear power plant, receiving 5 mrem per year.	2 days ("official" estimate); 24 days ("anti-nuclear" estimate)
One year's normal driving (10,000 miles)	2.8 days
Average coffee drinker	6 days
Fire	27 days
Exposure to prompt radiation from bomb blast, 1 mile from ground zero at Hiroshima	39 days
Drown	41 days
Be among 24,000 people receiving most radiation from Chernobyl accident	43 days
Never use seat belts	50 days
Get murdered	90 days
Get killed by drunk driver	105 days
Die of pneumonia or influenza	141 days
Job involving radiation exposure	160 days
Be a coal miner	3 years
Be 30% overweight	3.6 years
Average cigarette smoker	4.2 years

data source: B. L. Cohen and I. Lee, "A Catalog of Risks," *Health Physics* 36 (1979), pp. 707–722.
* "Official" estimates are based on 1990 cancer dose of 2,000 rem per fatal cancer; "anti-nuclear" ones are based on 167 rads, given by J. Gofman on p. 294 of *Radiation and Human Health* (Sierra Club Books, 1981).
† Assumes 20% of electricity is generated by nuclear power.

7 days—more than triple the risk of staying next to the nuclear plant.

Although the risk most of us face from radiation may be small, the number of individuals affected need not be. Even the very low chance of your personally contracting fatal cancer from nuclear power operations implies that between 25 and 600 people (depending on whether you use pro- or anti-nuclear estimates) in the United States will contract fatal cancer each year because of nuclear power.[4] Is that acceptable? Or do we demand that nuclear technology produce no excess cancers? And if we demand that of nuclear technology, why not of coal-burning technology, or of pesticides, or of artificial sweeteners? Nuclear choices, like those involving low-level radiation, cannot be divorced from a broader technological context.

Although the average individual may have little chance of contracting cancer from low-level radiation, not everyone is the average individual. In particular, the young and the unborn are at much higher risk of cancer and other radiation effects. Because cancer often takes decades to develop, those with more life ahead of them are intrinsically at greater risk. And because radiation's effects are greatest on cells that are actively dividing, young, growing individuals with many dividing cells are more likely to be harmed. John Gofman suggests that the cancer risk increases enormously with decreasing age (table 4.2). For people beyond middle age, Gofman's cancer doses are far greater than our "official" estimate of 2,000 rems per fatal cancer, indicating that older people are not very susceptible to radiation-induced cancer. But at 64 rems per

Table 4.2
Cancer dose by age.

Age when irradiated	Cancer dose (rems)
0	64
10	96
20	224
30	259
40	587
50	14,025

Data source: John Gofman, *Radiation and Human Health* (Sierra Club Books, 1981), pp. 285–288.

fatal cancer, Gofman suggests that newborns are 30 times more susceptible than the "official" figure would suggest. In fact, Gofman believes, half the cancer deaths in a population exposed to radiation will occur in those who were under 10 years old at the time of exposure.

Even more susceptible than newborns are human embryos in the first few months of gestation. In addition to the long-term risk of cancer, embryos often face radiation damage to the nervous system. When pregnant women are exposed to enough radiation to cause even mild radiation sickness, studies suggest that their babies are more likely than not to be aborted spontaneously, still-born, or afflicted with mental retardation and other major defects. These studies involve doses of 100 rads and more, but the extreme sensitivity of the fetus makes it prudent for pregnant women to avoid radiation exposure as much as possible. And the link between cancer and parental radiation exposure may begin even earlier; a 1990 British study suggests that children may develop leukemia if their fathers are exposed to radiation *before* conception.[5]

Medical Uses of Radiation

As we have just seen, radiation can cause a host of health problems, from radiation sickness and cancer to genetic abnormalities. Yet, surprisingly, radiation is one of modern medicine's most valuable tools for diagnosing and treating diseases, especially cancer. As figure 3.15 showed, medical procedures expose the average American to far more radiation than other artificial radiation sources, such as nuclear power or the testing of nuclear weapons. Whether to use radiation for medical purposes is yet another nuclear choice we face. In most cases, the health benefits of medical radiation exposure outweigh the possible harmful effects.

Cancer Treatment

The same properties that make radiation harmful to living cells make it a valuable tool for fighting cancer. The cells most suscep-tible to radiation damage are those that divide rapidly; that is why, for example, the bone-marrow cells that produce red blood cells

Nuclear News: The Aftermath of Chernobyl

The 1986 accident at Chernobyl, in the Soviet Ukraine, ranks as the world's worst disaster involving a nuclear power plant. Chernobyl released about a million times the radiation emitted in the 1979 nuclear accident at Three Mile Island, in Pennsylvania. What are the health implications of Chernobyl?

Radiation sickness took its toll among the operators of the plant, the fire crews, the relief workers, and a few local residents. There were more than 200 cases of radiation sickness, and at least 29 people died. Many radiation victims were spared death through the efforts of an American physician, Robert Gale, who traveled to Chernobyl to perform bone marrow transplants that allowed some of the victims' bodies to replenish their red blood cells.

About 24,000 people received high enough radiation doses (average: 43 rem) that they can expect more than double the usual number of leukemia cases in the decade following the accident. Still, that's only about 26 additional cases over a decade. In the rest of the world's population, the total number of excess cancer deaths is estimated to be about 17,000 for the 50 years following the accident. This number is so small relative to the 500 *million* cancer deaths normally expected over the same period that there is almost no chance of detecting Chernobyl-induced cancers by statistical methods.

Was the Chernobyl accident serious? In the United States alone, drunk drivers will kill 250,000 people in the next decade. Coal-burning power plants will kill another 100,000. Chernobyl's worldwide toll of 17,000 cancer deaths pales in that kind of comparison. But in absolute terms, 17,000 fatal cancers is a big number. In fact, it is very nearly the number of days in 50 years. So you could say that accepting 17,000 deaths from Chernobyl amounts to permitting the killing one person at random each day. We are back to fundamental questions of the nuclear age: How safe do we require nuclear technology to be? Will we accept *any* deaths attributable to nuclear technology?

News source: "No Detectable Health Risk Is Found Outside Chernobyl Vicinity," *New York Times*, December 27, 1988; "4 Years Later, Chernobyl Still Claims Victims," *New York Times*, April 26, 1990.

and the cells that replenish the intestinal lining are preferentially destroyed by high radiation doses. Since they divide rapidly, cancer cells are more easily damaged by radiation than most normal body cells. Directing a beam of radiation at a tumor can destroy enough cancer cells to shrink or eliminate the tumor, while doing less drastic damage to surrounding tissue.

A variety of techniques are used in radiation treatment of cancer. X-ray beams were turned to cancer therapy only a few years after Wilhelm Roentgen's 1895 discovery of the rays. In modern x-ray treatment, high-energy accelerators hurl electrons at metal targets, producing penetrating beams of x rays. Gamma rays, more energetic than x rays, were widely used in early cancer treatment. A typical gamma-ray unit consists of a chunk of radioactive cobalt-60 surrounded by heavy lead shielding; a channel in the shielding allows a well-defined beam of gamma rays to emerge. Cobalt-60 treatment units are still in use in some developing countries, but because of the bulky lead shielding and the lack of control over gamma-ray energy they have fallen out of favor in technologically advanced areas. Increasingly, x-ray and particle beams from high-energy accelerators have replaced gamma rays from nuclear materials. Accelerators permit the beam's energy to be controlled precisely, for maximum effectiveness against a given patient's tumor. Finally, some tumors call for implantation of radioactive material at the tumor site; radiation exposure is then limited to a small region surrounding the tumor. Gold-198, which undergoes beta decay with a 65-hour half-life, is often used for this purpose. For blood cancers, such as leukemia, injection of phosphorus-32 into the bloodstream brings cancer-fighting radiation right to the diseased blood cells.

Radiation therapy is most effective against individual tumors, although it is also used to prolong life and relieve pain in cases of more widespread cancer. Radiation is often combined with surgery and chemotherapy. Approximately half of all cancer patients receive radiation treatment at some point during their therapy.

Very high doses of radiation are used in cancer therapy. In some instances, localized areas may receive cumulative doses measured in thousands of rads. Average doses to the whole body are

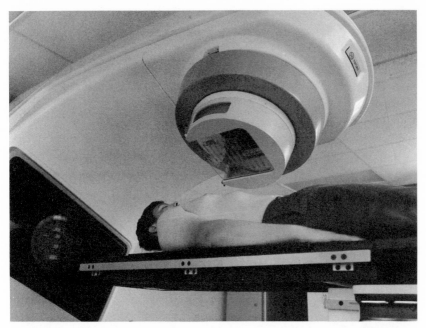

Figure 4.3 This radiation therapy machine produces high-energy beams of electrons or x rays for treatment of cancer. (Varian Medical Group)

lower, yet still in the range where harmful effects can be expected. But radiation therapy usually prevents or delays death, which would otherwise follow from uncontrolled cancerous growth.

Medical Diagnosis Using Radiation

Nearly all of us have had diagnostic x-ray photographs taken of our teeth or bones. X rays are like nuclear gamma rays, but with lower energy. They are produced by slamming high-energy electrons into metallic targets. With appropriate energy, x rays penetrate soft tissues but not hard bones and teeth, so the hard tissues stand out in x-ray photographs. Applying x-ray-opaque materials to soft tissues allows them to be imaged as well. X-ray photos of the stomach, for example, are made after the patient swallows a solution of x-ray-opaque barium. The barium coats the walls of the esophagus, the stomach, and the intestine, so these organs show up in the x-ray photograph (figure 4.4). X-ray imaging combined with

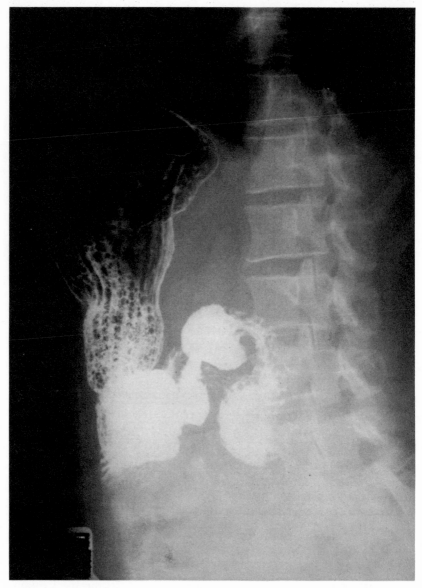

Figure 4.4 An x-ray photograph of a human stomach, made possible by having the patient swallow a solution of the x-ray-absorbing element barium. (Holm Radiology)

sophisticated computer analysis is called *computerized axial tomography,* or CAT. CAT scans give vividly detailed images through selected sections of the body .

The techniques of **nuclear medicine** make use of radiation from radioactive isotopes. Administering radioactive versions of biologically important chemicals allows doctors to follow these chemicals through the body and into particular organs. Iodine, for example, is necessary for the functioning of the thyroid gland. If a patient drinks a solution containing a radioactive isotope of iodine, subsequent measurement of radiation from the thyroid gives an indication of whether or not the gland is functioning normally. In general, a radioactive substance is called a **radioactive tracer** when it is used to trace the flow of material in some system (in this case, the human body).

Many radioisotopes are in wide use as tracers. Radioactive substances can be chosen that are readily absorbed by particular organs, or by tumors. The radiation emitted then gives an image of the region under study. This technique can locate tumors deep within the brain (figure 4.5). Similar scanning techniques are used to search for cancers of the liver, the bones, and other organs. In some cases, isotopes with very short half-lives are used in medical diagnosis. Oxygen-15, for example, has a 2-minute half-life and is used in studies of blood flow and lung ventilation. The very short half-life makes this isotope especially safe to handle and to administer, since its radioactivity decays in such a short time. But a hospital using O-15 must produce it "in house," since there would be no time to transport it from elsewhere. Some major hospitals have cyclotrons or other particle accelerators to produce short-lived radioisotopes for medical diagnostics.

Nonmedical Uses of Radiation

Radiation serves a host of applications, including industrial uses, scientific research, archaeology, insect control, food preservation, and security. Here we will sample just a few of radiation's myriad uses.

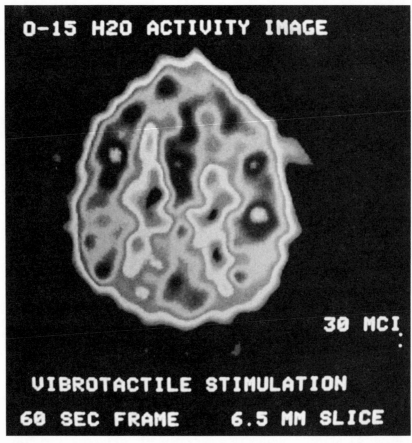

Figure 4.5 This image of a human brain was taken with radiation produced after the patient was administered the short-lived radioisotope oxygen-15. (GE Medical Systems/Scanditronix)

Radioactive Tracers

We have just seen how radioisotopes can be used as tracers in medical diagnosis. The same idea works in other applications. Suppose, for example, that a botanist wants to know the rate at which a given plant absorbs carbon dioxide from the air. (In this age of concern about global warming due to a buildup of carbon dioxide in the atmosphere, that is a very relevant question.) He puts the plant in a closed environment whose atmosphere contains some carbon dioxide made with radioactive carbon-14. The plant absorbs the radioactive carbon dioxide. Using radiation detectors to measure the plant's radioactivity, the botanist can then infer the rate at which the plant absorbs carbon dioxide. Or suppose an engineer develops a new material for use in automotive engine bearings. To study how well the material wears, she can introduce radioisotopes into a sample bearing. As the bearing wears, radioactive material accumulates in the engine oil. Measuring the radioactivity of the oil allows the engineer to infer the rate of wear.

Insect Control

Since radiation preferentially damages rapidly dividing cells, large radiation doses to the reproductive organs may result in sterility. This effect helps eradicate insect pests. A large quantity of insects are irradiated with massive doses. Released in the wild, these sterile insects mate with normal insects. No offspring result from the matings, and the population of offending insects declines. The screw-worm fly, a livestock pest prevalent in the southeastern United States, was eradicated in the 1950s in the first large-scale application of insect control through radiation sterilization. More recently the medfly, a serious pest of citrus crops, has been partially controlled in the same way.

Food Preservation

High doses of radiation preserve food by destroying bacteria and enzymes that cause spoilage. At very high doses—millions of rads—bacteria are completely destroyed, and sealed food then keeps for years, just as canned food does. At lower doses, bacterial populations fall to the point where the shelf life of food extends for many months. Finally, irradiation of fruits and vegetables inhibits sprouting and overripening, again greatly extending useful shelf life (figure 4.6).

Is irradiated food safe? Proponents of irradiation point to a method that is less destructive of food values than the heat treatment used in traditional canning, and which may reduce the incidence of salmonella and other diseases. They are also quick to point out that irradiation does not make food radioactive, since the gamma radiation does not induce nuclear reactions. Opponents agree that there is no radiation hazard to the consumer, but they note that ionizing radiation causes chemical changes that may result in the formation of toxic or carcinogenic substances. And some opponents of food irradiation see an insidious connection between food and nuclear weapons. They argue that the quantities of radioisotopes needed to support large-scale food irradiation must come from by-products of the production of nuclear weapons, and that a nation hooked on irradiated food will therefore be more inclined to continue building nuclear weapons.

Radioisotope Dating

As they decay, radioactive isotopes act as "clocks," enabling archaeologists, geologists, and art historians to determine the ages of ancient objects. For things that were once alive—bones, charcoal from ancient campfires, vegetable matter, textiles, and the like—the isotope carbon-14 is used. Carbon-14 forms continuously in the atmosphere as cosmic rays interact with nitrogen. Chemically identical to ordinary carbon, this carbon-14 is incorporated into plants and then into plant-eating animals. You are mildly radioactive, in part because of the carbon-14 in your body. Carbon-14 decays with a half-life of 5,730 years, but the level of carbon-14 in a living organism remains fairly constant, as the intake of newly formed carbon-14 balances the radioactive decay. When the organism dies, it stops taking in C-14. But radioactive decay continues, so the level of C-14 drops. By measuring the ratio of C-14 to stable isotopes of carbon in an ancient sample, scientists can determine how much time has passed since the sample was alive. Suppose, for example, that archaeologists excavate an ancient village and find that the skeletons they unearth have only 1/8 the carbon-14 fraction

Figure 4.6 Food preservation by radiation. The potato at top is 8 months old, and was untreated. The potato at bottom is the same age, but was irradiated with 20,000 rads of gamma radiation. (Brookhaven National Laboratory)

Nuclear News: Food Irradiation

Food irradiation is not a new idea. Since the 1950s, America's astronauts have taken irradiated food into space. As early as 1964, the Food and Drug Administration approved irradiation to preserve potatoes and wheat, although irradiation did not prove economical for those products. In 1986, the FDA approved irradiation of fruits, vegetables, and pork; poultry joined the list in 1990. In approving irradiation, the FDA cited tests in which laboratory animals fed large amounts of highly irradiated foods showed no ill effects. But environmental and anti-nuclear groups argued that laboratory animals' diets hardly duplicated human patterns of eating and food preparation, and that dangerous by-products of the irradiation process might still be expected in human food. In response to these concerns, several states—including New York, New Jersey, and Maine—have passed laws banning the sale of irradiated food. New Jersey's 1989 act was particularly significant, since the state hosts seven irradiation plants. Built primarily for radiation sterilization of medical supplies, some of these plants were eager to move into food irradiation. The nuclear choice made by New Jersey legislators may be a significant setback for food irradiation.

News source: "Jersey Passes Bill Banning Sale of Irradiated Foods," *New York Times,* January 8, 1988; "F.D.A. Approves the Irradiation of Poultry," *New York Times,* May 2, 1990.

of present-day bones. To get to 1/8 of its original level, the C-14 must have diminished by one-half, then done so again, then done so once more, for a total of three half-lives. Since carbon-14's half-life is 5,730 years, the charcoal sample must be just over 17,000 years old. This method, called **radiocarbon dating**, provides quite accurate results back to about 20,000 years, and is used as far back as 50,000 years. Figure 4.7 depicts the principle of radiocarbon dating.

For longer time spans, up to the several-billion-years ages of rocks, the ratios of different isotopes and their decay products are compared. For example, lead-206 is the stable end product of the uranium-238 decay chain considered in the preceding chapter. Comparing the amounts of lead-206 and uranium-238 tells how long ago the nuclei in a material formed. That is how we know that the Earth and the Moon are made from stuff of essentially the same

Figure 4.7 The principle of radiocarbon dating. (a) Carbon-14, formed in the atmosphere by the interaction of cosmic rays, is incorporated into living things through the food chain. While they live, organisms maintain a mild but steady level of C-14 radioactivity. (b) At death, C-14 uptake ceases. (c) Much later, C-14 activity has decayed considerably. (d) Archaeologists excavate the long-dead remains. By measuring C-14 radioactivity, they infer the time since death. Note that the archaeologists, with their active intake of C-14, are more radioactive than their ancient ancestor. In practice, radioactivity measurement is carried out not in the field but on samples taken to the laboratory. (artwork by Robin Brickman)

Nuclear News: The Shroud of Turin

For centuries the Shroud of Turin, a linen cloth bearing the imprint of a human body, has been revered by many Roman Catholics as the burial cloth of Jesus. In 1988 the shroud's claim on authenticity fell to radiocarbon dating, done independently by laboratories in England, Switzerland, and the United States. By measuring the ratio of carbon-14 to other carbon isotopes, all three laboratories agreed that the shroud dates from 1200–1300 a.d.

The dating of the shroud was done with sophisticated methods that are beyond the reach of most radiocarbon-dating facilities. Normally, carbon-14 content is measured by detecting radioactive decay. But to date the shroud in this way would have required destruction of several samples, each as large as a handkerchief. The more advanced technique used postage-stamp-size pieces of the shroud (figure 4.8). Each laboratory extracted the carbon from its sample, then accelerated the individual carbon atoms to high energies. Passing the atoms through a magnetic field then separated them according to their masses. Effectively, the method counted all C-14 atoms instead of only those that happened to decay; thus, a much smaller sample could be used.

News sources: "Church Says Shroud of Turin Isn't Authentic," *New York Times,* October 14, 1988; "How Carbon 14 Was Used to Fix Date of Shroud," *New York Times,* October 14, 1988.

age. Indeed, much of our knowledge of our own past, of our planet, and of our solar system comes from radioisotope dating.

Activation Analysis

Is my water safe to drink, or does it contain toxic elements such as arsenic? What, exactly, is in this sample of air pollution I have collected on a filter? Is there a bomb in this airline passenger's luggage? What elements are present in this meteorite, which came from outer space? What kind of paint did Leonardo Da Vinci use for the Mona Lisa? These and similar questions are often answered by means of a radiation technique known as **activation analysis**.

In activation analysis, a sample to be analyzed is first bombarded with neutrons, protons, or gamma rays. A few of these cause nuclear reactions in the sample, resulting in radioactive nuclei. The activated nuclei decay, each type of nucleus giving off its own

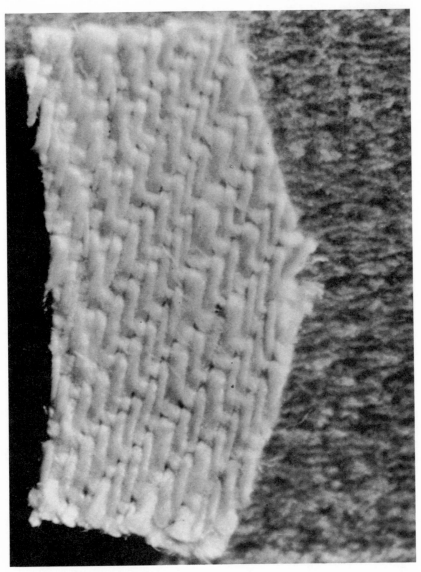

Figure 4.8   This piece of the Shroud of Turin, smaller than a postage stamp, was used in the radiocarbon dating of the shroud's origin to 1200–1300 A.D. (Accelerator Mass Spectrometry Laboratory, University of Arizona)

characteristic radiation. Measuring the intensities and energies of those radiations gives the quantities and types of radioactive nuclei present, and from this the numbers and types of the original non-radioactive nuclei can be determined.

Activation analysis has many uses in archaeology, art history, geology, chemistry, and environmental science. Its advantages include the ability to detect elements present only in minute quantities and its nondestructive nature. A work of art, for example, can be analyzed without the need to remove any material. With the recent surge in terrorist bombings of aircraft, activation analysis has emerged as a leading candidate for detection of explosives in airline luggage.

Summary

The high-energy particles that constitute nuclear radiation can do grievous damage as they tear through biological systems, disrupting cell functions and damaging genetic material. The results may include acute radiation sickness, initiation of cancer, and undesirable genetic traits that are passed on to future generations. Fortunately, relatively large radiation doses are needed to produce significant effects. Fatal radiation sickness occurs at about 400 rems. The natural mutation rate doubles at doses of 100 rems or more. There is controversy about the average dose needed to cause cancer; antinuclear estimates run as low as 120 rems per fatal cancer, and the most recent official estimates put the cancer dose at about 2,000 rems. Either of these doses is much higher than one individual would receive in any but the most serious nuclear accident. On the other hand, even very low doses to large populations may result in significant incidences of cancer.

The penetrating radiation emitted in nuclear processes need not be altogether harmful. The same disruptive effects that damage living things help preserve food. Medical applications of radiation have saved countless lives through diagnosis and treatment. Radioisotopes are used to trace the flow of material in biological and industrial processes, as their radiation emission provides an easily detectable indication of where the material goes and how much is

Nuclear News: Sniffing Out Explosives

On December 21, 1988, a Pan American 747 jet was blown up over Scotland by less than a pound of plastic explosive, which terrorists had concealed in a cassette player. Plastic explosives, a favorite of terrorists, are not detected by most present-day airport security devices. But future passengers may find their luggage subject to nuclear activation analysis aimed particularly at explosives.

All explosives are characterized by high densities of both nitrogen and oxygen. In one explosive-detection scheme, luggage is scanned with neutron beams (figure 4.9). Collisions of neutrons with nitrogen and oxygen nuclei in the luggage excite the nuclei, which decay almost immediately by gamma radiation. The precise energies of these gamma rays are known, so their numbers provide a measure of the quantity of nitrogen and oxygen present. Sophisticated gamma-ray detectors coupled with computer imaging systems then show where high nitrogen and oxygen content might indicate the presence of explosives.

Can such detection systems work? Are they reliable? Are they safe? Can we afford them? Certainly we can build systems that distinguish explosives from almost all other materials. They do require luggage to become radioactive—but only for a tiny fraction of a second. Safety to the public is not an issue. But can we build sophisticated particle accelerators that can be operated for long periods by people who aren't PhD physicists? And can the flying public afford this new level of security? These nuclear questions remain unanswered.

News sources: "New Machines Can Detect Terrorists' Bombs, Usually," *New York Times*, September 12, 1989; "Questions Are Raised on New Bomb Detector," *New York Times*, February 4, 1990; "Terror Creates a Need: Better Bomb Detectors," *New York Times*, May 23, 1990.

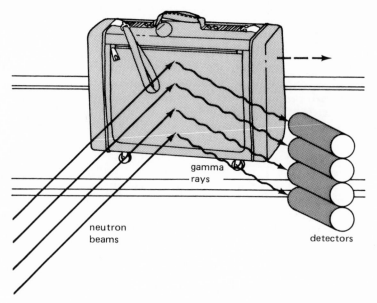

Figure 4.9 Neutron activation could be used to check airline baggage for explosives. Here, a suitcase is scanned with neutron beams that excite nitrogen and oxygen nuclei within it. The excited nuclei emit gamma rays, which are detected and analyzed to determine if the luggage has the high nitrogen and oxygen content characteristic of explosives. (adapted, with permission, from a drawing by Lee Grodzins)

there. Naturally occurring radioisotopes act as "clocks"; their half-lives allow us to date materials from thousands to billions of years old. And the creation of short-lived radioisotopes in materials of uncertain composition gives rise to radiation whose characteristics reveal that composition.

Is radiation bad or good? The answer has to be that it is both, and that each potential use or occurrence of radiation requires a nuclear choice that weighs the bad against the good.

Notes

1. See W. J. Schull, M. Otake, and J. Neel, "Genetic Effects of the Atomic Bombs: A Reappraisal," *Science* 213 (September 11, 1981), pp. 1220–1228.

2. Recall that we are treating the rad and the rem as essentially interchangeable measures of radiation exposure.

3. National Academy of Sciences, *Health Effects of Exposure to Low Levels of Ionizing Radiation (BEIR V)* (National Academy Press, 1990). There is considerable uncertainty in this cancer-dose estimate, and it could vary by a factor of 2 either way.

4. These figures assume a cancer dose of 2,000 rems per fatal cancer. The antinuclear figure is based on individual doses of 5 millirems per year (somewhat more than the typical dose to persons living near a nuclear power plant) to the entire population; the pro-nuclear figure assumes an average dose of 0.2 mrem per year.

5. See "British Radiation Study Throws Experts into Tizzy," *Science* 248 (April 6, 1990), pp. 24–25.

Further Reading

Bulletin of the Atomic Scientists 46 (September 1990). This issue is devoted to a special report entitled "Radiation and Health."

Charles E. Cobb and Karen Kasmauski, "Living with Radiation," *National Geographic* 175 (April 1989), p. 403. Essays and color photos survey harmful and beneficial effects of nuclear radiation.

Bernard L. Cohen, *Before It's Too Late: A Scientist's Case for Nuclear Power* (Plenum, 1983). Chapters 2 ("How Dangerous Is Radiation?") and 4 ("Understanding Risk") underscore the author's contention that radiation's dangers are grossly exaggerated in the public mind.

John W. Gofman, *Radiation and Human Health* (Sierra Club Books, 1981). A leading anti-nuclear physician-scientist argues that radiation's effects are more serious than "official" studies suggest. Includes detailed statistics and analyses.

Eric J. Hall, *Radiation and Life* (Pergamon, 1976). A brief survey of radiation's effects on living things. Somewhat dated but well illustrated.

Michio Kaku and Jennifer Trainer, *Nuclear Power: Both Sides* (Norton, 1982). A collection of essays on opposite sides of nuclear issues. Chapter 2 deals with radiation effects, with four convincing essays arguing contradictory points of view.

David W. Lillie, *Our Radiant World* (Iowa State University Press, 1986).

National Academy of Sciences, *Health Effects of Exposure to Low Levels of Ionizing Radiation (BEIR V)* (National Academy Press, 1990). This latest "official" study of radiation's health effects increases the estimates of radiation-induced cancer by a factor of 3–4. Thorough, detailed, and mathematical.

Harvey Wasserman and Norman Solomon, *Killing Our Own: The Disaster of America's Experience with Atomic Radiation* (Dell, 1982). Outspokenly anti-nuclear writers detail incidents involving radiation exposure in the U.S. population.

Glossary

activation analysis Determination of the elemental content of materials through bombardment with radiation and subsequent analysis of radioactive decays.

cancer dose The radiation dose that, when spread over a population, is likely to cause one excess cancer death. The cancer dose is believed to be about 2,000 rems.

genetic effects Effects on an organism that are passed on to future generations.

ionization The removal of an electron from an atom. Radiation is one possible cause of ionization.

ionizing radiation Radiation of sufficient energy to knock electrons from atoms. X rays and nuclear radiation are ionizing; light, infrared, microwaves, and radio waves are not.

mutation A change in genetic characteristics brought about by alteration of DNA, the genetic material. Radiation can cause mutations.

nuclear medicine Medical techniques making use of radioactive isotopes.

radiation sickness Illness—characterized by nausea, vomiting, intestinal bleeding, anemia, and other symptoms—caused by high-level radiation exposure (over 100 rems in humans).

radiation therapy Use of radiation to fight cancer or other diseases.

radioactive tracer A radioactive substance used to trace the movement of material in biological or other systems.

radiocarbon dating Radioisotope dating using the isotope carbon-14; useful for dating once-living materials as much as 50,000 years old.

radioisotope dating Use of radioisotopes to date ancient materials by measuring the extent to which unstable isotopes in those materials have decayed.

repair enzymes Substances in living cells that repair damage to DNA, possibly reducing the effects of low-level radiation. At high radiation doses, the repair enzymes themselves may be destroyed.

somatic effects Effects on an individual organism not passed on to future generations.

Energy from the Nucleus

5

We have spent two chapters considering radiation, which occurs when unstable nuclei emit energetic particles. Although radiation is energetic, far larger amounts of energy come from wholesale rearrangement of the nucleus. It is this larger energy that powers nuclear reactors and nuclear weapons. In this chapter we will explore several approaches to the liberation of nuclear energy.

The Curve of Binding Energy

Chapter 2 stressed the nuclear difference—the millionfold increase in energy from chemical to nuclear reactions. We found that this huge increase is associated with the extreme strength of the nuclear force that binds protons and neutrons into the atomic nucleus. Now let us ask a more detailed question: Just how tightly bound are the nucleons in a given nucleus?

A more precise way to put this question is this: How much energy is required to remove one nucleon from the nucleus? This quantity is called the **binding energy**. In a more familiar context, imagine a tennis ball and a bowling ball sitting on the floor. You want to lift each to a tabletop. Clearly, it takes more energy to lift the bowling ball. Why? Because it is more tightly bound to the Earth, and that is because the gravitational force on it is stronger. The energy required to lift each ball is its binding energy, associated in this case with the gravitational force that binds it to the Earth.

Why is binding energy important? Because it tells not only how much energy is *required* to take a nucleus apart, but also how much energy is *released* when the nucleus forms. Back to our gravitational analogy. To lift the bowling ball to the tabletop, you have to supply energy; but if you drop the bowling ball from the tabletop, you get that energy back—in the form of a loud sound, a slight heating of the floor, and perhaps permanent damage to the ball and the floor. In nuclear terms, binding energy is therefore a measure of the energy released when a nucleus is formed; to be precise, it is the energy released per nucleon. Ultimately, binding energy tells us how much energy we can get out of nuclear reactions.

We found in chapter 2 that the makeup and the stability of a nucleus are determined by the interplay between the strong but short-range nuclear force and the weak but long-range electrical force. For small nuclei, the nucleons are close together, and the short-range nuclear force dominates. On the other hand, there aren't many nucleons in a small nucleus, so each nucleon is bound to only a few others. As the size of the nucleus increases, each nucleon feels the nuclear attraction of the additional nucleons. The nucleus becomes more tightly bound, so the binding energy increases. For very large nuclei, however, some nucleons are so far apart that the nuclear force between them is insignificant. If they are protons, those nucleons still experience an electrical repulsion. The result is a nucleus that is less tightly bound. In the preceding chapters, we saw that this effect could make nuclei unstable (that is, subject to radioactive decay). Now, in the context of nuclear energy, the same effect reduces the binding energy per nucleon for very large nuclei.

In light of these considerations of nuclear binding, we might expect binding energy at first to increase with increasing nuclear size, then at some point to decrease as electrical repulsion becomes important. Measuring the binding energies of different nuclei confirms this trend. Figure 5.1 shows the **curve of binding energy**, a graph of binding energy per nucleon versus size of the nucleus. (Size, in this context, is determined by the total number of nucleons, which was introduced in chapter 2 as the nuclear mass number.) The curve shows that for light nuclei the binding energy per nucleon increases very rapidly with increasing nuclear size. The binding energy eventually peaks at the common iron isotope iron-56,

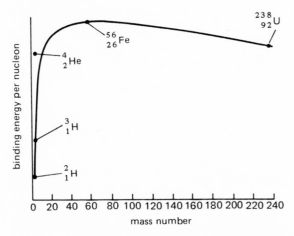

Figure 5.1 The curve of binding energy. This graph plots binding energy per nucleon versus nuclear size, as measured by the nuclear mass number. The most tightly bound nucleus, at the top of the curve, is iron-56 ($^{56}_{26}$Fe). Several other common isotopes are are also labeled.

then begins a gradual decline toward the heaviest naturally occurring isotope, uranium-238.

Nuclear Fusion

The increase in binding energy for light nuclei means that combining two such nuclei results in a more tightly bound nucleus, and therefore in the release of nuclear energy. Helium-4 ($^{4}_{2}$He), for example, is much higher up the binding-energy curve than deuterium ($^{2}_{1}$H). Combining two deuterium nuclei to make a single helium-4 nucleus would therefore release a substantial amount of nuclear energy. This process of combining light nuclei is called nuclear **fusion**. Figure 5.2 shows two important fusion reactions. The energy released in fusion is so huge that fusion of even the minuscule amount of deuterium in water would make a gallon of ordinary water the energy equivalent of nearly 400 gallons of gasoline.

Nuclear fusion is vital to our existence; it makes the Sun shine, and therefore it sustains life on Earth. And we ourselves are made from the products of fusion. Carbon, oxygen, nitrogen, and other elements necessary for life were formed through nuclear fusion in

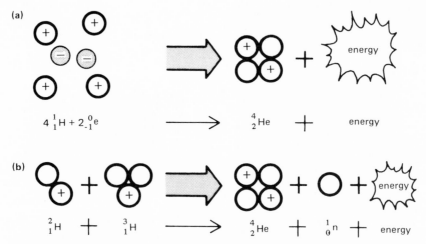

(a)

$$4\,_1^1\text{H} + 2\,_{-1}^0\text{e} \longrightarrow \,_2^4\text{He} + \text{energy}$$

(b)

$$_1^2\text{H} + \,_1^3\text{H} \longrightarrow \,_2^4\text{He} + \,_0^1\text{n} + \text{energy}$$

Figure 5.2 Two important fusion reactions. (a) In the Sun, a sequence of several fusion reactions has, as its net result, the conversion of four protons and two electrons into a helium-4 nucleus. Energy released in this reaction is what makes the Sun shine. (b) In thermonuclear weapons—"hydrogen bombs"—deuterium ($_1^2\text{H}$) and tritium ($_1^3\text{H}$) fuse to form helium-4, and a neutron is released. This reaction may someday find peaceful use in electric power plants.

the cores of ancient stars. Some of those stars exploded in violent nuclear blasts called supernovas, and spewed their material into interstellar space. Much later, the Sun, the Earth, and other constituents of our solar system formed out of this cosmic debris.

With the exception of hydrogen and helium, which formed shortly after the universe began, essentially all the other elements up to iron were "cooked" by fusion in the interiors of stars. The most massive stars have cores of iron, surrounded by an "onion-skin" structure of lighter elements. Fusion in each layer gives rise to the heavier nuclei in the next layer inward.

Iron is the most tightly bound nucleus, so once iron forms it is not possible to release further energy by nuclear fusion. So where did the elements heavier than iron come from? Astrophysicists believe that they formed only in the cataclysmic supernova explosions themselves, when enough excess energy was available to form these less-tightly-bound nuclei. We and virtually everything around us are therefore products of the stars, with our constituent elements formed in nuclear reactions that occurred billions of years ago.

It is uplifting to think of ourselves as "star children," tracing our ultimate origins to fusion events inside ancient suns. But fusion also presents a sobering prospect. Fusion energy released in our thermonuclear weapons has the potential to destroy us. Although fusion powers the stars, we humans have not yet learned to harness it for any purpose but destruction.

Fusion is a simple nuclear process, but getting fusion reactions started is anything but simple. To fuse nuclei, you must get them close together—so close that the strong but short-range nuclear force can bind them tightly. But all nuclei carry positive electric charge, so they repel one another. The repulsive electric force acts over a much longer range than the attractive nuclear force, so electrical repulsion must be overcome before nuclei can get close enough to fuse. Getting one nucleus to fuse with another is like dropping a ball into a deep hole that is surrounded by a barrier. A lot of energy will be released when the ball drops into the hole, but to get it there you first have to supply energy to get it over the barrier (figure 5.3).

If the ball in figure 5.3 moves very rapidly toward the hole, it may have enough energy to overcome the barrier. Similarly, if two

---

Figure 5.3   A gravitational analogy for fusion. The energy released in fusion is like the energy released when a ball falls into a deep hole. But to get to the hole, the ball first has to overcome a high barrier, analogous to the electrical repulsion between fusing nuclei. If the ball moves very rapidly toward the hole, it may have enough energy to overcome the barrier.

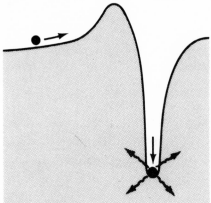

Nuclear News: Cold Fusion

One way to get the ball in figure 5.3 over the barrier is to push it slowly, but with a large enough force to get it over the top. That, basically, is the approach that was attempted in the "cold fusion" schemes that made headlines in 1989. In cold fusion experiments, deuterium was forced into the spaces between atoms of a metallic crystal. According to the experimenters, enough deuterium could be crammed into these spaces, and enough pressure exerted by the metal atoms, that some deuterium nuclei were forced close enough to fuse. After careful study, most physicists have concluded that cold fusion does not work, and that the experiments involve instead some unknown chemical reaction.

News source: "Panel Rejects Fusion Claim, Urging No Federal Spending," *New York Times,* July 13, 1989.

nuclei are moving fast enough toward each other, they may have enough energy to surmount the barrier of their mutual electrical repulsion. In the traditional "hot" approach to fusion, heating of the fusing material brings the nuclei to the required speeds. But very high temperatures are needed. For the deuterium-tritium fusion reaction illustrated in figure 5.2b, the mixture of deuterium and tritium must be heated to about 100 million degrees! How can you get something that hot? And what can you keep it in? Those two questions—heating and containment—are the essential challenges facing fusion scientists.

The stars conveniently meet both of fusion's challenges with their immense gravity. A star forms when a cloud of gas collapses under its own gravity, compressing and heating as it does so. A big enough cloud gets hot enough to initiate fusion, and a star is born. Once fusion starts, the energy generated in the fusion reactions keeps the gas hot enough to sustain more fusion, and the star's immense gravity confines the hot gas. On Earth, we don't have access to a star's strong gravity. Terrestrial scientists have had to devise other ways to heat and confine fusing material. In later chapters we will see how this is done in thermonuclear weapons, and will explore several approaches to controlling nuclear fusion for peaceful energy production.

Nuclear Fission

We have seen how nuclear fusion releases energy by moving nuclei up the curve of binding energy, fusing lighter nuclei into heavier, more tightly bound ones. Now look again at the curve of binding energy, repeated in figure 5.4. You can see that it is also possible to release energy by moving up the curve from the heaviest nuclei to the more tightly bound, lighter nuclei in the vicinity of iron. How can you get from heavier to lighter nuclei? By splitting heavier nuclei into lighter pieces—a process known as **nuclear fission**.

Some of the heaviest and therefore least stable nuclei undergo **spontaneous fission**, in which a heavy nucleus spontaneously splits into two lighter pieces and a few neutrons. In naturally occurring isotopes, the half-lives for spontaneous fission are so long (often greater than the age of the universe) that the process is very infrequent and the rate of energy release therefore insignificant.

Of greater significance is **neutron-induced fission**, in which a heavy nucleus splits after being struck by a neutron. Nuclei that can undergo neutron-induced fission are said to be **fissionable**. Figure 5.5 shows the neutron-induced fission of a uranium-235 nucleus. The U-235 nucleus absorbs the neutron and becomes a highly excited nucleus of uranium-236. This excited nucleus under-

Figure 5.4 The curve of binding energy, showing that energy can be released by fusion of light nuclei or by fission of heavy nuclei.

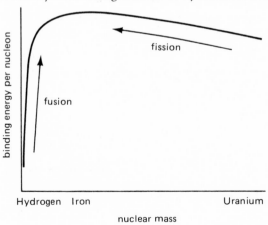

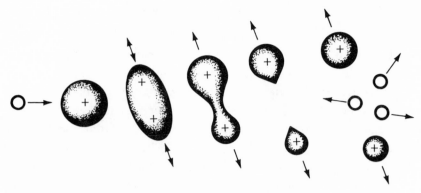

Figure 5.5 Neutron-induced fission of uranium-235. At left, a neutron strikes the U-235 nucleus, forming U-236. The highly excited U-236 nucleus oscillates violently, taking on a dumbbell shape. Electrical repulsion drives the two ends of the dumbbell apart, resulting in two smaller nuclei. Several neutrons are released in the process.

goes violent vibrations that distort it into a dumbbell shape. With this elongated shape, the short-range nuclear force weakens further, and the electric repulsion is then strong enough to drive the two pieces of the nucleus apart. The entire process takes only about a trillionth of a second.

Many heavy nuclei are fissionable, including the uranium isotopes U-235 and U-238. Most fissionable isotopes, including U-238, require that the incoming neutron have considerable energy. However, for three isotopes—uranium-233, uranium-235, and plutonium-239—fission occurs with neutrons of arbitrarily low energy. These three are called **fissile** isotopes. Of the three, only U-235 occurs naturally, and it constitutes less than 1 percent of natural uranium. As we will soon see, U-233 and Pu-239 can be produced in nuclear reactors, and Pu-239 has become a very significant nuclear material.

The Nuclear Chain Reaction

Neutrons induce fission, and the fission process itself releases neutrons. Those neutrons may go on to induce more fission reactions, giving rise to still more neutrons and more fission. The resulting self-sustaining process (figure 5.7) is called a **chain reaction**.

What is required to sustain a chain reaction? On the average, at least one neutron from each fission event must cause another fission; otherwise the reaction will soon fizzle to a halt. A chain reaction in which each fission event gives rise, on average, to exactly one other fission event is said to be **critical**. A critical chain reaction releases energy at a steady rate, and is therefore the desirable state in a nuclear reactor. If the number of fission events arising from each event is smaller than one, the reaction is **subcritical**. In a subcritical reaction, energy release decreases and the reaction soon halts. If, on the other hand, each fission event triggers more than one additional fission, then the reaction is **supercritical**. Energy release grows rapidly, and may lead quickly to a nuclear explosion. A nuclear fission bomb employs a supercritical chain reaction. Supercriticality in a nuclear reactor can be disastrous.

A typical fission reaction releases two or three neutrons (in commonly used nuclear reactors, the average is 2.47 neutrons per fission). It might seem easy to produce a critical or a supercritical chain reaction, since each fission event provides one or two more neutrons than are needed to sustain the reaction. But what happens to these neutrons? Imagine a chunk of uranium, consisting of a mixture of U-235 and U-238. Several things could happen to a neutron released in a fission reaction within this mass:

- It might leave the mass altogether, causing no further fission.
- It might be absorbed by a U-238 nucleus. Since the neutron energy required for fission of U-238 is very high, fission in this case is unlikely. (However, as we will soon see, neutron absorption in U-238 is significant for another reason.)
- It might be absorbed by a U-235 nucleus, resulting in fission and the production of two or three additional neutrons.

Other outcomes are possible, too. For example, neutron absorption in U-235 does not always result in fission. The presence of other materials in or around the uranium mass may result in more neutron absorption without fission. But considering only the three outcomes listed will help you understand the problems you would face in trying to achieve a fission chain reaction.

To sustain a chain reaction requires that at least one of the two or three neutrons released in a fission event be absorbed by U-235, inducing another fission event and giving rise to more neutrons.

Nuclear News: The Discovery of Fission

The discovery of nuclear fission reads like a fascinating detective story, laced with international scientific intrigue and spiced with the approaching cloud of World War II. The first hints came in 1934, when the Italian physicist Enrico Fermi and his collaborators bombarded uranium with neutrons. They found that a number of new radioactive isotopes resulted. When she heard of their results, the German chemist Ida Noddack speculated on a new possibility: "It is conceivable that in the bombardment of heavy nuclei with neutrons, these nuclei break up into several large fragments. . . ." Noddack's was the first known suggestion of nuclear fission, but scientists working with uranium did not follow up on her suggestion.

Later, in 1938, the German chemists Otto Hahn and Fritz Strassmann were astonished to find a radioactive isotope of the medium-weight element barium in uranium that had been bombarded with neutrons. How did the barium get there? Hahn and Strassmann expected neutron absorption to give rise to heavier nuclei than uranium, and indeed they had found such nuclei. But the lighter-weight barium was totally unexpected, given what was then known of nuclear physics. Nevertheless, Hahn and Strassmann came close to guessing its origin, exclaiming that the masses of barium and another nucleus that they also suspected was present summed that of uranium-238 and a neutron.

The clearest realization that fission had occurred came in December 1938, when the Austrian physicist Lise Meitner and her nephew Otto Frisch discussed Hahn and Strassmann's results. On a walk through the Swedish countryside, where Meitner had fled to escape Hitler, she and Frisch sketched a diagram not unlike our figure 5.5, suggesting that uranium nuclei had undergone fission to produce the barium that so puzzled Hahn and Strassmann. Meitner and Frisch also calculated the enormous energy that would be released in the process. Shortly thereafter, they were the first to use the word *fission,* in a paper entitled "Disintegration of Uranium by Neutrons: A New Type of Nuclear Reaction."

Word of the fission discovery spread rapidly throughout the world's physics community, and with it the realization of the enormous energy potential of a fission chain reaction. The public, too, soon learned of fission. On January 29, 1939, the *New York Times* declared "Atomic Explosion Frees 200,000,000 Volts," and an editorial the following week spoke of "utopias where whole cities are illuminated by energy in a little matter—this time uranium." But soon there were more ominous possibilities: On April 30, the *Times* headlined "Vision Earth Rocked by Isotope Blast . . . Scientists say bit of uranium could wreck New York." On the eve of World War II, the military implications were obvious.

The result, less than seven years after the recognition of fission, was the detonation of the first nuclear explosion in the New Mexico desert, followed only weeks later by the nuclear devastation of Hiroshima and Nagasaki.

News sources: "Atomic Explosion Frees 200,000,000 Volts," *New York Times,* January 29, 1939, p. 2; "Revolution in Physics," *New York Times,* February 3, 1939, p. 14; "Vision Earth Rocked by Isotope Blast," *New York Times,* April 30, 1939, p. 35. For details of the fission story, see Richard Rhodes, *The Making of the Atomic Bomb* (Simon and Schuster, 1986), chapter 10. Ida Noddack quoted in Samuel Glasstone, *Sourcebook on Atomic Energy* (Van Nostrand Reinhold, 1967), p. 475.

But this will not happen if too many neutrons leave the mass or are absorbed in U-238. How can these neutron losses be minimized? Loss by absorption in U-238 can be reduced by **enriching** the uranium; that is, by increasing the proportion of the fissile U-235 relative to the nonfissile U-238. Later we will see how this enrichment is accomplished; for now, note that enrichment to about 3 percent U-235 is common in U.S. nuclear power reactors, with much higher enrichment in some research reactors and in weapons. Enrichment technology is extremely sensitive, for a nation possessing it can produce nuclear weapons material from relatively abundant natural uranium.

Enrichment reduces losses by absorption in U-238, but it doesn't prevent neutrons from leaving the uranium mass. The easiest way to prevent that loss is to make the mass bigger; the bigger the mass, the more likely a neutron is to undergo absorption rather than to escape. Imagine starting with a very small chunk of enriched uranium. Most of the neutrons arising from fission in the mass escape without causing more fission. But as the size is increased, more of the neutrons are likely to hit a U-235 nucleus, causing additional fission. Eventually a size is reached at which, on average, exactly one of the neutrons from each fission event causes another fission. At that point, the chunk of uranium constitutes a **critical mass**; it has the minimum size needed to sustain a chain reaction. A larger mass is supercritical; in it, neutrons and fission reactions

Figure 5.6 Lise Meitner and Otto Hahn. Meitner and her nephew Otto Frisch interpreted the experiments of Hahn and Strassmann as evidence for neutron-induced fission of uranium. Meitner, an Austrian physicist, fled to Sweden to escape Hitler. Earlier, as a woman in a male-dominated field, she had braved sexist policies that denied her access to laboratories when men were present. By the 1930s she had become one of the world's most prominent and respected nuclear physicists. (from Otto Hahn, *A Scientific Biography* [Scribner, 1966], courtesy of AIP Niels Bohr Library)

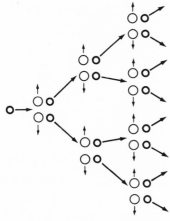

Figure 5.7 A fission chain reaction. In this case the reaction is supercritical, each fission giving rise to two others. The rate of fission grows rapidly, causing an explosive release of nuclear energy.

multiply rapidly, resulting in catastrophic energy release. The result is a nuclear explosion.

How big is a critical mass? That depends on the proportion of fissile material and on its shape and configuration. Surrounding the mass with a neutron-reflecting substance further reduces the size needed for criticality, by returning neutrons that would otherwise escape. So how big is a critical mass? In a world where fissile materials are increasingly available, it is alarmingly small. For pure uranium-235, somewhere around 30 pounds could constitute a critical mass. For plutonium-239 the critical mass is even smaller—roughly 5 pounds, an amount about the size of a tennis ball. These numbers are for weapons configurations, whose details we will discuss later; the critical mass in nuclear power reactors is much larger because of low enrichment, a vastly different configuration of fissile fuel, and the presence of cooling water and other materials.

We have seen how to sustain a chain reaction, but how does the reaction get started? That, it turns out, is not difficult. Cosmic rays interacting with the atmosphere produce stray neutrons that can initiate fission, as do occasional spontaneous fission events. If a critical mass of fissile material is formed, stray neutrons will ensure the start of a chain reaction. Greater reliability (especially in sophisticated nuclear weapons, where timing is crucial) can be

Nuclear News: Uranium Enrichment and Weapons Proliferation

In July 1987, Toronto businessman Arshad Pervez was arrested in Pennsylvania as he worked on a deal to purchase a special high-strength steel called Maraging 350 from the Carpenter Steel Corporation. Why should that seemingly routine business be a crime? Because Maraging 350 is an expensive steel with very limited uses, because Pervez was acting as export agent for a Pakistani firm headed by a retired army general, and because Pervez had told an undercover U.S. customs agent that the material was bound for Pakistan's Kahuta nuclear facility. In December, Pervez was convicted of conspiring to make illegal exports of sensitive material.

Pakistan wanted the super-strength Maraging 350 to build centrifuges that would whirl gaseous uranium around at high enough speed to separate fissile uranium-235 from the slightly heavier U-238. Maraging 350 steel is not itself a nuclear material, but its potential use in uranium enrichment makes it a sensitive material in a world concerned with the proliferation of nuclear weapons.

The Maraging 350 deal is not an isolated incident. Pakistan and several nations are vigorously seeking uranium-enrichment facilities or other sensitive items for the purpose of building nuclear weapons. Intelligence reports suggest that Pakistan has assembled a complete uranium-enrichment plant from components purchased individually throughout the industrialized world. There is some ambiguity about enrichment technology, since it can produce mildly enriched uranium for civilian power reactors or highly enriched uranium for weapons. But in the cases of Pakistan, and its adversary India, and Iraq, and South Africa, and Israel, and perhaps other countries, it is fairly evident that sensitive technology goes to nuclear weapons programs. Do we want the next century to see a world seething with nuclear-armed nations? That is a nuclear choice which the technologically advanced nations will have to make when considering attempts to control the spread of uranium-enrichment technology.

News sources: "A Bomb Ticks in Pakistan," *New York Times,* March 6, 1986, Section 6; "German Concern Said to Aid Pakistan A-Weapons," *New York Times,* January 29, 1989; "CIA Chief Wary of Pakistani Nuclear Program," *New York Times,* May 19, 1989.

Nuclear News: Missing Plutonium

The small quantity of plutonium required for a critical mass calls for stringent security and accurate record keeping wherever plutonium is handled. Workers at plutonium installations must pass through radiation monitors as they exit sensitive areas. This procedure not only warns workers of possible contamination but also guards against intentional diversion of fissile plutonium for unauthorized purposes. The bookkeeping and inventory procedures at nuclear weapons facilities and fuel-reprocessing plants are among the strictest in any industry.

Despite these precautions, accounts of missing nuclear materials have surfaced throughout the nuclear age. A 1989 report by the U.S. House of Representatives' Energy and Commerce Subcommittee on Oversight and Investigation cited inadequate security in plutonium-handling areas of the Lawrence Livermore National Laboratory in California, resulting in uncertainty regarding the whereabouts of plutonium quantities in excess of a critical mass. More serious is the 1990 discovery of more than 60 pounds—perhaps seven bombs' worth—of plutonium in air vents at the Rocky Flats nuclear weapons plant in Colorado. If the system can't account for that much plutonium, how can we be sure similar amounts haven't been diverted to terrorists or hostile would-be nuclear nations? In fact, there is good evidence that clandestine diversion of fissile material has occurred.

News source: "Congress Report Tells of Security Lapses at Nuclear Weapons Plant," *New York Times,* December 4, 1989; "30-Year Plutonium Loss at Plant Equals 7 Bombs," *New York Times,* March 29, 1990.

achieved using specially designed neutron sources. But that is not essential in a crude chain reaction; in fact, stray neutrons cause the opposite problem: the reaction in a fission weapon may "preignite," blowing the device apart before fission is complete.

The Products of Fission

In fission, a nucleus of uranium or plutonium splits into two lighter nuclei. Those lighter nuclei are called **fission products**. What are they?

The most likely outcome of a single fission reaction is the formation of two nuclei of unequal mass, one with mass number roughly in the range 85–105 and the other in the range 130–145. A common U-235 fission reaction, for example, gives rise to molybdenum-102 ($^{102}_{42}$Mo) and tin-131 ($^{131}_{50}$Sn), along with three neutrons:

$$^{1}_{0}n + {}^{235}_{92}U \rightarrow {}^{102}_{42}Mo + {}^{131}_{50}Sn + 3{}^{1}_{0}n.$$

This reaction is particularly significant because the tin-131 quickly beta-decays to radioactive iodine-131, a dangerous environmental contaminant that lodges in the thyroid gland. Other important isotopes among fission products are strontium-90 (a radioisotope that mimics calcium and is therefore incorporated into bone) and cesium-137.

The products of nuclear fission have one thing in common: They are highly radioactive. Figure 5.8 shows why. As was discussed in chapter 2, the heaviest nuclei have a considerable abundance of neutrons over protons, in order to overcome the electrical repulsion of protons with the nuclear force exerted by neutrons. When a heavy nucleus undergoes fission, the fission products preserve the same ratio of neutrons to protons. But, as figure 5.8 shows, that leaves them with an excess of neutrons, which renders them unstable. Some of those neutrons "boil off" almost instantaneously, giving the two or three neutrons produced in each fission event. But the remaining isotopes are still rich in neutrons. They decay, through a sequence of beta emissions, until they achieve stability. So the fission products are inherently radioactive. The half-lives of typical fission products range from less than a second to hundreds of years; since those times are much less than the roughly billion-year half-lives of uranium isotopes, the fission prod-

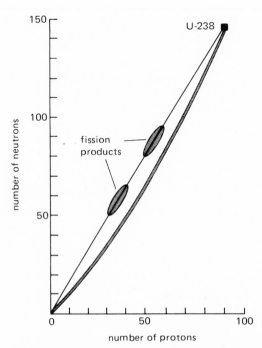

Figure 5.8 A chart of the nuclei, simplified from figure 2.13. The thick curve is the region of stability, extended to include the long-lived unstable isotopes through uranium. The thin straight line indicates the neutron-to-proton ratio for uranium and thus for the products of uranium fission. It shows that fission products necessarily lie above the stable region, and are therefore radioactive. Gray areas mark the most probable fission products.

ucts are much more intensely radioactive than the original uranium. During the refueling of a nuclear reactor, for example, fresh uranium fuel is handled without the need for elaborate radiation protection. But spent fuel, rich in fission products, is kept underwater to shield workers from its intense radiation (figure 5.9).

The high level of radioactivity in fission products is what makes the waste from nuclear reactors difficult to deal with. It is also what produces fallout from nuclear weapons. Unlike a reactor, in which every attempt is made to contain fission products, a nuclear weapon disperses its fission products directly into the environment. Those who where young children in the period from the mid-1950s to the early 1960s had their bones form during the time when the United States and the Soviet Union were vigorously engaged in the atmospheric testing of nuclear weapons; as a result,

Figure 5.9 A fuel bundle being moved underwater into the spent-fuel storage area during refueling of the Vermont Yankee nuclear power plant. The fuzzy glow around the bottom of the fuel bundle is caused by the intense beta emission from the radioactive fission products in the spent fuel. (Vermont Yankee Nuclear Power Corporation)

---

they have higher-than-normal levels of the fission product stron-
tium-90 in their bones, and can be expected to develop bone cancer
at a somewhat higher-than-average rate.

Middleweight fission products are not the only radioactive
isotopes formed when a chain reaction occurs in uranium. Neutrons
absorbed by uranium-238—the heaviest naturally occurring isotope—give rise to still heavier nuclei called **transuranics** (meaning
"beyond uranium"). Many transuranics have half-lives that, although short enough that these isotopes are no longer present in
nature, are long compared with those of fission products. The most
important transuranic isotope is plutonium-239. The process leading to Pu-239 begins when U-238 absorbs a neutron, producing
highly unstable U-239. U-239 decays by beta emission with a half-
life of only 24 minutes, forming neptunium-239. Np-239 again beta
decays, now with a half-life of 2.4 days, to form plutonium-239
(figure 5.10).

Figure 5.10   Formation of plutonium-239 occurs when uranium-238 absorbs
a neutron. Two subsequent beta decays occur, giving first neptunium-239
and then plutonium-239.

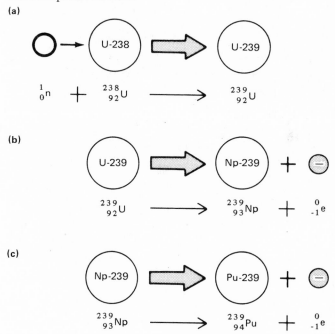

(a)

$$^{1}_{0}n \;+\; ^{238}_{92}U \;\longrightarrow\; ^{239}_{92}U$$

(b)

$$^{239}_{92}U \;\longrightarrow\; ^{239}_{93}Np \;+\; ^{0}_{-1}e$$

(c)

$$^{239}_{93}Np \;\longrightarrow\; ^{239}_{94}Pu \;+\; ^{0}_{-1}e$$

With its half-life of about 24,000 years, Pu-239 stays around for a long time, yet its radioactivity is still considerable. Most important, Pu-239 is more virulently fissile than U-235, fissioning more readily and requiring a smaller critical mass. This means that neutron absorption by U-238, although antithetical to the fission chain reaction, ultimately results in a more fissile product. This "**breeding**" of plutonium in uranium fission reactors is a widely used method of producing fissile Pu-239 for weapons. In power reactors, plutonium breeding may extend the world's uranium supplies, by converting some of the 99.3 percent of natural uranium that is nonfissile U-238 to fissile Pu-239.

$E=mc^2$ and All That

There has been been no mention so far of Albert Einstein's famous formula $E=mc^2$, which for most people is inextricably linked with nuclear energy. $E=mc^2$ does apply to nuclear energy, and in 1939 Lise Meitner used the formula to calculate the energy released in uranium fission. But it is a common misconception to think that $E=mc^2$ is exclusively about nuclear physics, and therefore to make Einstein ultimately responsible for nuclear technology.

$E=mc^2$ expresses a remarkable unity, which Einstein was the first to recognize: that matter and energy are really manifestations of the same basic stuff that makes up the universe. $E=mc^2$ is the key to this interchangeability of matter and energy. It says that a piece of matter with mass m is equivalent to a quantity of energy given by the product of m times the square of the speed of light, c. Since the speed of light is a huge number, the energy equivalent of even a small mass is very large. If we knew how to convert it entirely to energy, a single raisin could power a large city for a day. And $E=mc^2$ works both ways. A bundle of pure energy, such as a gamma ray, can suddenly turn itself into a pair of elementary particles of matter.

The equivalence of matter and energy manifests itself in nuclear reactions. If you weigh a uranium nucleus, then fission it, then weigh all the fission products, you will find they weigh slightly less than the original uranium. The "missing mass" has been converted to energy in the amount given by $E=mc^2$. Similarly, weigh two deuterium nuclei and then fuse them to make helium. The helium weighs less than the two deuterium nuclei you started with, and $E=mc^2$ relates the mass difference to the energy released. Or weigh a nuclear bomb before it goes off, and then collect and weigh the debris after it explodes. You will find a difference, and multiplying that difference by the square of the speed of light gives the energy released in the explosion. On a larger scale, the Sun, through fusion, converts 4 million tons of matter to energy every second.

But $E=mc^2$ is a *universal* statement. It applies to *all* events that release energy. Weigh a candle and all the oxygen around it. Burn the candle and weigh the smoke and carbon dioxide that come out, along with what is left of the candle. If you could weigh with enough precision, you would find that after burning the total weight is less. Again, matter has been converted to energy. Again, $E=mc^2$ relates the decrease in mass to the energy released. The only difference between the burning candle and the nuclear bomb is that the latter converts proportionately more of its mass to energy. That is the nuclear difference, expressed in Einstein's terms. It isn't that nuclear processes convert mass to energy whereas other processes don't; rather, nuclear processes convert proportionately more mass—about a million times more than chemical processes. And even the nuclear processes of fission and fusion are not all that good at converting mass to energy; in both, less than 1 percent of the mass gets converted.

So is Einstein's formula at the heart of nuclear energy? Yes, but only in the sense that it is at the heart of every energy-conversion process—burning gasoline, metabolizing food, stretching a rubber band, running a nuclear reactor. There is nothing uniquely "nuclear" about it. Was Einstein's work an essential step on the path to nuclear technology? No more than it was essential to the harnessing of fire or the development of the steam engine.

Einstein did play a role in instigating the nuclear weapons effort of World War II. But Einstein and his famous formula are widely misunderstood; neither the man nor the equation is responsible for our nuclear dilemmas.

Summary

Nuclear energy is released when atomic nuclei rearrange to form more tightly bound structures. The curve of binding energy shows that there are two kinds of energy-releasing nuclear reactions. In fusion, light nuclei join to form a heavier nucleus. Fusion is the process that powers the stars, but the electrical repulsion of nuclei makes makes fusion difficult to achieve under terrestrial conditions.

For the heaviest nuclei, energy is released when the nuclei split, or undergo fission. Fission occurs most readily when a nucleus is struck by a neutron, and fission itself releases more neutrons. In the few so-called fissile isotopes—notably uranium-235 and plutonium-239—a self-sustaining fission chain reaction is possible, as neutrons from one fission event cause additional fission. A critical mass is necessary to ensure a self-sustained reaction.

The middleweight nuclei that result from fission contain excess neutrons, and are therefore highly radioactive. Absorption of neutrons by nonfissile U-238 creates other radioactive materials, the heavy transuranic elements. The most important of these is the fissile isotope plutonium-239, which, like uranium-235, can fuel nuclear reactors and weapons.

Nuclear fusion and fission convert matter to energy according to Einstein's formula $E=mc^2$. But nuclear processes share that equation with every other energy-releasing process, including commonplace chemical reactions. The nuclear difference lies in the millionfold increase in the amount of matter converted to energy, not in the fact of that conversion.

You now know about the basic nuclear processes involved in radiation and energy production, and are familiar with some of the materials important in those processes. In the remainder of this book we will look at the applications of nuclear processes, first in power production and then in nuclear weapons.

Further Reading

Samuel Glasstone, *Sourcebook on Atomic Energy* (Van Nostrand Reinhold, 1967). Published under the auspices of the former U.S. Atomic Energy Commission, this book is a comprehensive compendium of basic nuclear knowledge.

Paul G. Hewitt, *Conceptual Physics* (Scott, Foresman, 1989).

Raymond Murray, *Nuclear Energy* (Pergamon, 1988).

Glossary

binding energy   The amount of energy needed to remove a nucleon from the nucleus; equivalently, the energy per nucleon released when the nucleus forms.

breeding   Production of plutonium-239 (or other useful isotopes) by neutron absorption.

chain reaction   A self-sustaining fission reaction in which neutrons from one fission event trigger subsequent fission.

critical chain reaction   A chain reaction in which, on average, exactly one neutron from each fission event triggers another fission. A critical reaction releases nuclear energy at a steady rate.

critical mass   A mass of fissile material large enough to sustain a fission chain reaction. Minimum critical masses for plutonium-239 and uranium-235 are about 5 pounds and 30 pounds, respectively.

curve of binding energy   A graph showing the binding energy per nucleon versus nuclear size. The graph peaks at iron, showing that all other nuclei are less tightly bound.

enrichment   A process whereby the proportion of fissile U-235 is increased above its naturally occurring value of 0.7 percent. Enrichment to 3 percent is typical of fuel for power reactors. Weapons-grade uranium may be enriched to 90 percent or more.

fissile nucleus   A fissionable nucleus that will undergo fission when struck by a neutron of arbitrarily low energy. The significant fissile nuclei are uranium-233, uranium-235, and plutonium-239. Of these, only U-235 occurs naturally, and it constitutes just 0.7 percent of natural uranium.

fission   The splitting of a heavy nucleus to form two medium-weight nuclei, accompanied by the release of nuclear energy.

fission product   Any medium-weight nucleus formed during fission of a heavy nucleus. Fission products are inherently radioactive.

fissionable nucleus   A nucleus that can undergo fission when struck by a neutron. Most fissionable nuclei require a certain minimum neutron energy to cause fission.

fusion   The combining of two light nuclei to form a heavier nucleus, accompanied by the release of nuclear energy.

neutron-induced fission   Fission that occurs when a heavy nucleus is struck by a neutron.

plutonium (Pu)   A transuranic element usually formed by neutron absorption in uranium-238. The isotope plutonium-239 is highly fissile, and is widely used in nuclear weapons.

spontaneous fission   Fission that occurs spontaneously, without any initiating event. Spontaneous fission occurs infrequently.

subcritical chain reaction   A chain reaction in which, on average, fewer than one neutron per fission event trigger another fission. A subcritical reaction soon fizzles to a halt.

supercritical chain reaction   A chain reaction in which, on average, more than one neutron per fission trigger additional fission. Energy release in a supercritical reaction grows explosively.

transuranic element   Any element heavier than uranium. The half-lives of these elements are short enough that they exist only when produced artificially, often by neutron absorption in fission chain reactions.

uranium (U)   The heaviest naturally occurring element. Its two most common isotopes, U-235 (0.7 percent of natural uranium) and U-238 (99.3 percent) are fissionable, but only U-235 is fissile.

Nuclear Power

II

Energy and People

6

What *is* energy? As we have seen, nuclear reactions release a million times the energy of their chemical counterparts, but what is it that gets released? A physicist might define energy with mathematical formulas, or might mumble something about "ability to do work." But in essence, energy is what makes everything happen. All motion entails energy, whether it is the motion of a planet, a car, or a neutron. The random dance of the molecules in matter involves the energy we associate with the term *heat*. Take away energy, and everything would stop. No change could occur. Nothing would happen.

Motion is not the only manifestation of energy. Stretch a rubber band and you have stored energy. A gasoline molecule is a miniature version of the stretched rubber band, with energy stored in the tension of the electric forces between its constituent atoms. So, too, is a uranium nucleus; its enormous stored energy is associated with the interplay of electric and nuclear forces. In all these cases, release of the stored energy makes things happen.

Other forms of energy include light and its cousins: radio waves, microwaves, infrared and ultraviolet rays, x rays, and gamma rays. These arise as energy is released in atomic, nuclear, and other processes. In turn, they can transform into other forms of energy. In a microwave oven, microwave energy enters the food and jostles the water molecules into motion that manifests itself as heat. A plant spreads its leaves in the sunshine, capturing light energy for storage as the chemical energy of sugar. In radiation therapy, gamma rays bombard a tumor, their energy breaking apart the molecules within cancer cells.

Ultimately, energy is a fundamental constituent of our universe. It takes many forms, and it readily changes from one form to another. Energy is measurable, quantifiable stuff with the potential to make things happen. We humans, especially those of us in the industrialized societies, use lots of energy to make lots of things happen. This chapter explores our enormous energy appetite.

Exercise!

Put down this book, stand up, put your hands on your hips, and start doing knee bends—up, down; up, down, about once a second. How much energy are you expending? That depends, in part, on how long you continue to do the knee bends; in 2 minutes you expend twice the energy you expend in 1 minute. So a better question might be this: At what *rate* do you expend energy? The answer to that question—the rate of energy use or supply—is called **power**.

There are many units for measuring power. Two you have probably heard of are the **watt** and the horsepower. To say that you have a 1,000-watt hair dryer is to tell the rate at which the hair dryer uses energy. To say that a car has a 200-horsepower engine is to tell the rate at which the engine can produce energy of motion from the energy stored in gasoline. In this book, we will use the **watt** (W), the **kilowatt** (kW, equal to 1,000 watts), and the **megawatt** (MW, equal to 1,000,000 W or 1,000 kW) as units of power. Electric light bulbs and appliances have their power consumption specified in watts, so these units should be familiar to you. A typical desk lamp might have a 60-watt bulb. The bulb in a slide projector uses energy at the rate of several hundred watts. A burner on an electric stove might use 2,000 watts, or 2 kilowatts. The stove burner takes energy at 20 times the rate of a 100-watt (0.1-kilowatt) light bulb. One horsepower is 746 watts, or roughly 1 kilowatt, so a 100-horsepower car could expend energy at about the same rate as 75 hair dryers, or 750 100-watt light bulbs.

Although you may think of watts in the context of electrical devices, this unit of power applies to anything that supplies or uses energy. The Sun, for example, supplies energy at the stupendous rate of 400,000,000,000,000,000,000,000,000,000 watts. At noon, solar energy reaches the Earth's surface at the rate of about 1 kilowatt for each square yard. If you could convert all that energy to electricity, you could run your hair dryer with a square yard's worth of noonday sun. Gasoline, burned at the rate of a gallon an hour, releases energy at the rate of about 40 kilowatts. The electrical power output of a typical large power plant is 1 billion watts, or 1,000 megawatts. A clothes dryer and a water heater each use energy at the rate of 5 kilowatts. A night light needs a few watts, and a calculator or a digital watch uses just thousandths of a watt.

Now back to the knee bends. At what rate are you expending energy? The answer, for an average-size person, is about 100 watts.[1] You're working pretty hard for those 100 watts; you wouldn't want to keep it up all day. Think of that next time you leave a 100-watt light on when you don't need it! Back at the power plant, some process must expend energy at the same rate as you with your knee bends. If the power plant consisted of human slaves turning hand-cranked electric generators, one slave would be completely occupied just keeping your light bulb burning.

The human body does other things than knee bends. Even when you are sitting still, your heart is pumping, your lungs breathing, and your brain thinking—all of which require energy. Maintaining your 98.6°F body temperature also requires energy. All in all, the human body expends energy at an average rate of about 100 watts. Strenuous exercise adds to that basic figure; some champion athletes can sustain several hundred watts (about half a horsepower) for extended periods and up to 5 kilowatts (7 horsepower) in brief bursts. But 100 watts (0.1 kilowatt, and the equivalent of a typical light bulb) is a good rough figure for the power of a human being. Through the next few chapters, we'll use the basic 100-watt human power as a standard for comparing other energetic processes. Table 6.1 lists some typical values for power.

Energy and Power

Power is the *rate* of energy use or supply. If you burn a 100-watt lamp for 2 hours, you use twice the energy you would use in 1

Table 6.1
Rates of energy use (power).

Energy user	Power
Human body	0.1 kW (100 W)
Flashlight	0.002 kW (2 W)
60-watt light bulb	0.06 kW
Car (60 mph, 30 mpg)	80 kW
Home heating, cold winter day	10 kW
Stove burner	2 kW
Microwave oven (power supplied to food)	0.6 kW
Large power plant	1,000,000 kW =1,000 MW (coal: 400 tons per hour; nuclear: 8 ounces U-235 per hour)
Sunlight on 1 square meter	1 kW
Hot shower	40 kW
U.S. oil imports, 1988	500,000,000 kW =500,000 MW
1 horsepower	0.746 kW

hour. Energy is therefore equal to the product *power* × *time*. And if the power is in kilowatts and the time in hours, the energy is in **kilowatt-hours** (kWh).[2] Example: Run a 2-kW stove burner for 1 hour, and you use 2 kWh of energy. Run it for 2 hours, and you use 4 kWh. Keep it on a full 24-hour day, and you use 48 kWh. At the typical price of 10¢ per kilowatt-hour of electrical energy, that would cost you $4.80. Suppose you did your knee bends for 15 minutes, or 0.25 hour. You are expending energy at the rate of 100 watts, or 0.1 kW. So you expend only 0.025 kWh in that 15 minutes. On the other hand, a 1,000-MW power plant puts out 1,000 megawatt-hours (MWh), or 1 million kWh, of electrical energy each hour, for a total of 24 million kWh in a day.

You can go the other way, too. A gallon of gasoline contains about 40 kWh of energy. Suppose you burn that gallon of gasoline in 2 hours. Then you use energy at the rate of 40 kWh per 2 hours, or 20 kWh per hour. And what is a kilowatt-hour per hour? Simply a kilowatt—the unit expressing *rate* of energy use. So you are using gasoline energy at the rate of 20 kW. Energy (measured in kilowatt-hours) is the "stuff," and power (measured in kilowatts) is the rate

Table 6.2

	Distance/Speed	Energy/Power
Quantity: unit	distance: miles	energy: kilowatt-hours
Rate: unit	speed: miles per hour	power: kilowatts (kilowatt-hours per hour)
Relationship	distance = speed × time	energy = power×time

at which that "stuff" is used or supplied. The relation between energy and power is like the relation between distance and speed. Distance (measured in miles) is the "stuff" a car traverses; speed (measured in miles per hour) is the rate at which the car traverses that "stuff." Table 6.2 illustrates this analogy.

People get very confused about the difference between energy and power. Part of the confusion comes from the fact that our unit of rate—power—is the single word *kilowatt,* while the unit of "stuff"—energy—is the compound word *kilowatt-hour.* In contrast, speed—a rate—has the compound unit *miles per hour,* whereas distance is simply *miles.* We will almost always be talking about power, since questions about actual amounts of energy are often irrelevant or meaningless. Consider the following:

- How much energy does a certain light bulb use? I can't answer that unless you tell me how long you are going to leave it on. What I can tell you is the *rate* at which it uses energy—and that rate is measured in watts or kilowatts.

- How much energy does a given nuclear power plant supply? Do you mean in a second, or in a year, or over its 40-year lifetime? I can tell you its rate of energy supply—for example, 1,000 megawatts—and from that you can determine how much energy it will supply in a given time.

- How much oil does the United States import? Do you mean each year? That is a rate, expressed, say, in gallons per year. If you ask about the energy content of that oil, then it is again a rate, now in kilowatt-hours per year. You could equally well express that rate in kilowatt-hours per hour, which is the same as kilowatts. That is why the entry for U.S. oil imports appears in table 6.1 as a rate, expressed in kW and MW. Incidentally, it is a big rate—

500,000 MW—equal to the electrical power output of 500 large power plants.

A common response to statements like "This light bulb uses 100 watts" or "This power plant produces 1,000 megawatts" is to ask "Is that 100 watts in an hour?" or "Is that 1,000 MW each day?" Such a response reflects confusion between energy and power: 100 W and 1,000 MW are already *rates* of energy use and supply; they don't need an additional time qualifier. Avoid such confusion. Power (watts, kilowatts, megawatts) always refers to a *rate*, and energy (watt-hours, kilowatt-hours, megawatt-hours, megawatt-years, or whatever) to the actual "stuff" being used or supplied.

Energy, rather than power, is the useful quantity when we ask about the energy stored in fuels or the energy released in a single event such as the formation of a carbon dioxide molecule, the fission of a uranium nucleus, or the explosion of a bomb. How much energy is in a gallon of gasoline? There is a definite answer: about 40 kWh. It makes no sense to ask how much power is in a gallon of gas. I could burn it in an engine for 1 hour, releasing energy at the rate of 40 kW, or I could light it with a match and see the whole 40 kWh released in 1 second. In the latter case, the power would be enormous,[3] but that enormous power output would last only a brief time. Table 6.3 gives the energy contents of some important substances. Note how the nuclear difference, introduced in chapter 2, manifests itself in the huge disparity between the energy contents of chemical and nuclear fuels.

Energy Slaves

The industrialized nations are prodigious energy consumers, especially the United States. What, exactly, does "prodigious" mean? What is our actual rate of energy consumption? Instead of answering with a cold, abstract number of kilowatts, let us put the question this way: Suppose our energy were not supplied by coal, oil, uranium, and so forth, but by slaves each working at the human body's power output of 100 watts. How many such "energy slaves" would a typical U.S. citizen need to meet his or her energy demand?

For the first humans, the answer to such a question would

Table 6.3
Approximate energy contents of various substances.

coal	7,300 kWh/ton
oil, gasoline	40 kWh/gallon
natural gas	30 kWh/100 cubic feet
ice cream	5 kWh/gallon
sugar	0.02 kWh/teaspoon
cow manure	4,300 kWh/ton
uranium (fission)	
natural abundance	850 kWh/ounce
pure U-235	570,000 kWh/ounce
deuterium (fusion)	
natural water	13,000 kWh/gallon
pure deuterium	2,600,000 kWh/ounce
Hiroshima bomb	15,000,000 kWh

have been "none." But as soon as people harnessed fire and domesticated animals, they began using more energy than their own bodies could supply. The number of equivalent "energy slaves" began to rise, and it has risen more or less steadily ever since.

You can begin to guess at the number of your "energy slaves" by thinking about your own life. Whenever you have a 100-watt light burning, that is one slave. Your car may burn an average of 1 gallon of gasoline each day; that is 40 kWh over 24 hours, or an average of 1.7 kW—17 more slaves. Suppose it is winter and your furnace is consuming energy at the rate of 10 kW. If there are four people in your family, you account for 2.5 kW, or another 25 slaves. You cook supper, picking up another few slaves. A load of laundry, washed in hot water and run through the dryer, adds a few more slaves. Occasionally you travel by plane; you must include that energy use, too. And what about the energy it took to make the plane, and your car, and to process your food, and to run the refrigerator that kept it fresh for you? And the gasoline to run the tractor to plow the field that grew the grain that fed the cow that got ground into the hamburger you cooked? Then there is the store where you shopped: banks of bright lights, open freezer cases, meat grinders, huge trucks delivering food with their engines idling as they unload. All these things use energy, and some of it in your name. Without further details, you can see that the number of your "energy slaves" is large.

You could get the average number of "energy slaves" for a U.S. citizen by adding up all the energy used in the country each year, then dividing by the number of hours in a year. The result is the total rate of energy consumption in the United States, measured in kilowatts. Divide by the population, and you have the rate of energy consumption per person. Taking each slave as 100 W or 0.1 kW, you can then find the number of "energy slaves."

The answer, for U.S. citizens in the late twentieth century, is very nearly 100 (see figure 6.1). If the energy used in our names were supplied by human slaves, each of us would have 100 slaves working round the clock, for an average power of 10,000 W or 10 kW. In other words, we use energy at 100 times the rate our own bodies can supply it. This is what it means to say we live in a "high-energy society."

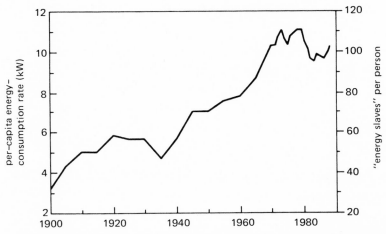

Figure 6.1 Per-capita energy-consumption rate for the United States in the twentieth century, given both in kilowatts and in equivalent "energy slaves" of 100 watts each. The per-capita energy-use rate peaked at nearly 12 kW— 120 "energy slaves"—just before the energy crisis of the 1970s, then dropped to about 10 kW with modest energy-conservation measures. The Great Depression shows as the obvious dip in the 1930s, hinting a possible relation between energy consumption and economic prosperity. (data sources: Romer, *Energy Facts and Figures* [Springside, 1985]; U.S. Department of Energy, *Annual Energy Review 1988*)

Energy and Prosperity

What does our high energy use buy us? Energy for transportation brings freedom of movement unparalleled in earlier times or in less developed societies. Our lives are easier and more comfortable thanks to a host of energy-using devices. Burgeoning electronic and optical technologies, coupled with communication satellites, use energy to spread information and entertainment at ever-increasing rates. Energy for manufacturing and distribution gives rise to unprecedented material well-being.

One might well expect energy use to correlate directly with material prosperity; in fact, one might argue that energy causes that prosperity. A way to test this expectation is to look at energy use in relation to gross national product (GNP), a traditional indicator of economic prosperity. Figure 6.1 hinted at that correlation, showing a dip in energy consumption corresponding to the Great

Depression of the 1930s. Figure 6.2 is a more direct comparison, graphing the per-capita GNPs of selected countries versus their per-capita energy-consumption rates. The most obvious feature of this graph is a strong correlation between prosperity and energy consumption: The greater your rate of energy consumption, the greater your material prosperity, as measured by your share of your country's GNP. Ethiopians, among the poorest of our planet's citizens, consume nonfood energy at the rate of only 29 watts—less than one "energy slave." Brazilians average 1.1 kW—just over ten slaves. Most of the industrialized nations are in the range of 5–7 kW, or 50–70 slaves. And the United States, among the richest and most energy-intensive of nations, averages 10 kW, or 100 slaves.

Despite the obvious correlation between energy and material prosperity, a significant question remains: What is driving what? Is enhanced GNP a result of energy use, or are the citizens of prosperous nations simply able to afford more energy and more energy-intensive technologies? Looking back at figure 6.1, do we think the Great Depression was caused by a drop in energy consumption, or was it the other way round? Certainly a high living standard in the twentieth century requires substantial energy consumption. But

Figure 6.2 Gross national product versus energy-consumption rate for selected countries. Countries above the line use energy more efficiently than average; those below, less efficiently. Multiplying the number on the horizontal axis by 10 gives the number of "energy slaves" per person in the various countries. (data source: World Bank, *World Development Report 1989*)

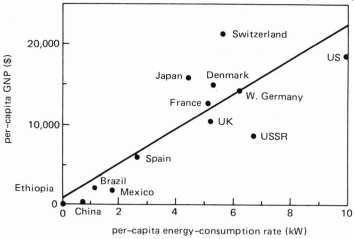

does it follow that further advances in living standards require greater energy consumption? That question must be answered as we plan for our energy future. It may well be a nuclear question, too: If we decide we must continue with increasing energy consumption, then we will need to weigh the costs and benefits of nuclear energy sources against the alternatives.

But look again at figure 6.2. What about Switzerland, Japan, and Denmark? They lie well above the line describing the average relation between GNP and energy consumption. Somehow, they are managing to produce greater GNP than their energy use would suggest; in that sense, they use energy more efficiently. The United States, the Soviet Union, and the United Kingdom, on the other hand, have lower GNPs than their energy use might suggest; they appear to be less efficient.

The Slaves' Work

What do our 100 "energy slaves" do for us? Figure 6.3 gives the answer for the United States. Energy use divides among four broad

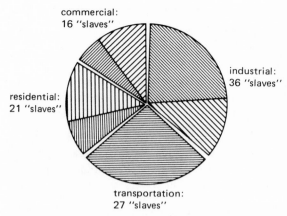

commercial:
16 "slaves"

industrial:
36 "slaves"

residential:
21 "slaves"

transportation:
27 "slaves"

Figure 6.3 Energy consumption in the United States for 1988. Coarser shading indicates energy use to generate the electricity used in each sector. The entire pie represents the average per-capita energy-consumption rate of about 10 kW, or 100 "energy slaves." (data source: U.S. Department of Energy, *Annual Energy Review 1988*)

areas: transportation, residential, commercial, and industrial. In the United States, we use just over one-fourth of our energy on transportation. Twenty-seven of your slaves are busy moving you and your goods around. Of those 27 transportation slaves, over half drive your private car; the rest are at work in public transportation or in moving goods that are ultimately for your use. Half of your residential and commercial energy slaves work to heat buildings, while others run water heaters, air conditioners, refrigerators, and other appliances. Your 36 industrial slaves are engaged in a variety of activities. Nearly half of them synthesize basic chemicals and refine metals, while the remainder process food and produce the vast array of goods available in our industrial society.

Figure 6.3 indicates the part of each sector's energy that goes to generating electricity—a total for all sectors of 36 slaves, or over one-third of all U.S. energy consumption. We are particularly interested in electricity because the only large-scale peaceful use of nuclear energy today is for electric power. As we narrow our discussion toward nuclear energy, we need to focus on electricity and, specifically, on nuclear-generated electricity.

Electrical Slaves

Of the first electric power plants, built before 1900, about half were powered by fossil fuels (coal, oil, gas) and about half were hydro-electric (water-powered) units. The abundance of cheap fossil fuels, and the convenience of siting fossil-fuel power plants, soon made fossil fuels the dominant source of electric power in most parts of the United States. Coal has always been the staple fuel for most electric-power generation, although oil made a more significant contribution before the price increases of the 1970s sent utilities scrambling to convert from oil to coal. Nuclear fission came on the commercial scene in the late 1950s, and its share of U.S. and world energy production has risen more or less steadily since then. By 1990, nuclear power plants supplied about 20 percent of the United States' electricity. Figure 6.4 traces the history of electrical energy in the United States since 1950, and figure 6.5 shows the mix of electrical-energy sources in 1988.

Figure 6.3 showed that about 36 percent of your average energy consumption—36 slaves—is due to the generation of electricity. Figure 6.5 shows that today most of those are coal slaves; only 7 of your slaves (the nuclear 20 percent of your 36 electric slaves) are in the nuclear business. (That number varies considerably; the Pacific Northwest, for example, has abundant hydroelectric resources, and gets only 9 percent of its electricity from nuclear sources. New England, on the other hand, averages 35 percent nuclear electricity.) But the nuclear share has been growing, and it will continue to grow if we as a society decide that we need more energy and that nuclear fission is environmentally or economically preferable to other sources. This is a big nuclear choice we face in the coming years.

We have looked mostly at energy use in the United States. As figure 6.2 showed, however, other industrialized nations are also substantial energy users. For some, less rich in fossil fuels than the United States and unambivalently committed to a nuclear future, the nuclear portion is higher than in the United States. France leads the world, with 70 percent of its electricity generated by nuclear fission in 1988. Figure 6.6 shows the proportions of nuclear electricity in several industrialized nations.

Nuclear News: Electricity Is Cheap

The news media are full of stories about increases in the price of electricity. Some examples, all from recent editions of the *Wall Street Journal,* follow:

April 7, 1987: "El Paso Electric Company Seeks Rate Increase"
April 29, 1987: "Ohio Edison Smooths Path for 9% Rate Boost"
June 11, 1987: "Detroit Edison Seeks Rate Increase of 6.5%"
September 29, 1988: "Power Unit Will Request $96.5 Million Rate Boost"
November 14, 1988: "Tucson Electric Seeks 22% in Rate Increases"
December 19, 1988: "Electric Utilities Unit in Mass Seeks Rate Increase"
August 21, 1989: "Central Vermont P.S. Seeks Rate Increase"

Consumer groups and state regulatory boards often complain and contest the utility companies' rate-increase requests:

Wall Street Journal, May 5, 1987: "Missouri Officials Ask Reduction in Rates"
Washington Post, May 13, 1987: "Pepco Told to Cut Rates 5.2%"
Wall Street Journal, February 17, 1988: "Rate Cut Proposed for Commonwealth Edison"
Wall Street Journal, September 21, 1987: "Electric Bills Hinder Chicago's Rebound"
New York Times, November 6, 1987: "State Utilities Panel Denies Lilco Rate Increase"
Los Angeles Times, July 29, 1989: "After 16 Years of Fighting for Lower Utility Rates . . . Outspoken Advocate . . . Will Be Missed"

Is electricity really costly? It *is* the most expensive form of energy available on a large scale, for good technical and economic reasons. In the United States, the actual cost of electricity in 1990 averaged roughly 10¢ per kilowatt-hour. Suppose you had to hire people to make your electricity, using the power of their own bodies. What would their wages be if they produced electricity costing 10¢/kWh?

The average human power output is 100 watts, or 1/10 kW. At that rate, your energy servants would each have to work for 10 hours to produce one kilowatt-hour of electrical energy. With 10¢ worth of electricity to show for 10 hours' labor, their wages would be only 1¢ per hour. Electricity is cheap!

Of course, the low cost of electricity and other forms of energy is what makes our energy-intensive society possible. If we had to pay as much for energy as it would cost to hire workers at minimum wage to produce it for us, we would surely not each be using energy at the equivalent rate of 100 human bodies.

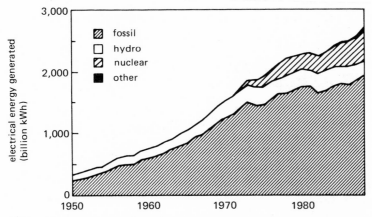

Figure 6.4 Sources of electrical energy generated in the United States from 1950 to 1988. The category "other" (thick line at top) includes geothermal, solar, wood, wind, and power plants fueled by waste products. (data source: U.S. Department of Energy, *Annual Energy Review 1988*)

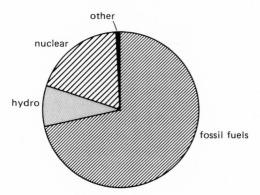

Figure 6.5 Sources of electrical energy generated in the United States in 1988. The bulk of the energy is from fossil fuels, predominantly coal. Nuclear fission supplied 20 percent; sources in the category "other," including geothermal, solar, wood, wind, and power plants fueled by waste products, supplied only 0.4 percent. (data source: U.S. Department of Energy, *Annual Energy Review 1988*)

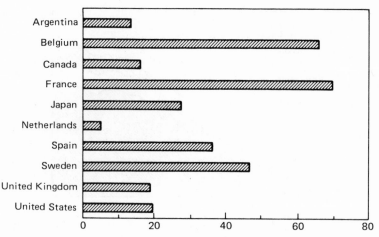

Figure 6.6 Percentage of electricity generated by nuclear sources in selected countries. France, with a strong commitment to an independent nuclear program, leads the world in its dependence on nuclear power. (data source: U.S. Department of Energy, *Commercial Nuclear Power 1989*)

Summary

Citizens of modern industrialized societies use energy at enormous rates—many times the 100 watts a human body can supply. That statement—like many others about energy—is expressed in terms of power, or *rate* of energy use. Power is measured in watts (and kilowatts, and megawatts), while energy itself is measured in watt-hours (and kilowatt-hours, and megawatt-hours). For the United States, the per-capita energy-consumption rate is about 10 kW, or 100 times the power of the human body. About 36 percent of that energy goes to make electricity, and about 20 percent of the electric share is provided by nuclear fission. In countries more dependent on nuclear power than the United States, the nuclear share is as high as 70 percent.

In the remaining chapters of part II, we wil look in detail at the nuclear power plants used to generate electricity throughout the world, and will consider the safety, environmental, and other issues that surround their operation. Those issues become increasingly important as concern for the global environment makes fossil fuels decidedly unattractive, and as a world population approaching 6 billion strives toward the higher material standards of the energy-intensive industrial societies.

Notes

1. If you have had a high school physics course, you can easily estimate this power. As I do knee bends, I raise the upper part of my body about 10 inches, or about 0.25 meters. Suppose about 2/3 of my 145 pounds (about 66 kilograms) is involved. Then the energy involved in raising that mass m a distance $h=0.25$ meters against the $g=9.8$ newtons/kilogram force of gravity is mgh, or

$(2/3) \times (66 \text{ kg}) \times (9.8 \text{ N/kg}) \times (0.25 \text{ m}) = 108 \text{ joules}.$

I do this once a second, and one watt is one joule per second, so my power output is 108 watts.

2. Other energy units you may know of include the calorie, used to describe the energy in foods and chemical reactions; the joule, official energy unit of the international scientific community; the electron-volt (eV), used in atomic and nuclear physics; the British thermal unit (Btu), used by the U.S. engineering community and building trades; and the megaton, a unit for the explosive energy of nuclear weapons. All these units measure the same thing, namely energy.

3. One second is 1/3,600 of an hour, so it would be

$40 \text{ kWh}/(1/3{,}600 \text{ hour}) = 144{,}000 \text{ kW}$, or 144 MW.

That is about 15 percent of the output of a large power plant!

Further Reading

Gordon Aubrecht, *Energy* (Merrill, 1989). Textbook for a general introduction to energy, with chapter 3 particularly relevant to basic concepts of power and energy.

Robert Romer, *Energy Facts and Figures* (Spring Street Press, 1985). A wealth of data on energy use and supply, energy content of fuels, Earth's energy balance, and energy resources.

Robert Romer, *Energy: An Introduction to Physics* (Freeman, 1976). Unfortunately out of print, this book provides an excellent introduction to the concept of energy in relation to energy use in industrialized society. The voluminous tables of energy data have been extracted and updated in Romer's *Energy Facts and Figures*.

U.S. Department of Energy, Energy Information Administration, *Annual Energy Review*. Detailed statistics, in both graphic and tabular form, on U.S. energy consumption by source and by end use.

U.S. Department of Energy, Energy Information Administration, *International Energy Annual*. Statistics on world energy consumption and supply.

Glossary

kilowatt (kW)   A unit of power, equal to 1,000 watts.

kilowatt-hour (kWh)   A unit of energy. A device using energy at the rate of 1 kilowatt uses 1 kilowatt-hour of energy each hour.

megawatt (MW)   A unit of power, equal to 1,000,000 watts or 1,000 kilowatts.

power   The rate at which energy is used or supplied; measured in watts or kilowatts.

watt   A unit of power.

Making Electricity

7

Electrical energy today is produced almost exclusively in large, centralized power plants that send their product over a complex network of transmission lines to homes, industries, and other users. Every outlet in your house is connected to that network, and to all its implications. Flip a light switch and you may be asking coal miners to dig a little deeper and maybe to die a little sooner from "black lung" disease. Turn on your electric stove and you may contribute to the acid-rain-causing emissions from coal-burning power plants. If you have an electric water heater, your morning shower may result in the formation of radioactive waste and in the production of the bomb material plutonium. Install an air conditioner and you may accelerate the damming of rivers and the displacement of native populations in Canada's northern wilderness.

As a citizen of a high-energy society, you and the electrical devices you use are directly responsible for the unpleasant aspects of electric-power generation. Does it have to be this way? Are large, centralized power plants the most efficient and the most environmentally sound way to make electricity? Must those plants generate air pollution or radioactive waste, or wreak havoc with our rivers? And do we really need all that energy, anyway? We will confront those questions in chapter 11. For now, let us examine the generation of electricity as it is done today: with large power plants.

Electric Generators

The electrical energy that flows along wires to power our homes and industries is only one of the many forms energy can take. A variety of processes produce electricity from other forms of energy. In a battery the energy stored in chemical compounds is transformed into electrical energy. Solar cells convert the energy of sunlight directly into electricity. Heat energy can also be converted directly, albeit inefficiently, into electricity. Spacecraft traveling to the outer planets, where sunlight is too weak for power, carry radioactive materials whose decay heat produces electricity. Squeezing certain crystals generates electricity, an effect put to use in "matchless" fire starters. But in large-scale electric-power generation, one method dominates: the conversion of mechanical to electrical energy in devices known as **electric generators**.

The principle of the electric generator is a simple one, based on a fundamental law of physics. That law links electricity and magnetism, stating that electric current arises when an electrical conductor and a nearby magnet move relative to each other. (It doesn't matter whether the magnet or the conductor moves—a manifestation of Einstein's theory that only relative motion matters.) In principle, you could connect the two ends of a piece of wire to a light bulb and wave a magnet around near the wire, and the bulb would light. If you did the experiment with a sensitive meter instead of a light bulb, you would see the meter's needle move, indicating an electric current (figure 7.1). To get a bulb to light, you would have to move the magnet very rapidly indeed, but in principle you could do so.

A magnet waved around near a wire is a simple electric generator, but a more effective generator consists of a wire loop rotated between the poles of a strong magnet. In practice, the loop is a coil made up of many turns of wire. Brushes contact the rotating structure and convey the electric current to stationary wires, which carry it where it is needed. Figure 7.2 shows the essential features of an electric generator.

The generator principle, incidentally, is used not only to produce power. A video or audio cassette contains a thin ribbon of magnetized material that moves past a wire coil, the so-called head

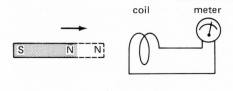

Figure 7.1 The principle of the electric generator. (above) Moving a magnet near coil of wire produces an electric current in the wire, indicated by deflection of a meter. (below) It is the motion that is important; holding the wire still near the magnet gives no current.

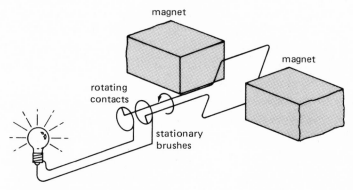

Figure 7.2 An electric generator. A wire loop rotates between magnetic poles, giving rise to an electric current. Conducting brushes contacting the circular rings convey the current from the rotating loop to wires that carry it where it is needed—in this case, to a light bulb.

of the VCR or cassette player. This motion generates minute electric currents that correspond to the information stored in magnetic patterns on the tape; the current is then amplified to drive a video display or loudspeakers. Computer disks work in much the same way, spinning under a wire-coil head whose electric currents mimic the magnetization pattern that holds the disk's information. So does the magnetic strip on your credit card, scanned by a merchant's verification device to see if your credit is good.

To make electricity with the generator of figure 7.2, you have to spin the coil. And that may not be easy. Why not? Because electrical energy is leaving the generator, in this case to light a lamp. And where does that energy come from? It comes from whomever or whatever is turning the generator. You can't get something for nothing! If the generator is producing electrical energy, it must be supplied with mechanical energy to keep it turning. The generator does not *make* energy; it only converts energy from mechanical to electrical form.

If you have ever had the opportunity to turn a hand-cranked electric generator, then you've felt with your muscles the need to supply the generator with energy. First turn the generator when it is not connected to anything; it spins easily. Then throw the switch that connects it to a 100-watt light bulb. Ugh! It suddenly gets very hard to turn. The generator "knows" it is being made to light the lamp, as evidenced by its being hard to turn.[1] You can see this same effect by switching on your car's headlights as the engine is idling slowly. You will notice a slight drop in engine speed (the engine may even stall if it is idling very slowly). The car's electric generator, which is turned by the engine, suddenly has to produce energy to run the headlights, so it gets harder to turn, and the engine slows down.

The same effect applies to the large generators in power plants. If it didn't, the power companies could set their generators spinning and they would happily produce power forever without burning any coal or fissioning any uranium. But the generators in power plants are hard to turn—very hard, since they power entire cities. And they get harder to turn as they are called upon to produce more power. When you flip on a 100-watt lamp, a generator on your power network gets a little harder to turn. (How much harder? As much harder as it was for you to turn that hand-cranked gen-

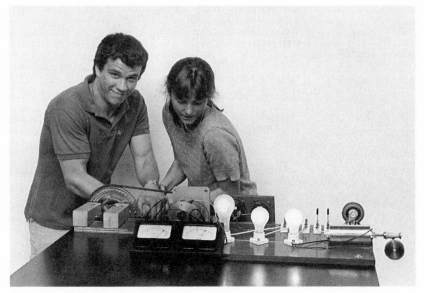

Figure 7.3 A hand-cranked generator. With 200 watts of light bulbs connected, two people work hard to keep the generator turning and the bulbs lit. (photo by Erik Borg)

erator, or to do your 100-watt knee bends.) To keep the generators turning steadily, the power companies have to burn more coal, or fission more uranium, or let more water over a dam. You can't turn on your light without something like that happening. (No one actually makes the decision but you; automatic mechanisms keep the generators turning at constant speed, and regulate the burning, fissioning, or water flow to maintain that speed.) The fourteen 110-car trainloads of coal delivered each week to the power plant in figure 2.1 show just how hard it can be to keep a big generator turning.

So how are the generators made to turn? The simplest way is through direct application of mechanical energy, most commonly the energy of moving water. Hydroelectric power plants, built at natural waterfalls or artificial dams, use the power of moving water to spin fan-like turbine blades (figure 7.5). The turbines connect directly to the rotating-coil structures of electric generators. Hydroelectric power plants include some of the smallest and some of the largest facilities in the electric-power industry, with power outputs ranging from a few hundred kilowatts (a few thousand

Figure 7.4 This 678-megawatt turbine-generator at Boston Edison's Pilgrim nuclear power plant works on the same physical principles as the generators shown in figures 7.2 and 7.3. (Boston Edison)

"slaves") to thousands of megawatts (tens of millions of "slaves"). In Norway, Sweden, Canada, and many developing nations, hydropower is the dominant source of electricity. In the United States, however, hydropower supplies only about 8 percent of our electricity—less than half of nuclear power's share.

Thermal Power Plants

Of the 92 percent of the United States' electrical energy that is not hydroelectric, nearly all is produced in **thermal power plants**. These are essentially huge steam engines in which a heat source boils water to make high-pressure steam. The steam spins a turbine connected to an electric generator. The heat source can be burning coal, oil, or gas, or it can be nuclear fission. The basic operation of a thermal power plant, and even many of the details, are the same whether the plant is fossil-fuel or nuclear.

Figure 7.5   A hydroturbine being installed at Wheeler Dam in Alabama. (Tennessee Valley Authority)

Nuclear News: Converting a Power Plant

In 1984, Consumers Power Company of Michigan found itself facing financial disaster. The company had sunk $4.1 billion into a nuclear power plant, which was 85 percent complete. But the plant's reactors, built on weak foundations, were sinking into the ground. Regulatory and financial difficulties compounded the situation. In June 1984, construction on the nuclear plant was halted.

Then, in 1986, the troubled plant got a new lease on life. A number of companies joined in a consortium to convert the facility to a gas-fired power plant. Clever financial arrangements helped ensure profitability. But so did the nature of thermal power plants: Much of the former nuclear plant's equipment—$1.5 billion worth, including the turbine-generators—would work in a fossil-fuel plant. In the reincarnated plant, gas-heated steam spins turbines originally meant to be driven by steam from nuclear fission. And the plant's owners enjoy financial success that stems in part from the similarities between nuclear and fossil-fuel power plants.

News source: "Nuclear Write-Off to Success Story," *New York Times*, September 25, 1989.

Figure 7.6   A simplified diagram of a thermal electric power plant.

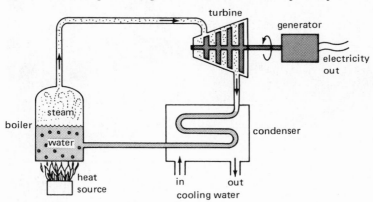

Figure 7.6 shows the essential features of a thermal power plant. The boiler might be a large vessel with an outlet pipe for steam and an inlet pipe for water, as depicted; or it might consist of a great many thin pipes carrying water through the heated region. Steam leaves the boiler at high pressure. It hits the small-diameter turbine blades, then expands, cools, and slows through a sequence of larger blades engineered to extract the maximum energy from the steam (figure 7.7). The turbine turns the generator, and out goes electricity.

What happens to the steam after it has given up as much energy as it can? It eventually returns to the boiler, but first it must be converted back to water. That is the job of the **condenser**, in which steam pipes come into contact with cool water (usually pumped in

Figure 7.7 Turbine assembly removed from a 820-megawatt turbine-generator. Note the smaller blade diameter at center, where the high-pressure steam enters. (Pennsylvania Power & Light Company)

from a river, a lake, or a bay). The steam condenses, giving up energy to the cooling water. To prevent ecological disturbance, the heated water is cooled before being returned to its source. Often the cooling is done in the enormous cooling towers that, for many people, symbolize nuclear power. Those towers are as likely to be found at fossil-fuel plants, since any thermal power plant needs a condenser and a cooling system.

The cooling system dumps a lot of energy into the environment—energy wrung from expensive coal, oil, gas, or uranium. Isn't that a huge waste? It is. In fact, it is an enormous waste. *A typical thermal power plant dumps about two-thirds of the energy released from the fuel into the environment as waste heat.* Think about that! It means that only one-third of the energy in the fuel ends up as useful electricity. A 1,000-MW power plant is really a 3,000-MW plant; however, 2,000 MW go up the cooling tower as waste heat.[2] So

---

Figure 7.8 Pennsylvania Power & Light Company's Martins Creek power plant. Clouds rise from the 414-foot-high cooling towers, where water from the condenser is cooled by updrafts of air before returning to the Delaware River. This plant burns coal and oil to produce 1,940 megawatts of electric power. (Pennsylvania Power & Light Company)

---

when you turn on your 100-watt lamp, it isn't really one slave you press into service, it is three—one to light the lamp and two to make waste heat. Of your 36 electrical slaves, only about 12 actually supply you with electricity; the other 24 just produce waste heat.

What is wrong? Are our engineers so dumb they can't make an efficient power plant? Is there too much friction in the generators? The problem lies deeper than that. Even a perfectly constructed thermal power plant can't approach 100 percent efficiency at converting heat to electricity. To understand why, we need to consider energy's quality as well as its quantity.

Energy Quality

If I offer you a kilowatt-hour of energy, would you rather have that energy in the form of electricity or as a tank of hot water? If you want to take a bath, either form will do; you can turn 1 kWh of electricity into 1 kWh of heat simply and without any loss of energy, or you can use the hot water directly. But suppose you want to run the electric motor in your washing machine. You can do that with electricity, but not with hot water. In that sense, electricity is a more useful and therefore a higher-quality form of energy.

This notion of energy quality is the subject of one of the most famous laws of physics: the **second law of thermodynamics**. The name sounds complicated, but the second law is really very simple, and it governs much of what happens in our lives. Fundamentally, the second law of thermodynamics says this: Left to themselves, things get more chaotic, not more organized.[3] Scrambling an egg is a manifestation of the second law. You beat the egg until its white and its yolk disappear in a uniform yellow blend. Keep beating, even run the beater in reverse, and you will never see the white and the yolk reemerge. The scrambled state wins out over the original ordered state. Or take glasses of cold and hot water, and put them next to each other. Come back later and both are lukewarm. A long time later, they're still lukewarm. They will never spontaneously return to the original hot/cold state, even though the total energy is the same either way.

The example of the water glasses further elucidates the notion of energy quality. Suppose you need two cups of lukewarm water. You could draw two cups, lukewarm, from the faucet, or you could mix hot and cold water. But if you need hot water, you can't get it from lukewarm without supplying additional energy. High temperatures represent higher energy quality, in that you can go readily from hotter to cooler but you cannot efficiently go the other way. So there is a hierarchy of energy qualities, with electricity (and motion) at the top, progressing downward through ever lower temperatures (figure 7.9).

What does all this have to do with power plants? A thermal power plant starts with heat energy, released from burning or fissioning fuel, and ends up with electricity. Doesn't that violate the second law of thermodynamics? It would, *if* all the relatively low-quality heat ended up as highest-quality electricity. But the second law doesn't say you *can't* go from lower-quality to higher-quality energy; it just says you can't do it with 100 percent efficiency. You can go from low to high quality, but you lose energy in the process. That is exactly what happens in a power plant, where a great deal of the heat energy is lost as waste heat, unable to spin the turbines and make electricity.

But need we lose fully two-thirds of the energy? The second law speaks to that, too. It says that the maximum possible efficiency of a power plant depends on the highest and lowest temperatures

Figure 7.9 Higher-quality energy is more valuable, since it can be transformed without loss into energy of lower quality, whereas lower-quality energy cannot be transformed with 100 percent efficiency to energy of higher quality.

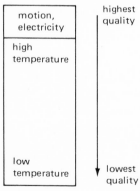

available. The lowest temperature is set by the environment, so we can't do much about that. But we could raise the highest temperature—the temperature in the boiler—and thereby raise the efficiency. Unfortunately, limitations on the materials used in power-plant construction set a maximum temperature of about 1,000°F for the thin-tube boilers used in fossil-fuel power plants, and about 600°F for the large pressure vessels in most U.S. nuclear plants. Coupled with less fundamental inefficiencies, these maximum temperatures lead to typical power-plant efficiencies between 30 and 35 percent. So our statement that thermal power plants convert only one-third of their fuel energy to electricity is approximately correct, although nuclear plants are usually slightly less efficient than their fossil-fuel counterparts.

Isn't there anything that could be done with the waste heat, instead of simply dumping it into the environment? Indeed, waste heat from power plants constitutes a valuable source of energy that could be used wherever the lesser quality of low-temperature heat is adequate. For example, waste heat could be circulated through city buildings to provide central heating, and spent steam from turbines is still useful in a number of industrial processes. In the United States, we rarely make such use of the waste heat from power plants. In many European countries, though, power plants are often designed and sited so their waste heat can be used. That accounts in part for the differences in energy efficiency that were discussed in the preceding chapter.

The process of generating electricity in a thermal power plant and simultaneously using the waste heat is called **cogeneration**. Although few large power plants in the United States practice cogeneration, many industries have begun to build their own smaller units to provide both electricity and heat. Michigan's nuclear-to-gas conversion, discussed earlier in this chapter, is an exception for the United States: a large power plant that produces electricity for a commercial utility while supplying steam for industrial uses at the Dow Chemical Company.

Differences in energy quality have significant implications for wise use of energy. Suppose, for example, that you are shopping for a new water heater. Gas and electric heaters cost about the same, and their operating expenses may not be too different. And both seem efficient: When you burn gas, nearly all the energy stored in the gas ends up as heat in the water; when you run an electric water heater, nearly all the electrical energy ends up in the water, too. But that is not the whole story. Back at the electric-power plant, with its roughly 33 percent efficiency, the amount of fuel consumed was 3 times what it would have taken to heat the water directly. It is a waste to convert that hard-won, highest-quality electrical energy into low-grade heat; where heat is needed, it makes more sense to burn a fuel directly and thereby put essentially all its energy content to good use.

Summary

In the United States, about 36 of each individual's 100 "energy slaves" work to produce electricity. Nuclear fission accounts for 20 percent of that electricity, so each of us has only 7 nuclear slaves. Nuclear electricity, like that made from fossil fuels, is generated in thermal power plants. The fuel in these plants boils water to steam, and the steam turns a turbine attached to an electric generator. By spinning a wire coil in the presence of a magnet, the generator produces electric current. How difficult it is to turn the generator depends on how much electric power it is called upon to produce. Every time you turn on an electric appliance, a generator somewhere gets harder to turn, and more energy must be supplied to keep it turning.

The second law of thermodynamics shows that electricity is energy of the highest quality, and that it is not possible to convert lower-quality heat into electricity with 100 percent efficiency. The efficiency of thermal power plants now in operation is only about 33 percent. This means that two-thirds of all the energy released from the fuel is dumped to the environment as waste heat. That energy could be recovered for applications requiring low-temperature steam or hot water, but it cannot be used to produce more electricity.

Notes

1. How does the generator "know"? That involves another link between electricity and magnetism. The electric current in the generator coil makes the

coil behave like a magnet, and repulsive forces between the coil "magnet" and the stationary magnet make it hard to move the coil. The greater the power being produced, the greater the electric current—and therefore the greater the force and the harder it becomes to turn the generator.

2. Sometimes you will encounter the abbreviations MW_e and MW_{th}, for "megawatts electric" and "megawatts thermal." The former gives the electric-power output of a power plant, the latter the total power released from the fuel as thermal energy (including both the energy that ends up as electricity and the waste heat).

3. You may have heard of *entropy,* a physicist's word for disorganization. The second law says that the entropy of a closed system—such as the entire universe—can never decrease. If you do succeed in making hot and cold from lukewarm, you may decrease the entropy in your kitchen; but back at the power plant that made the electricity that ran your refrigerator, there was a more-than-compensating increase in entropy.

Further Reading

Christopher Flavin, *Electricity's Future: The Shift to Efficiency and Small-Scale Power* (Worldwatch Institute, 1984). This brief report argues that large-scale electricity-generating stations are a thing of the past, and that the future will bring a greater diversity of smaller generating units.

Marc Messing, Paul Friesema, and David Morell, *Centralized Power* (Environmental Policy Institute, 1979). A thorough study of the history of electric-power generation and the structure of the electric-power industry in the United States. Case studies compare large and small power plants and alternative energy sources. Tables and charts cover many aspects of electric power.

U.S. Department of Energy, Energy Information Administration, *Inventory of Power Plants in the United States* (Government Printing Office, published annually). A listing, by state, of all commercial electric-power generating facilities in the United States, including all energy sources: nuclear, fossil fuels, garbage, hydropower, wind, and solar. Skimming the listings gives a fascinating profile of the U.S. electric utility industry.

Glossary

cogeneration   Generating electricity while utilizing the waste heat.

electric generator   A device that converts mechanical into electrical energy by moving electrical conductors in the presence of a magnet.

energy quality   A characteristic of energy that determines how useful it is for performing a variety of tasks. Electricity and motion, the highest-quality forms of energy, can be used fully for any task requiring energy. High-temperature heat energy is lower in quality, and low-temperature heat is lowest. Lower-quality energy cannot be converted to higher-quality with 100 percent efficiency.

second law of thermodynamics The statement that things tend toward more chaotic states. Its implications include the inability of energy to flow from lower-quality (cooler) to higher-quality (hotter) forms, with the highest energy quality in the form of electricity or motion. The second law sets fundamental limits on the efficiency of thermal power plants.

thermal power plant A power plant in which a heat source—usually burning fuel or nuclear fission—boils water to steam that turns a turbine connected to an electric generator.

Nuclear Reactors

8

In a nuclear power plant, fission provides the heat that boils water to drive the turbine-generator system. We examined the basic process of nuclear fission in chapter 5, where we saw how the splitting of a fissile nucleus—uranium-235 or plutonium-239—gives rise to two highly radioactive fission product nuclei, several neutrons, and a great deal of energy. Most of the energy takes the form of rapid motion of the fission products. These collide with particles in the surrounding material, producing the heat that ultimately turns the generator.

We saw in chapter 5 how fission can be sustained by means of a chain reaction, in which neutrons released in fission trigger additional fission. In a power plant, we want the chain reaction to be just critical, meaning that an average of exactly one neutron from each fission causes another fission; that way, the rate of energy release remains steady. A supercritical chain reaction, illustrated in figure 5.7, would be disastrous in a power plant. With more than one neutron from each fission causing a subsequent fission, the energy release would grow rapidly and would soon outstrip the power plant's ability to utilize the energy. The result would be violent disruption of the power plant. That is exactly what happened in the Chernobyl accident.

The essential goal in running a nuclear power plant, then, is to maintain a steady chain reaction. At the heart of the power plant is its **nuclear reactor**, a system engineered to sustain a controllable chain reaction. Understanding the many controversies surrounding nuclear power requires a close look at the workings of the several varieties of nuclear reactors.

Neutrons, Fast and Slow

When a neutron hits a uranium nucleus, will it always cause fission? We have already seen part of the answer: for low-energy neutrons (also called **slow neutrons**), only the isotope U-235 will undergo fission. Nuclei of the more common isotope U-238 may absorb a neutron, eventually becoming plutonium-239. High-energy neutrons (**fast neutrons**) will also fission U-235 and, if fast enough, may even fission U-238. But the likelihood of fission varies drastically with neutron energy. For U-235, slow neutrons are several thousand times more likely than fast neutrons to cause fission. Thus it is much easier to achieve a chain reaction if the high-speed neutrons emitted in fission can be slowed. In contrast, U-238 fission requires extremely fast neutrons—even the fast neutrons emitted in fission aren't fast enough. This is why it is not possible to sustain a chain reaction with U-238. Table 8.1 summarizes the likelihood of fission with different neutron speeds.

Because slow-neutron fission of U-235 is so much more likely, a chain reaction employing slow neutrons can be sustained in uranium even when a great deal of nonfissile U-238 is present. Even though the U-238 absorbs many of the neutrons, those slow neutrons that do strike U-235 nuclei are very effective in causing

Table 8.1
Fission probability versus neutron speed.

	Neutron speed		
Isotope	Slow	Fast (as emitted in fission)	Very fast (faster than emitted in fission)
U-235	Extremely likely	Possible	Possible
U-238	Impossible	Extremely unlikely	Possible

fission. A slow-neutron chain reaction may therefore work in uranium with U-235 at its natural abundance of only 0.7 percent, although in many reactors slight enrichment with U-235 is necessary. The use of natural or slightly enriched uranium in power-plant reactors is important for two reasons: First, enrichment is an expensive process. Second, natural or slightly enriched uranium cannot be used directly to make nuclear weapons.

The fact that most power reactors use slow neutrons provides an answer to an often-asked question: Can a nuclear reactor blow up like a bomb? The answer, for a slow-neutron reactor, is a definitive No. A bomb gets its destructive power not only from the sheer amount of energy released but also from the suddenness of that release; a typical nuclear bomb explosion is over in a millionth of a second. In a bomb, the chain reaction is sustained by fast neutrons that spend only about ten billionths of a second between being released in one fission event and striking a nucleus to cause another fission event. In a reactor, neutrons take 10,000 times longer between fission events, so even a reactor that was badly out of control could not undergo a bomb-like explosion. That is not to say that a nuclear reactor could not fail catastrophically or even explosively; however, even the most violent reactor disaster would not have an explosive effect approaching that of a nuclear weapon.

In a nuclear power reactor, slow neutrons sustain the fission chain reaction. But the neutrons released in fission have high energy—they are fast neutrons. Slowing them down is the job of the **moderator**, a substance that absorbs neutron energy. At the microscopic level, collisions between neutrons and nuclei in the moderator are what slow the neutrons.

When objects collide—whether they be subatomic particles, billiard balls, or cars—the most effective energy transfer occurs if the objects have similar masses. A good moderator, then, is a substance whose nuclei have about the same mass that neutrons have. The single proton in a hydrogen nucleus has essentially the same mass as the neutron, so hydrogen is a good moderator. That makes water (H_2O) an appropriate moderator substance, and indeed water is the moderator in nearly all U.S. power reactors. But hydrogen has one failing: It often absorbs neutrons. A moderator should absorb neutron *energy*, but not the neutrons themselves; otherwise it reduces the number of neutrons available to sustain the chain reaction. This loss of neutrons means that water-moderated reactors cannot run on unenriched, natural uranium; there simply aren't enough U-235 nuclei around to sustain a chain reaction with the reduced number of neutrons. So fuel for U.S. power reactors must be enriched, to about 3 percent U-235, to overcome the effect of neutron absorption in the moderator. The expense of enrichment is balanced by the abundance of the moderator substance, water.

After ordinary hydrogen (^1_1H), the next heaviest nucleus is deuterium (^2_1H, also called D), the hydrogen isotope whose nucleus contains a proton and a neutron. Since its mass is twice that of the neutron, deuterium isn't as efficient at absorbing neutron energy. But it makes up for that inefficiency by having a very low probability of absorbing neutrons themselves. Neutrons in a deuterium-moderated reactor are therefore so abundant that the reactor can run on natural, unenriched uranium despite its low (0.7 percent) proportion of fissile U-235. Since it is chemically similar to ordinary hydrogen, deuterium also combines with oxygen to make water—in this case **heavy water** (^2H_2O, or D_2O). Heavy water is present in minute quantities in ordinary water, and can be separated and used as a reactor moderator. The Canadian nuclear program emphasizes **heavy-water reactors**, or HWRs. (U.S. reactors, in contrast, are called **light-water reactors**, or LWRs.) Heavy water has been considered a "sensitive material" since the dawn of the nuclear age, since a nation possessing it can make plutonium for bombs without needing enriched uranium.

In some reactor designs, solid graphite—the common form of carbon used in pencil "lead"—serves as the moderator. The carbon nuclei in graphite are much more massive than neutrons but, like deuterium, exhibit very little neutron absorption. Graphite-moderated reactors are used in a number of European power plants;

Figure 8.1 Locations of commercial nuclear power reactors in operation in the United States as of 1990. (U.S. Department of Energy/Energy Information Administration)

most graphite reactors in the United States produce plutonium for nuclear weapons.

People are often confused about the role of the moderator in a nuclear reactor. The name suggests that it moderates—that is, controls or tempers—the nuclear reaction. In fact, the opposite is true: The moderator makes the nuclear reaction go. Without the moderator, only fast neutrons would be present. In reactor fuel, with its low proportion of fissile U-235, those fast neutrons could not sustain a chain reaction. What the moderator does moderate is the speed of the neutrons; the slow neutrons that result are more effective at causing fission.

Controlling the Chain Reaction

To keep the chain reaction running steadily, something must maintain the delicate average of exactly one neutron from each fission causing another fission. If that number—called the **multiplication factor**—drops below 1, the reaction will fizzle to a halt; if it goes over 1, the reaction will quickly go out of control.

To grasp the disastrous consequences of a rise in the multipli-

Nuclear News: Heavy Water

In the winter of 1943, six Norwegians working with the British army staged a daring commando raid that destroyed what was then the world's only commercial heavy-water production facility, in Nazi-occupied Norway. The Germans had chosen heavy water to moderate the reactors of their fledgling nuclear weapons program, and the raid set that program back at least a year.

Half a century after the Norwegian raid, heavy water remains a sensitive material. In 1983, 15 tons of Norwegian heavy water disappeared from a shipment intended for West Germany. Norwegian officials claim the material was diverted to India. Although it denies the allegations, India is suspected of using the heavy water in reactors producing plutonium for nuclear weapons. Indeed, India's sole nuclear explosion, in 1974, used plutonium from a Canadian-made research reactor whose heavy-water moderator had come from the United States. Because of heavy water's potential use in weapons production, the Nonproliferation Treaty requires that the International Atomic Energy Agency (IAEA) be notified of all heavy-water shipments in excess of a ton. But India has not signed the treaty, and the alleged diversion of Norwegian heavy water was not reported to the IAEA. The 1983 heavy-water episode appears not to be an isolated case: In 1990 Romania revealed that its former Communist government had assisted in the diversion of a 1986 Norwegian shipment of heavy water to India.

News sources: "Norway Is Missing Atom Arms Water," *New York Times,* May 4, 1988; "Norway Details How Heavy Water Went to India," *New York Times,* May 7, 1989; "Romania Is Reported in Nuclear Deal with India," *New York Times,* April 30, 1990.

cation factor, consider a nuclear reactor running at its rated power level of 3,000 MW. Suppose the multiplication factor changes from 1 to 2. Then, in the time—called the **generation time**—that it takes neutrons from one fission to cause another, the rate of fissioning will double, and the reactor's power will increase to 6,000 MW. Another generation time and it will be 12,000 MW; one more, and it will be 24,000 MW—8 times what the power plant is designed to handle, and sure to cause violent destruction. A multiplication factor of 2 is unrealistic, but suppose it gets just a little above 1— say, 1.01. Then after one generation the power level will be 3,000 MW × 1.01, or 3,030 MW. After two generations, it will be 3,030 MW × 1.01, or 3,060 MW. Carrying this procedure through 10 generations gives a power level of 3,314 MW (still tolerable) but after 100 generations the power has nearly tripled, exceeding 8,000 MW. And after 500 generations, the reactor power has risen to nearly 150 times its design value. Since the generation time is only a fraction of a second, a multiplication factor even a little over 1 can quickly lead to disaster.

Controlling a nuclear reactor means keeping the multiplication factor—the number of neutrons from each fission that cause subsequent fission—at *exactly* 1. This is done by inserting neutron-absorbing material, in the form of **control rods**, between the uranium fuel rods. If the multiplication factor rises above 1, the control rods are inserted further to absorb more neutrons and thereby lower the multiplication factor. Should the multiplication factor drop below 1, the rods are withdrawn to provide more fission-causing neutrons and thus maintain the chain reaction. Control rods also adjust the power output of the reactor, and can be inserted to halt the chain reaction. An emergency shutdown, or **scram**, occurs when the control rods are inserted fully into the reactor. Reactors are generally designed to scram automatically whenever a potentially dangerous situation arises (see figure 8.2).

What if a reactor does get out of control—can we be sure the control rods will have time to react? Even a slow neutron takes only 1/10,000 second from its release in one fission event until it causes another fission. At that rate, the 500 generations of neutrons that blew up our 3,000-MW reactor a few paragraphs ago would be over in less than 1/10 second. No mechanical system can possibly respond in that time. Fortunately, reactor control *is* possible, thanks

Figure 8.2 A reactor scram occurs when all control rods are inserted fully in response to a potentially dangerous condition. The operating mechanisms for the rods are therefore very sensitive, as this sign in the Vermont Yankee reactor building suggests. (photo by Dexter Mahaffey)

to the fact that a very few neutrons are released not in the fission event itself but in the subsequent decay of fission products. The effect of these **delayed neutrons** is to make the average generation time about 1/10 second. That means it takes a relatively long time for a reactor's power to increase significantly if the multiplication factor rises slightly above 1, so there is enough time for mechanical control rods to operate.

Even with delayed neutrons, care in reactor engineering and operation is essential. Control is maintained as long as the reactor has its multiplication factor so close to 1 that the chain reaction wouldn't continue without the delayed neutrons. Then slight variations result in slow changes in reactor power, and the control rods can respond to these changes. But if the multiplication factor goes high enough that the so-called **prompt neutrons** alone can sustain the reaction, the generation time becomes much shorter and the reaction may go rapidly out of control. This is precisely what happened at Chernobyl, where operators inadvertently put the reactor into this situation of *prompt criticality*. The reactor power then soared to 500 times its design value in only 5 seconds.

Keeping Cool

Fission in a nuclear reactor produces heat. Unless that heat is removed, the reactor's temperature will continually rise, possibly resulting in a **meltdown**. In a power reactor, of course, we want to get the heat out of the reactor in the form of high-energy steam to turn the turbine-generator. Removing heat from the reactor is the job of the **coolant**, a fluid that circulates through the reactor to pick up heat energy and transport it where it can be useful.

Coolants, like moderators, vary with reactor design. In U.S. light-water reactors, ordinary water is the coolant. The same water, in fact, serves as both coolant and moderator. This is an important safety factor, since a loss of coolant also means a loss of moderator. With no moderator, the fission chain reaction comes to a halt. There is still the danger of meltdown, since the decaying fission products generate considerable heat, but the power level is well below that of the chain reaction itself. In addition to ordinary water, coolants used in commercial slow-neutron reactors include heavy water and the gases helium and carbon dioxide.

Reactor Designs

All slow-neutron reactors share the three common features that have just been discussed: moderator, control rods, and coolant. Working with those three elements, nuclear engineers around the world have developed very different reactor designs. The details those designs are not just for engineers, since they have direct bearing on the public debate over the safety of nuclear power. A brief overview of the reactor designs commonly used in nuclear power plants follows.

Light-Water Reactors

Asked to design a nuclear reactor, you might imagine putting pieces of uranium fuel in a container of water. Circulating around the fuel, the water would slow neutrons to maintain a chain reaction, and would simultaneously absorb heat. The water would boil, and the resulting steam could run a turbine-generator. That, in a nutshell, is a **boiling-water reactor**, or BWR—in many ways, the simplest design for a power reactor. Your BWR-based power plant might look something like figure 8.3.

About one-third of U.S. nuclear power reactors are BWRs. The uranium fuel in these reactors is sealed in long tubes of cor-rosion-resistant metal, with many tubes joined to form fuel bundles (figure 8.4). These bundles, mounted inside the heavy steel pressure vessel, make up the reactor's **core**. The water that serves as coolant and moderator circulates among the fuel rods. In normal operation, most of the dangerously radioactive fission products remain encased in the fuel. But some gaseous fission products work their way into the water, and, in addition, neutron absorption in the water gives rise to radioactive tritium. For these reasons, the water and the steam circulating through the turbine are radioactive. This means that a break in a steam pipe will release radioactivity, and it also requires radiation protection for workers servicing the turbine (figure 8.5).

The **pressurized-water reactor**, or PWR, is a close cousin of

Nuclear News: A Natural Reactor

In 1972 a worker at a French nuclear-fuel plant discovered a curious thing: Samples of uranium arriving from a mine at Oklo in the West African Gabon Republic contained even less fissile uranium-235 than the normally low 0.7 percent. This result was particularly baffling because the ratio of U-235 to U-238 is believed the same throughout the solar system, as confirmed by measurements on meteorites and moon rocks.

What could be the cause of the U-235 depletion? The clue emerged in further analysis of the Oklo samples: Not only were they depleted in U-235, but the samples also contained an unusual blend of isotopes that would normally be expected among the stable "offspring" formed in the decay of nuclear fission products. The conclusion was inescapable: A natural fission chain reaction had occurred at Oklo some 2 billion years ago. Humans did not invent the fission reactor.

We have just seen the difficult technological steps required to sustain a chain reaction, including the enrichment of uranium or the procurement of heavy water and the construction of carefully engineered reactor systems, including a moderator to slow the neutrons. How could random natural events put a reactor together? Several circumstances conspired to make the Oklo chain reaction possible. First, the ore at Oklo is rich in uranium; that is why it was developed for a mine. Second, the ore body at the time of the reaction was saturated with groundwater that could serve as a moderator, and the rich uranium vein was thick enough that fission neutrons were unlikely to escape. But how could ordinary light water moderate a chain reaction in natural uranium? We have seen how the neutron-absorbing properties of light water make that impossible in today's reactors. But 2 billion years ago, things were different. The half-life of uranium-235 is 700 million years; that of U-238 is 4.5 billion years. U-235 has decayed more rapidly than U-238, and that means there was a greater proportion of U-235 in the past. 2 billion years ago, in fact, the proportion of U-235 in natural uranium was about 3 percent—its value in today's enriched light-water reactor fuels.

Eventually six separate natural reactor zones were identified at Oklo. The reactors probably ran for several hundred thousand years, with a total power output between 10 and 100 kW. The chain reactions were probably kept under control by their need for moderating water: If the reaction ran too fast, water boiled away and the reaction slowed. The very low power level in the reactor zones precluded meltdown.

The fossil reactors at Oklo are more than scientific curiosities. They have served as natural laboratories for studying the long-term behavior of nuclear fission products. Analysis shows very modest migration of fission products from uranium-bearing regions into adjacent clay; plutonium decay products, on the other hand, show no migration—an indication that plutonium remained fixed at the sites where it formed for at least its 24,000-year half-life. These results are encouraging to those who advocate underground storage of nuclear wastes.

News source: "A Natural Fission Reactor," *Scientific American,* July 1976, p. 36.

Figure 8.3 Diagram of a nuclear power plant using a boiling-water reactor. The turbine is turned by steam generated in the reactor vessel itself.

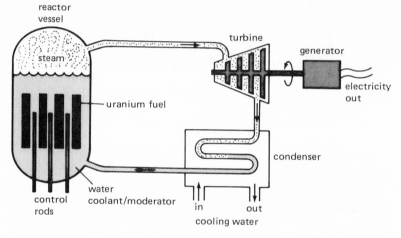

Figure 8.4   A pair of fuel bundles, shown before being lowered into the reactor core. (courtesy of GE Nuclear Energy)

the BWR. The main difference is that water in the PWR's reactor vessel is held under such high pressure that it is not able to boil, despite its high temperature. This superheated water goes from the reactor vessel to one or more separate **steam generators**, where it flows through pipes in contact with water that is allowed to boil. The resulting steam turns the turbine-generator, is condensed back to water, and then returns to the steam generator to complete the cycle. The turbine, the condenser, and the boiling-water part of the steam generator constitute the **secondary loop** of the cooling system. Water in the secondary loop never comes in contact with the reactor core, so it is not radioactive. Radioactive water is limited to the **primary loop**, consisting of the reactor vessel and the primary coolant pipes that flow through the steam generators. Steam generators are mounted close to the reactor vessel, within a thick containment structure; this makes the release of radioactivity unlikely even in the event of a failure in the primary coolant system.

Figure 8.5 Maintenance being performed on the turbine at the Vermont Yankee nuclear power plant. Since Vermont Yankee has a boiling-water reactor, the turbine assembly becomes contaminated with radiation. Note the workers' protective hats and clothing. The large concrete wall in the foreground blocks gamma radiation to keep off-site exposures below Nuclear Regulatory Commission limits. (Vermont Yankee Nuclear Power Corporation)

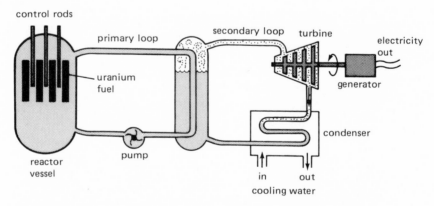

Figure 8.6 Diagram of a power plant using a pressurized-water reactor, showing the primary and secondary coolant loops that share the steam generator.

Figure 8.6 shows the essential features of a pressurized-water reactor.

About two-thirds of the commercial nuclear power plants in the United States use PWRs. They have proved more reliable than BWRs, with nine of the top ten long-term performance records going to PWRs. On the other hand, the most significant reactor accident in the United States—that at Three Mile Island—involved a PWR.

PWRs and BWRs share a number of important features. In both, the reactor core is enclosed in a heavy steel pressure vessel. To refuel the reactor, it must be shut down and the lid of the pressure vessel must be removed (figure 8.7). Spent fuel rods are removed—a process that must be carried out under water because of the intense radioactivity—and replaced with fresh ones. Typically, a reactor is refueled about every 12–18 months, one-third of the fuel bundles being replaced. Refueling lasts 6–8 weeks, a time utilities seek to minimize because of the high cost of replacement power. The long time interval between refuelings has a positive aspect, though: It makes spent fuel from light-water reactors a less-than-ideal source of weapons-grade plutonium.

As we have seen, the use of a common moderator and coolant is a safety feature of light-water reactors, a category that includes U.S. PWRs and BWRs. Loss of coolant means loss of moderator, which necessarily halts the fission chain reaction. The use of water as the moderator has an additional safety effect that occurs because the density of water decreases as its temperature rises: A slight increase in the reactor's power raises the water's temperature, lowering its density and therefore—since there are fewer hydrogen nuclei in a given volume—reducing the water's effectiveness as a moderator. As a result, fission slows and the temperature again drops. This feature of light-water reactors provides some stability and some protection against runaway chain reactions.

Heavy-Water Reactors

Several countries, particularly Canada and Great Britain, have developed commercial power reactors moderated with heavy water. Since heavy water absorbs very few neutrons, these reactors can operate with unenriched uranium. The most popular HWR design

---

Figure 8.7 Removing the lid of the pressure vessel before refueling a boiling-water reactor. (Pennsylvania Power & Light Company)

is the **CANDU** (for **Can**adian **D**euterium-**U**ranium). In the current generation of CANDU reactors, bundles of unenriched uranium fuel are mounted in individual coolant channels, through which pressurized heavy-water coolant circulates. The entire assembly of fuel bundles and coolant channels is immersed in a tank of unpressurized heavy water, which serves as the neutron moderator. Thus the CANDU design is cooled and moderated by heavy water, but with the two functions kept physically distinct. The pressurized coolant carries heat to steam generators, where it boils ordinary light water, which drives the turbine-generators as in a U.S. pressurized-water reactor.

Because the CANDU's coolant flows within individual pressurized channels, there is no need for the large, heavy pressure vessel found in U.S. light-water reactors. The lack of a single pressure vessel makes catastrophic failure of the high-pressure coolant system less likely than in a U.S. reactor. Furthermore, individual channels can be removed and refueled without shutting down the reactor. In practice, CANDU refueling is done almost continuously, with an average of fifteen fuel bundles replaced each day. Continuous refueling has several important implications. First, it makes controlling the reactor easier, because conditions within the reactor core remain essentially the same over time. (In contrast, the depletion of U-235 and the buildup of fission products over the long refueling interval make for substantial changes in the core of a U.S. light-water reactor.) As a result the CANDU design requires fewer control rods.

A second implication of continuous refueling entails the ever-present connection between nuclear power and nuclear weapons. Fuel that has spent a long time in a reactor—as happens in light-water reactors—is a poor source of bomb-grade plutonium. But in a reactor that can be refueled frequently, it is easier to "cook" fuel just long enough to optimize plutonium for use in nuclear weapons. With a continuously refueled reactor, it is also less practical for international inspection teams to monitor diversion of nuclear materials.

Which is safer, the American light-water design or the Canadian CANDU? In some of its design features, the CANDU has a safety edge over U.S. LWRs. And the CANDU avoids the need for uranium enrichment, a process that can be carried further to make bomb-grade uranium. On the other hand, the CANDU design shares with other continuously refueled reactors the ability to make bomb-grade plutonium, and its continuous refueling makes international inspection more difficult. In fact, a CANDU research reactor provided India with its first bomb plutonium. So which reactor is safer? That nuclear choice requires the weighing of complex considerations involving physics, engineering, and politics.

Graphite-Moderated Reactors

The first human-made nuclear reactor was a 400-ton pile of graphite embedded with nearly 50 tons of uranium compounds. Built under the stands at the University of Chicago stadium, it first "went critical" on December 2, 1942 (figure 8.8). The successful demonstration at Chicago led quickly to the construction of huge graphite-moderated reactors at Hanford, Washington, which turned out plutonium for early nuclear weapons.

Graphite has also proved workable as a moderator in commercial power reactors. Graphite-moderated reactors in Great Britain and the United States use gases for cooling, with hot gas piped to a steam generator where it boils water to turn a turbine. In the United States, the **high-temperature gas-cooled reactor** (HTGR) uses helium gas coolant and a highly enriched fuel consisting of 93 percent uranium-235. Its high temperature makes the HTGR more efficient at converting heat energy to electricity (recall the discussion of energy quality in the preceding chapter). Its high level of U-235 requires enrichment facilities capable of producing weapons-grade uranium. Its helium coolant is less capable than water of handling excess heat in an emergency; on the other hand, helium cannot boil and does not react chemically with other substances. Steam or hydrogen explosions (as occurred at Three Mile Island) are therefore impossible in a helium-cooled reactor.

The current generation of U.S. HTGRs has not met with unequivocal success. The single commercial HTGR power plant, at Fort St. Vrain in Colorado, was retired in 1989 after years of mixed performance. Newer HTGR designs are benefiting from the experience of the Fort St. Vrain plant, and some nuclear engineers anticipate that advanced gas-cooled reactors will be intrinsically

Figure 8.8 A sketch of Chicago Pile 1, the first reactor to achieve a self-sustaining chain reaction. The reactor consists of a pile of graphite blocks embedded with uranium. The man standing on the floor in front of the pile is manually removing a control rod to start the chain reaction. Because of wartime secrecy, no photographs were taken of the completed reactor. (Argonne National Laboratory)

safer and more reliable than the light-water reactors now in wide use. We will look at these advanced designs in the next chapter.

A graphite-moderated reactor need not be cooled with gas. The Soviet Union's **RBMK** reactor design consists of a large cylinder of graphite blocks, pierced with channels containing uranium fuel enriched to 2 percent U-235. The coolant, ordinary light water, circulates through the channels and boils to steam, which runs the turbine-generator (figure 8.9). As in the Canadian CANDU, the RBMK's arrangement of individual fuel channels allows for continuous refueling; in contrast with the CANDU, some RBMKs are intentionally dual-purpose reactors, producing both electric power and plutonium for weapons.

The choice of graphite as a moderator has important safety implications. First, loss of coolant—whether gas or liquid—does not mean loss of moderator. Thus, the fission chain reaction does not automatically halt in the event of a cooling-system failure. In a water-cooled graphite-moderated reactor (the Soviet RBMK) there is an additional dangerous possibility: Because the hydrogen in light water absorbs neutrons, water in the RBMK reactor acts more like a control rod than like a moderator. Overheating and excessive

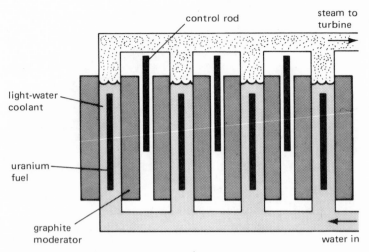

control rod

steam to turbine

light-water coolant

uranium fuel

graphite moderator

water in

Figure 8.9 Simplified diagram of the Soviet RBMK reactor, the design used at Chernobyl. Ordinary light-water coolant circulates through channels containing uranium fuel, and boils to drive the turbine. The coolant channels are set in a graphite moderator, as are channels for control rods.

boiling then enhance the rate of fission by reducing neutron absorption. The enhanced fission rate exacerbates the overheating, leading to a further increase in the fission rate. In contrast to the stabilizing effect of water in a light-water reactor, water in the RBMK design is actually destabilizing. Complex control systems are necessary to keep the RBMK reactor under control.

In April of 1986, a series of errors by its operators drove one of four RBMK reactors at the Chernobyl nuclear power plant in the Soviet Ukraine into a condition such that the control systems were ineffective. The reactor went supercritical, and in 5 seconds the power level surged to 500 times its design value. Cooling water flashed instantly to steam, blowing the heavy concrete cover off the reactor. The hot, flammable graphite moderator, normally protected from atmospheric oxygen, then caught fire. With the smoke went highly radioactive fission products.

Chernobyl was not the first major graphite-reactor accident. In 1957 a gas-cooled, graphite-moderated plutonium-production reactor at Windscale, England, caught fire, releasing 20,000 curies of radioactive iodine-131. (This compares with 10–25 curies at Three Mile Island, and 7 million curies at Chernobyl.) Milk contaminated by the iodine was dumped into the Irish Sea, and radioactive fallout spread across several European countries.

Are graphite-moderated reactors safe? Graphite, unlike water, is flammable—a circumstance that has exacerbated several reactor accidents. On the other hand, gas-cooled graphite reactors are immune to steam explosions, which might destroy water-cooled reactors. A graphite-moderated reactor will not automatically shut down when it loses its coolant. On the other hand, the massive block of graphite absorbs heat and may prevent a rapid temperature increase. So are graphite reactors safe? This is another complex nuclear question. There is no obvious "best" kind of nuclear reactor. Each has its pros and cons, involving not only safety but also economics and the ever-present issue of weapons proliferation.

Breeder Reactors

All the nuclear reactors we have considered so far have been slow-neutron reactors, employing a moderator to reduce neutron energy and enhance fission. With the exception of the high-temperature gas-cooled reactor, those slow-neutron reactors use natural or slightly enriched uranium at far below the enrichment necessary for weapons. Their relatively long generation time from one fission event to the next ensures that these reactors, even if wildly out of control, cannot blow up with the explosive power of a nuclear weapon.

Slow-neutron reactors have one great inefficiency: They have almost no use for the 99.3 percent of uranium that is U-238. Yet all that U-238 could be converted by neutron absorption to fissile plutonium-239, providing far more nuclear fuel than U-235 alone. Slow-neutron reactors do, in fact, produce Pu-239, but at a rate somewhat lower than one Pu-239 nucleus per fission event. The vast majority of the U-238 remains useless.

Although uranium is widely distributed throughout the Earth, its availability in commercially mineable quantities is limited. At the current rate of consumption, the known reserves of concentrated uranium ore may last only a few decades. Conversion of the U-238 in that uranium to plutonium-239 could prolong the availability of uranium to more than 1,000 years.

Enter the **breeder reactor**, a system in which neutrons from fission "breed" plutonium-239 from uranium-238. (Another breeding scheme involves making fissile U-233 from thorium-232.) Breeders have been around for a long time; in fact, the first nuclear-generated electricity came from an experimental breeder in 1951. The United States moved toward development of a commercial breeder with its Clinch River Breeder Reactor Project, but in 1977 President Jimmy Carter opposed Clinch River appropriations because of concerns over safety and weapons proliferation. Active breeder programs continue in Japan, the Soviet Union, West Germany, the United Kingdom, and especially France. The French "Superphénix" breeder produces electrical energy at the rate of 1,240 megawatts. However, its high cost and a serious liquid sodium leak in 1987 have dampened France's enthusiasm for breeder reactors.

A true breeder is distinguished from other reactors in that it produces more fissile fuel than it consumes; that is, for each fission event, more than one neutron gets absorbed in U-238 to produce Pu-239. Breeding therefore requires lots of neutrons. With one neutron needed to sustain fission, and more than one to breed new fuel, there isn't much leeway, since only two or three neutrons arise in each fission event. It turns out that more neutrons arise when fission is induced by fast neutrons, with the most neutrons obtained from fast-neutron fission in Pu-239. Successful breeders therefore use plutonium as fuel rather than uranium, and they have no moderator to slow the neutrons—hence the name **fast breeder**. Since fission is less likely with fast neutrons, fast breeders need more highly enriched fuel than ordinary, slow-neutron reactors.

Since it has no moderator, the core of a breeder can be very compact. A small core produces too much heat to permit the use of water as the coolant; furthermore, the need to be parsimonious with neutrons rules out neutron-absorbing coolants. And the coolant nuclei can't act as moderators, since the breeder requires fast neutrons. Current fast-breeder design favors the use of liquid sodium metal as the coolant. Sodium has enormous ability to carry heat without the need for high pressure, and it exhibits very low neutron absorption; on the other hand, sodium burns spontaneously on contact with air, and it reacts violently with water.

Contact between radioactive sodium and water would be especially disastrous; for this reason, a liquid-metal fast breeder re-

---

actor (LMFBR) has three separate cooling loops. The primary loop carries radioactive liquid sodium from the reactor core to a heat exchanger, where it heats a secondary loop of nonradioactive liquid sodium. That, in turn, boils water in a steam generator, and the steam drives a conventional turbine-generator. Figure 8.10 is a simplified diagram of a LMFBR.

The LMFBR is far more complex than a conventional reactor, and its operation poses new safety issues. Since there is no moderator, loss or even heating of the coolant only reduces the slight degree of neutron absorption and makes the fission reaction more vigorous. For this reason the fast breeder, like the RBMK, can be unstable; the fission reaction naturally tends to run out of control unless compensating effects are engineered into the system. Since a breeder uses *fast* neutrons, the power increase in case of an accident involving prompt-neutron criticality proceeds far more rapidly than in a slow-neutron reactor. And the higher concentration of fissile isotopes in breeder fuel makes an out-of-control breeder more violently explosive. Can a fast breeder blow up like a bomb? Probably not, but the answer isn't the unambiguous No that it was in the case of slow-neutron reactors.

For many of those who oppose breeder reactors, the most serious concern isn't the danger of reactor accidents; it is the ramifications of a plutonium-based energy economy. In an unstable world, do we really want plutonium to be a commercial commod-

Figure 8.10 A liquid-metal fast breeder reactor. The reactor core is surrounded by a blanket of uranium-238, which is bred to plutonium-239. Note the three-loop cooling system, designed to isolate radioactive liquid sodium from water.

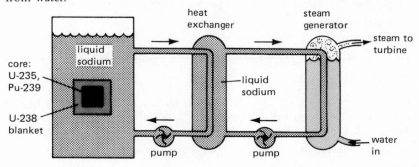

ity? What if terrorists hijack a plutonium truck? What if an airplane transporting plutonium crashes?

On the other hand, breeder reactors may reduce a nation's dependence on foreign oil and thereby lower the chance of international conflict. Fissioning plutonium instead of burning coal poses less danger of global climatic change. And, as we will see in chapter 10, breeder reactors produce less nuclear waste than nonbreeders.

Summary

Nuclear reactors come in many types. Common to all reactors are control rods that regulate the number of neutrons and therefore the rate of fission, and a coolant to carry off heat. Slow-neutron reactors make use of the fact that uranium-235 undergoes fission more readily when struck by slow neutrons, and therefore employ a moderator to slow neutrons from the high speeds they have when they emerge in fission events. The most common moderators are water, heavy water, and graphite. Because slow neutrons are highly efficient at causing fission, most slow-neutron reactors operate with natural or only slightly enriched uranium. Slow-neutron reactors exhibit a variety of pros and cons relating to safety, economics, and weapons proliferation.

Fast breeder reactors, in contrast, use fast neutrons to sustain the chain reaction. They require highly enriched fuel—usually plutonium—to overcome the fast neutrons' lower effectiveness in causing fission. But breeders produce more fuel than they consume, by converting nonfissile uranium-238 into fissile plutonium-239. Breeder reactors are more complex than conventional reactors, and they raise the specter of widespread plutonium trade in the international economy.

Further Reading

Jack Dennis, editor, *The Nuclear Almanac* (Addison-Wesley, 1984). In this collection of essays, members of the MIT faculty survey all aspects of nuclear issues. Part VII deals with nuclear power reactors.

Raymond Murray, *Nuclear Energy* (Pergamon, 1988). Chapters 11–13 detail reactor design and operation.

Nuclear News: International Plutonium Trade

In 1988 the United States authorized Japan to begin unlimited reprocessing of reactor fuel made from U.S.-supplied uranium. Initially at least, reprocessing of Japanese fuel will take place in Europe, with the recovered plutonium returned to Japan by air. Proponents of the agreement argue that without blanket permission to reprocess, Japan would simply buy its uranium elsewhere. That would hurt the United States' balance of trade and would leave the United States with no control whatsoever over Japanese plutonium use. Opponents see the agreement as a big step toward an international plutonium trade that will put staggering quantities of potential bomb material into commercial circulation. Over time, Japan alone stands to acquire more plutonium than is in the combined nuclear arsenals of the United States and the Soviet Union. And what about the safety of plutonium transport by air? After vigorous protests from the state of Alaska, originally proposed as a refueling stop for plutonium-carrying aircraft, the U.S.-Japan agreement was modified to ban flights over U.S. territory. Instead, the airplanes will fly over the North Pole, with permission for U.S. landings only in emergencies. And the United States retains the right to suspend the agreement if it sees a significant increase in the risk of weapons proliferation or threat to U.S. security. Does the plutonium agreement already pose a proliferation risk? Or is a nation bent on developing nuclear weapons going to get fissile material regardless of plutonium's commercial availability? These are nuclear questions to which U.S. government actions now give official answers. But are they the right answers? Are there right answers?

News sources: "Rising Nuclear Trade Stirs Fear of Terrorism," *New York Times,* November 5, 1987; "Alaska Seeks Halt to Plutonium Plan," *New York Times,* October 4, 1987; "U.S. to Allow Unrestricted Transfer of Plutonium," *Washington Post,* April 22, 1988; "Controls Added to Plutonium Pact With Japan," *Washington Post,* June 6, 1988.

Anthony Nero, *A Guidebook to Nuclear Reactors* (University of California Press, 1979). An authoritative treatment of reactor technology, designs, nuclear fuel and waste considerations.

U.S. Department of Energy, Energy Information Administration, *Commercial Nuclear Power* (Government Printing Office, published yearly). A wealth of data on nuclear power in the United States and throughout the world.

Glossary

boiling-water reactor (BWR)   A reactor in which water boils within the core to produce steam that drives a turbine-generator.

breeder reactor   A reactor designed to produce more fissile material than it consumes, by converting nonfissile uranium-238 into fissile plutonium-239.

CANDU   A Canadian reactor design using heavy water for moderator and coolant, and fueled by unenriched uranium. The acronym stands for **CAN**adian **D**euterium **U**ranium.

control rod   A rod made of neutron-absorbing material that is inserted into a nuclear reactor to control the fission rate and therefore the reactor's power.

coolant   A fluid that circulates through the core of a nuclear reactor to remove heat.

core   The heart of a nuclear reactor, containing the nuclear fuel, the moderator, and the control rods.

delayed neutrons   Neutrons that emerge in the decay of short-lived fission products, rather than in fission events themselves. Delayed neutrons help in the control of a reactor by increasing the time for significant changes in power.

fast breeder   A breeder reactor in which fast neutrons sustain the chain reaction.

fast neutrons   High-energy neutrons, with speeds comparable to the speed at which they are ejected in fission events.

generation time   The time between emission of a neutron in a fission event and that neutron's causing a subsequent fission event.

heavy water   Water whose hydrogen nuclei are those of the isotope deuterium ($^2_1$H).

heavy-water reactor (HWR)   A nuclear reactor that uses heavy water as its moderator. Heavy water absorbs fewer neutrons than ordinary (light) water, permitting the use of unenriched uranium in heavy-water reactors.

high-temperature gas-cooled reactor (HTGR)   A graphite-moderated reactor cooled by a gas, usually helium, and operating at much higher temperature than a water-cooled reactor.

light water   Ordinary water, whose hydrogen nuclei consist of a single proton ($^1_1$H).

light-water reactor (LWR)   A nuclear reactor using ordinary (light) water as its moderator. U.S. boiling-water and pressurized-water reactors are LWRs.

liquid-metal fast breeder reactor (LMFBR)   A fast breeder reactor using liquid sodium metal as its coolant.

meltdown   Melting of a reactor core due to loss of coolant.

moderator   A substance introduced into a nuclear reactor in order to slow neutrons. Common moderators include water, heavy water, and graphite.

multiplication factor   The average number of neutrons from each fission event that go on to cause another fission. In a reactor operating at steady power, the multiplication factor must be exactly 1.

nuclear reactor   A system, containing nuclear fuel and other materials, designed to sustain and control a fission chain reaction.

pressurized-water reactor (PWR)   A reactor in which cooling water circulating through the core is kept under enough pressure that it doesn't boil. This primary cooling water then transfers its heat to a secondary system in which water does boil, producing steam to drive a turbine-generator.

primary loop   The part of a reactor's cooling system that brings coolant into direct contact with the core.

prompt neutrons   Neutrons emitted during a fission event, rather than some time later. A reactor whose chain reaction is sustained by prompt neutrons alone is in danger of going rapidly out of control.

RBMK   A Soviet reactor design using a graphite moderator and water cooling. The Chernobyl reactor was of the RBMK type.

scram   Emergency shutdown of a reactor, accomplished by fully inserting the control rods.

secondary loop   A loop of coolant fluid that does not come into direct contact with the core, but picks up heat from the primary loop.

slow neutrons   Low-energy neutrons moving much more slowly than when ejected in fission events. Slow neutrons are far more effective than fast neutrons in causing uranium-235 to undergo fission.

steam generator   A device in which nonboiling coolant transfers its heat to water, which then boils to produce steam. Used in all reactor systems except boiling-water reactors.

Reactor Safety

9

Are nuclear reactors safe? We raised that question repeatedly in the preceding chapter, comparing the pros and cons of each reactor type. In this chapter, we will look further at reactor safety, considering both normal operation and the potential for nuclear accidents. The accidents at Three Mile Island and Chernobyl will serve to illustrate how sequences of minor errors or equipment failures can get a nuclear power plant into big trouble.

Reactor safety is far from an academic issue. Concern for public and environmental safety is the primary reason voters find nuclear questions on their ballots, and is what motivates the more militant anti-nuclear activists. Yet safety issues are often complex and subtle, and intelligent debate requires a modicum of knowledge about the operation of nuclear reactors and the safety implications of alternatives to nuclear power.

Normal Reactor Operation

Despite the keen interest of the news media in nuclear power plants, most plants operate day after day without anything newsworthy happening. What effect does this normal operation have on public health and safety?

Advocates of nuclear power argue that nuclear plants are among the cleanest of energy sources. In terms of sheer quantity of effluents, that is certainly true. A 1,000-megawatt coal-burning

Figure 9.1   Protesters near New Hampshire's Seabrook nuclear power plant after the NRC issued the plant's operating license, in 1990. (William Murphy, *Portsmouth Herald*)

power plant produces some 600 pounds of carbon dioxide gas, 30 pounds of ashes, and 10 pounds of acid-rain-causing sulfur dioxide *every second*. In contrast, most products of nuclear fission remain locked within the fuel rods of a properly operating nuclear power plant. A very few fission products—mostly radioisotopes of the gases krypton and xenon, and small amounts of radioactive iodine—work their way through the fuel-rod cladding material and into the cooling water, and neutrons absorbed in that water give rise to radioactive tritium. These substances are routinely released to the environment—the tritium in the plant's liquid discharges and the other substances to the atmosphere.

How bad are these normal radiation releases? Chapter 3 introduced the curie as the unit of radioactivity and noted that the amounts of radioactive iodine-131 released in the Three Mile Island and Chernobyl accidents were 15 curies and 7 million curies, respectively. These accidental releases occurred over a few days. In contrast, a 1,000-MW nuclear power plant normally releases less than 0.03 curies of I-131 per year. Much larger quantities of radioactive krypton and xenon are also released, but these gases are chemically inert and pose little health threat. The U.S. Nuclear Regulatory Commission limits the maximum public exposure from a nuclear power plant's effluents to 5 millirems per year for gases and 3 millirems per year for liquid discharges. In practice, individuals living in the immediate neighborhood of a nuclear plant typically receive about 1 millirem per year, well under 1 percent of their normal dose of background radiation. That 1 millirem is about the same as the radiation dose you would get from cosmic rays on a single flight between New York to Chicago. As was noted in chapter 4, 1 millirem might cause an average decrease in lifespan of 6 minutes, about half the risk of smoking one cigarette. Incidentally, coal-burning power plants also release radiation—in the form of uranium, thorium, and their decay products, which occur naturally in coal—and the radiation dose from a coal plant probably exceeds that from a nuclear plant of comparable size.[1]

It sounds as if normal emissions from nuclear power plants are pretty benign. On the other hand, we saw in chapter 4 how standards for exposure to low-level radiation have decreased over time, and how the cancer-causing effects of radiation underwent a major reassessment in 1990. And the greater effects of low-level radiation

Figure 9.2 This monitoring station samples airborne radioactivity at Massachusetts' Pilgrim Nuclear Power Station. (Boston Edison)

on children are often underemphasized in the official standards. While reports of bizarre deformities among creatures born near nuclear power plants are certainly exaggerated, nagging suggestions of increased human infant mortality, cancer, and other maladies in the vicinity of nuclear power plants surface occasionally. But the numbers in these cases are usually so small that it is hard to reach statistically valid conclusions. What is clear is that people aren't dying in droves from the effects of nuclear power plants—as they are from smoking, drunk driving, and handguns, or, for that matter, from coal-burning power plants, whose emissions probably cause 10,000 deaths each year in the United States.

When Things Go Wrong

A nuclear power plant is a complicated system, with a lot of things that can go wrong. Abnormal incidents occur frequently at nuclear plants, but most are minor and lead to no radiation release or other

Nuclear News: Pilgrim Cancer?

A 1989 report by Richard Clapp, former head of the Massachusetts State Cancer Registry, typifies the concern that keeps alive the debate over the safety of routine emissions from nuclear power plants. Clapp analyzed cancer rates in five Massachusetts towns near the Pilgrim nuclear power plant, and found a 60 percent increase in leukemia cases—27 cases versus 17 expected—for the period 1982–1984.

Although there is no proof that Pilgrim emissions caused the cancers, the increase in disease came 8–10 years after a period of excess radiation emission from the nuclear plant. That interval is equal to the expected latency period for radiation-induced leukemia. Clapp also reported evidence of an upturn in thyroid cancer in 1986; radiation-induced thyroid cancer has a latency period of 13–14 years. As a practical measure, Clapp urged that residents near Pilgrim be screened for common radiation-induced cancers. Pilgrim officials responded by noting that leukemia was below expected levels for 1985–86; that the 1982–1984 excess leukemias were nearly all confined to the town of Marshfield, the most distant of the five towns from the nuclear plant; that the excess radiation emissions resulted in public doses 100 times lower than would be expected to cause even one excess leukemia case; and that the numbers involved were small enough to be explained by random statistical variations.

If you are a resident of the Plymouth area, will you have yourself screened? That is a nuclear choice *you* need to make.

News sources: "Screening Sought in Cancer Link to Pilgrim," *Boston Globe,* September 19, 1989; *Leukemia Backgrounder—Update* (Pilgrim Nuclear Power Station, Rev. 1).

danger. Occasional errors or equipment failures do result in radiation emission exceeding NRC standards. Other failures—for example, in backup safety systems—raise the potential for serious accidents, but in most cases those accidents don't materialize. Only rarely does a sequence of failures compound into a major reactor accident.

The core of a nuclear reactor contains an enormous amount of highly radioactive material, whose dispersal to the environment is probably the worst thing that can happen at a nuclear power plant. Numerous safety systems exist to prevent accidents that could damage the core, and to contain radiation in the event that an accident does occur. How likely are such major accidents? How effective are the safety systems?

Overheating of the nuclear fuel is the most probable cause of a major reactor accident. That could occur either through an increase in the rate of nuclear fission or by a loss of coolant. We have already seen how different reactor designs differ in their potential for such occurrences. The Soviet RBMK reactor, for example, is prone to a runaway chain reaction—as happened at Chernobyl. On the other hand, the RBMK's massive graphite moderator absorbs heat, preventing a rapid temperature rise in the event of a loss of coolant. American light-water reactors are susceptible to very different safety problems. Since the coolant is also the moderator, the chain reaction in a light-water reactor stops automatically in the event of coolant loss. But even after shutdown, the reactor core continues to generate heat from the radioactive decay of fission products. Immediately after shutdown, decay heat amounts to nearly 10 percent of the reactor's normal power output; that figure drops to 1 percent only after several hours. Even a fraction of a percentage of full power represents tens of millions of watts, a rate of energy production that requires active cooling to prevent damage to the reactor core. For this reason, the worst-case accident scenario in a U.S. light-water reactor is a **loss-of-coolant accident** (LOCA).

Numerous events could lead to a LOCA. Pipes connected to the reactor core could break, dumping cooling water out of the system. Circulation pumps could fail, allowing the water in the reactor vessel to stagnate and boil away. Obstructions in the piping could block the flow of coolant. Less likely is a rupture of the

Nuclear News: Nuclear Incidents

The U.S. Nuclear Regulatory Commission classifies incidents at nuclear power plants into four categories: unusual event, alert, site area emergency, and general emergency. An *unusual event* has the potential to degrade plant safety, but with no abnormal off-site radiation expected. An *alert* represents an actual degradation of plant safety, but with consequent radiation releases far below Environmental Protection Agency guidelines for protective action. A *site area emergency* entails actual or expected major failures in a plant's safety systems, with significant radiation releases only near the plant boundary. In a *general emergency,* substantial core damage carries the potential for significant radiation release beyond the plant site.

Despite the abundance of media reports on nuclear reactor incidents, there have been only two site area emergencies and one general emergency since the NRC ratings began in the late 1970s. The first site area emergency occurred at Three Mile Island in 1979, and was soon upgraded to the only general emergency. The second site area emergency occurred in 1990 when a truck hit a utility pole supplying power to a nuclear plant in Georgia. A backup generator failed to start immediately, leaving the plant without emergency cooling. Power was restored in 17 minutes, and the situation was quickly downgraded to an alert. Even the number of alerts is small, typically under ten per year. Unusual events are more numerous, at several hundred per year. In 1988, for example, there were 212 unusual events and only six alerts. The latter included fires that compromised warning systems, inoperable containment spray systems for emergency pressure reduction, and power losses affecting control-panel indicators. None of these events threatened the public, and none required emergency procedures.

News sources: "Georgia A-Plant Accident Getting Sharp U.S. Scrutiny," *New York Times,* April 1, 1990; U.S. Nuclear Regulatory Commission, NUREG-1272, Section 5, p. 22.

Nuclear News: What Is In Those Pipes?

In March 1971, a pump inside the reactor vessel at Illinois' Dresden Unit 3 reactor showed inadequate water flow. Inspection turned up a 16-inch-diameter piece of plywood inside the reactor vessel. A welder had accidently left the plywood inside the emergency core-cooling pipes, and it had later washed into the reactor vessel, where it restricted coolant circulation. A year later a second Illinois reactor, Quad Cities Unit 2 at Cordova, experienced similar pump obstruction. This time an entire gas welding apparatus, including tanks and hoses, was found inside the reactor vessel!

These extreme cases are among many in which reactor coolant systems have been blocked through human error. Some other cases follow.

- October 1973 at Oconee Unit 2 in South Carolina: Valves for the emergency spray systems used to alleviate containment pressure buildup in the event of an accident were left closed; this could have led to containment breaching had an accident occurred.

- June 1975 at Maine Yankee: Valves that allow emergency core-cooling water to be pressurized so it can enter the reactor in the event of a coolant loss were locked in their closed position, rendering the ECCS inoperable. Under this condition, a coolant loss could have led to a meltdown.

- December 1975 at Millstone Unit 2 in Connecticut: Frozen pipes rendered inoperable instruments used to start the emergency core-cooling system in the event of an accident. A wiring error in the pipe heaters led to the freeze-up.

These and many other reactor incidents are documented in *The Nugget File*, which consists of formerly secret government accounts of safety-related incidents at nuclear facilities obtained by the Union of Concerned Scientists under the Freedom of Information Act.

News source: *The Nugget File* (Union of Concerned Scientists, 1979).

reactor vessel itself, accompanied by a catastrophic loss of coolant. These various failures might occur because of defective materials, sudden surges in pressure, human error, corrosion, or external factors such as earthquakes, tornadoes, airplane crashes, or acts of war.

In the event of a LOCA, all or part of the reactor core may lose contact with cooling water. The temperature of the uranium fuel then rises rapidly, melting the zirconium alloy cladding that surrounds it. At that point, radioactive fission products spread freely throughout the reactor vessel and the primary loop of the cooling system. They may also escape through whatever breach in the cooling system caused the loss of coolant in the first place. For just that reason, a heavy concrete **containment structure** surrounds the reactor vessel (and, in a PWR, the entire primary loop of the cooling system). (In a boiling-water reactor, the primary coolant loop necessarily extends to the turbine-generator, outside the containment structure.) The containment structure is designed to contain not only radioactive material but also pressure from hot gases and steam. On the other hand, today's containment structures cannot withstand the most violent events possible within a reactor, so there is a small chance that a LOCA could lead to breaching of the containment structure and dispersal of radioactive fission products.

In a serious LOCA, not only the fuel-rod cladding but also the uranium fuel itself may melt. In such a **meltdown**, 100 tons of molten uranium could melt its way through the reactor vessel and the floor of the containment structure, possibly resulting in a large steam explosion when the molten uranium reaches groundwater. The term "China syndrome"—popularized as the title of a 1979 film thriller about an accident at a U.S. nuclear power plant—describes such a meltdown, in which molten uranium melts its way into the ground, heading for China. In reality, the uranium would be cooled and contained within some tens of feet, and this type of containment failure might be less serious than an explosive breach of the containment with direct dispersal of radioactive material to the atmosphere. A more remote meltdown scenario has molten uranium regrouping in a critical configuration, starting an uncontrollable chain reaction.

A loss-of-coolant accident need not lead to a meltdown, provided replacement water can be supplied to the reactor core. That is the job of the **emergency core-cooling system** (ECCS). Emergency core cooling is the most important safety system in a nuclear power plant, and advocates of nuclear power point to the ECCS as a prime example of engineered safety. Critics, on the other hand, note that there has never been a full-scale test of an emergency core-cooling system. Advocates counter by pointing to ECCS tests at small experimental reactors, to computer simulations that indicate ECCS failure is unlikely, and to successful ECCS operation in a number of accidents that stopped short of meltdown. Critics point to accidents where the ECCS has been disabled, either through equipment damage or through human error. Operators at both Three Mile Island and Chernobyl turned off their emergency core-cooling systems, for reasons we will examine shortly.

The effects of a meltdown might be disastrous. Dispersal of radioactive material after the breaching of a containment structure could kill several thousand people immediately and nearly 50,000 with subsequent cancers, according to a 1974 Atomic Energy Commission study (the so-called WASH-1400 or Rasmussen Report). But the study found the probability of such an accident to be only

---

Nuclear News: Can a Candle Cause a Meltdown?

In March 1975, an electrician at the Brown's Ferry nuclear power plant in Alabama held a lighted candle to test for air leaks in an area of electrical cables beneath the power plant's control room. He accidentally started a fire that quickly destroyed cables controlling the safety systems of the plant's two operating reactors. The reactors scrammed, halting the chain reactions. But radioactive decay continued to produce heat in the reactor cores, and the fire had disabled the emergency core-cooling systems. In one reactor, water boiled away until the core was within 4 feet of being uncovered. But there was no meltdown, and no radiation release. Was the Brown's Ferry accident a near-disaster? Or did it affirm that nuclear plants are so safe they can withstand the loss of major safety systems?

News source: "Fire Raises Issue of Safe Reactors," *New York Times,* March 26, 1975.

---

once in a billion years' reactor operation. (With 100 reactors operating in the United States, that means a disaster of this magnitude would be expected once in 10 million years.) But an assessment of the Atomic Energy Commission report by the Union of Concerned Scientists suggested that the AEC underestimated accident probabilities by a factor of 20 and accident consequences by a factor of 6. If that is so, then the fatality rate from reactor accidents should be 120 times the AEC estimate, averaging 2–3 deaths per year for each operating reactor. Is that an acceptable price to pay for the energy from nuclear fission? And is a yearly death rate a reasonable way to quantify the large death tolls associated with extremely infrequent accidents?

The Big Ones: TMI and Chernobyl

The 1979 accident at the Three Mile Island nuclear power plant in Pennsylvania ranks as the most serious commercial nuclear power incident in the United States; the 1986 disaster at Chernobyl in the Soviet Union is far and away the worst reactor accident the world has yet seen. A close look at these accidents shows a complex interplay of basic nuclear physics, reactor design, safety systems, and human behavior. Studies show how these accidents could have been prevented, and how they might have been worse.

Three Mile Island

At 4 A.M. on March 28, 1979, valves malfunctioned in the condenser system of Metropolitan Edison's unit 2 reactor on Three Mile Island in Middletown, Pennsylvania. This minor failure stopped the flow of water in the secondary loop of the pressurized-water reactor's cooling system. Backup pumps started automatically to keep the cooling water flowing. Unfortunately, valves leading from those pumps had inadvertently been left closed, and a paper maintenance tag on the reactor's control panel obscured the lights indicating the status of those valves. Lack of cooling water caused the temperature and pressure in the reactor vessel to rise. What had been a minor failure in the condenser system now threatened the reactor itself.

Just 3 seconds after the start of the accident, pressure in the reactor rose enough to open a pressure-relief valve at the top of the

Nuclear News: Nuclear Insurance

Take a look at your homeowner's insurance policy. You will find a "Nuclear Hazard Clause," with a definition something like this: "'Nuclear Hazard' means any nuclear reaction, radiation, or radioactive contamination, all whether controlled or uncontrolled or however caused, or any consequence of any of these." A few lines further you will see "This policy does not apply . . . to loss caused directly or indirectly by nuclear hazard. . . ." That means your home is not covered in the event of damage from a nuclear reactor accident. If your house is contaminated beyond use when your local reactor has a meltdown, too bad; your insurance company isn't going to pay for a new house.

In the 1950s, nuclear power was so new and uncertain a technology that private insurance companies were unwilling to provide liability insurance to the nuclear industry. The U.S. government, eager to promote peaceful uses of nuclear energy, stepped in with the Price-Anderson Act, limiting total liability to $560 million in the event of a reactor accident. Initially, the government was to pay the lion's share of that figure. Subsequent amendments raised the ceiling to $700 million and reduced the government's share. But a major accident could cause billions of dollars in damage. Nuclear critics have viewed the Price-Anderson Act as unwarranted government protection of nuclear utilities against the true costs of their dangerous reactors. Proponents claim the act was necessary to get the fledgling nuclear industry started, and point out that utilities now carry the major burden of their Price-Anderson insurance.

In 1988, Price-Anderson liability was raised to $7 billion, a tenfold increase but still well below many estimates of damage in a major nuclear accident. Is Price-Anderson an unfair government subsidy to an inherently dangerous industry? Or is it a necessary part of doing business in the nuclear age? Your elected representatives grapple with that nuclear question as they debate Price-Anderson every 10 years.

News source: "Senate Votes to Raise Nuclear Industry's Liability," *Washington Post*, March 19, 1988.

Figure 9.3 The Three Mile Island nuclear power plant. The containment structures for reactor units 1 and 2 are the cylindrical buildings at right center. The accident occurred in the unit 2 reactor, at right. (Nuclear Regulatory Commission)

reactor vessel. Five seconds later the reactor scrammed, as control rods dropped automatically to stop the chain reaction. The pressure then fell. Already in these first 8 seconds, one minor equipment failure had occurred, one safety system had started as designed but had been foiled by a set of closed valves, and several other safety systems had acted perfectly to mitigate the buildup of pressure in the reactor. All would have been well but for another equipment failure.

As the reactor pressure dropped, the relief valve at the top of the reactor vessel was supposed to close. But it stuck open—despite a light in the control room indicating that it had closed. As a result, there was a breach in the primary coolant system through which water and steam could escape into the containment structure (figure 9.4). TMI was experiencing a loss-of-coolant accident, but the operators didn't realize it.

Less than a minute into the accident, the emergency core-cooling system came on automatically, as it was designed to do. But the operators believed, incorrectly, that there was too much

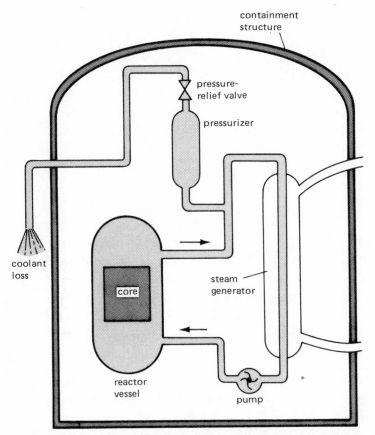

Figure 9.4 Simplified diagram of Three Mile Island unit 2 reactor, showing coolant escaping from pressure relief valve near the top of the containment structure. Normally the relief valve is closed, and the pressurizer is partially filled with gas to maintain fixed pressure in the primary coolant loop. But during the accident the valve stuck open, allowing radioactive water to escape into an adjacent building outside the containment structure.

water in the system. So they cut the ECCS water flow to almost nothing.

Confusion prevailed for the next hour as operators struggled with a situation they thought was caused by too much water. What was happening, in fact, was that steam in the reactor vessel was pushing up water levels in other parts of the system. But the control panel did not have a direct indication of water level in the reactor itself, so the operators were understandably misled. For more than 2 hours, they felt they had only a minor incident on their hands. All the while, though, the water level in the reactor vessel was dropping.

One clear indication of a LOCA would have been an alarm warning that radioactive materials had been detected in the containment structure. None of the operators remembered hearing this alarm. On the other hand, hundreds of alarms had sounded, overwhelming the abilities of the operators and the computer to process them. So the operators had turned many of them off. Either the crucial alarm had failed or the operators had missed it; in either

Figure 9.5 Discussions in the control room during the TMI accident. Normally four operators are on duty, but other personnel joined them as the accident progressed. (Nuclear Regulatory Commission)

event, they were not informed that radioactive water and steam were building up in the containment structure.

Nearly 2 hours into the accident, the operators, still unaware of the LOCA, shut off pumps that circulate water within the reactor's core. This act further reduced reactor cooling, and soon enough water had boiled to steam that the core became uncovered. Steam reacted chemically with the fuel-rod cladding to produce flammable hydrogen gas, and the hydrogen and steam together effectively blocked further water flow in the primary system. As the fuel cladding failed, radioactive fission products escaped through the stuck relief valve into the containment structure. Now, 2 hours into the accident, monitors in the containment structure began indicating abnormal radiation levels. In the next half-hour the operators finally diagnosed the stuck relief valve and blocked the flow through it. They had halted the loss of coolant, but they still had no idea that the reactor core was uncovered.

At 6:45 A.M., less than 3 hours into the accident, supervisory personnel declared a "site area emergency," based on the possibility of a release of radioactive material in the immediate vicinity of the plant. Forty-five minutes later the situation was elevated to "general emergency," since significant radiation might now escape the plant site. Still no one believed the core was uncovered, and most of the operators dismissed instrument readings suggesting abnormally high core temperatures.

By 9 A.M., 5 hours into the accident, members of the Nuclear Regulatory Commission had been notified of the accident, and by then at least some operators and NRC commissioners felt that the core was, in fact, uncovered. Gradually, through the morning, other federal and state agencies were notified of the emergency at TMI. But information was incomplete and often misleading. Local emergency officials complained that the information they received from the Pennsylvania Emergency Management Agency was either so vague or so technical as to be useless; often radio and television proved their best source of information. The mayor of Harrisburg got a call from a Boston radio station asking him about the nuclear emergency at nearby TMI, and replied: "What emergency?" No one really knew what had happened, and agencies responsible for public safety could not act effectively in the confusing situation.

Meanwhile, contaminated water overflowed holding tanks outside the containment structure, releasing low levels of radiation to the environment. Further releases accompanied attempts to relieve pressure inside the reactor. By noon—7 hours into the accident—officials began to discuss the possibility of evacuation. Confusion reigned again, as it was not clear what agency had the ultimate responsibility for evacuation.

Just before 2 P.M. instruments recorded a sudden, brief rise in pressure within the containment structure. Most in the control room thought this pressure "spike" was an instrument malfunction. Actually, there had been a minor explosion of hydrogen gas, not strong enough to damage the containment structure.

Within 10 hours of the accident the reactor was effectively under control. But releases of radioactive material, both intentional and inadvertent, continued for several days. Schools closed, people living near the plant were urged to stay indoors, and the general question of evacuation hung in the air. Actual measured radiation levels outside the plant were not high, and it is unlikely that any member of the public received more than 100 mrem, or one-third of a yearly background dose. But general fear and ignorance of radiation, coupled with confusing and often contradictory reports from the media, the utility company, and federal, state, and local officials, fostered a sense of panic and dread among the populace.

The accident began on a Wednesday. By Saturday a new worry occupied officials: that a further buildup of hydrogen in the containment structure might lead to an explosion that could rupture the structure and result in widespread dispersal of radioactive material. On Saturday afternoon the chairperson of the Nuclear Regulatory Commission told reporters that an evacuation of the nearly 1 million people living within 20 miles of TMI might prove necessary because of the danger of a hydrogen explosion. By Sunday afternoon, for reasons no one understood, the hydrogen bubble began to disperse. By late that day it was clear to the NRC that an explosion was impossible, but not until Monday morning was this good news made public. Although the danger was over, the damage had been done—damage to the reactor, damage to the public psyche, damage to the credibility of the Nuclear Regulatory Commission and other government agencies, and damage to the nuclear power industry in the United States. And the saga of TMI contin-

Nuclear News: TMI Gets Worse

Although the core of the Three Mile Island reactor was uncovered and sustained extensive damage from overheating, it was originally believed that the core temperature did not reach the melting point of the uranium fuel. The accident was therefore not viewed as a true meltdown. But when the reactor vessel was first examined in 1982, investigators found a 5-foot-deep cavern where the top of the core should have been. Subsequent examination showed clear evidence of fuel melting. And in 1989—more than 10 years after the accident—workers found evidence of cracks in the reactor vessel's stainless steel liner. When the last of the radioactive debris was removed in 1990, though, it became clear that the reactor vessel itself had come through virtually unscathed. The TMI cleanup showed that the partial meltdown within the reactor vessel was far worse than previously imagined, with some 20 tons of molten uranium at temperatures up to 5,000°F flowing into the bottom of the reactor vessel. At the same time, the integrity of the reactor vessel suggested that the danger to the public from even a meltdown might be less than previously anticipated.

News sources: "2 Possible Cracks in 3-Mile Reactor," *New York Times,* August 6, 1989; "After the Meltdown, Lessons From a Cleanup," *New York Times,* April 24, 1990.

ues, as ongoing cleanup operations reveal new and potentially serious aspects of the accident.

What caused the Three Mile Island accident? It started with a minor pump malfunction. But that was compounded by many factors: additional equipment failures, valves left in the wrong position, faulty instrument readings, operators' errors and confusion, miscommunication between government agencies and the public, a control panel and operating procedures that were poorly designed for emergency situations, and the failure of the operators to take seriously some indications of what was actually happening. There were broader causes, too: Nuclear Regulatory Commission licensing procedures that allowed plants with known safety problems to continue operation, inadequate training of plant operators, and communication failures within the nuclear industry. In short, the TMI accident had many causes—mechanical, human, and institutional. Improvements in any of these areas might have prevented it.

How serious was the TMI accident? In its consequences for public health, it was not at all serious. No one was killed outright, and most estimates suggest it is unlikely that any cancer deaths will result. The consequences for nuclear power in the United States were more serious. The accident dealt a devastating blow to an industry already plagued by public opposition and by declining orders for nuclear plants. In the years since this accident, no new commercial power reactors have been ordered in the United States, and 61 previously ordered reactors have been canceled (figure 9.6).

The TMI accident could have been much worse. Only a minuscule fraction of the radioactive fission products escaped to the environment. Breaching of the containment structure—which could have occurred had the operators not acted soon enough to halt the loss of coolant—might have spewed radioactive material over the surrounding countryside, resulting in death and illness. Nuclear critics argue that TMI came within 60 minutes of such a disaster; nuclear advocates counter by pointing to safety systems and operators' actions that ultimately worked to prevent a full meltdown and a containment breaching, despite numerous malfunctions and human errors. Critics say it is only a matter of time before a U.S.

Figure 9.6 Evolution of the nuclear power industry in the United States, showing reactor orders and cancellations from 1965 to 1988 grouped in 2-year intervals. No new reactor orders have been placed since 1978. (data source: U.S. Department of Energy, *Commercial Nuclear Power 1989*)

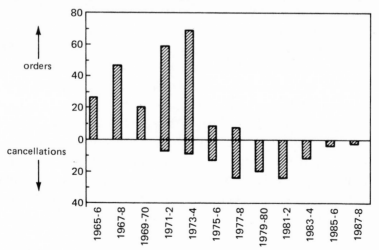

reactor experiences a full meltdown; proponents argue that safety standards adopted in the wake of TMI make that even less likely than before.

Chernobyl

Seven years after TMI, the world got its first taste of a really serious reactor accident. At the Chernobyl nuclear power plant in the Soviet Ukraine, a runaway nuclear reaction led to an explosion and fire that spread radioactive material over much of Europe.

Ostensibly, the Chernobyl accident occurred during a test of a reactor safety system. Officials wanted to be sure that the spinning turbine could provide temporary power to run the emergency core-cooling pumps in the event of a reactor shutdown. They were testing devices designed to maintain power to the ECCS in that circumstance.

A remarkably candid Soviet report shows how the test procedure, operators' errors, and the design of the reactor all contributed to the accident. Figure 9.7 shows the sequence of events. Starting on April 25, 1986, operators prepared for the test by reducing the reactor's thermal power from its normal 3,200 MW to 1,600 MW. They then turned off the emergency core-cooling system. Although the test procedure called for this action to prevent the ECCS from interfering with the test, it constituted a clear violation of the reactor's operating procedures.

As operators resumed lowering the reactor power, one of them neglected to set an automatic control that would have maintained power at the test level of around 700 MW. Instead, the reactor power plunged to a mere 30 MW.

It is very difficult to restart a reactor after a rapid decline in power. The reason has to do with a particular isotope, xenon-135, that absorbs neutrons voraciously. Xe-135 forms in the decay of the fission product iodine-135. In an operating reactor, Xe-135 is destroyed by neutron absorption as rapidly as it forms. But when the reactor's power drops, Xe-135 builds up because fewer neutrons are present. Attempts to increase power then fail because Xe-135 absorbs so many neutrons that increased fission is impossible; xenon "poisons" the reactor. The xenon-135 decays with a 9-hour half-life, so the reactor will restart easily after about a day.

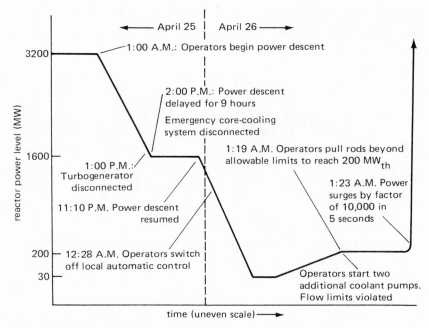

Figure 9.7 Sequence of events in the Chernobyl accident, shown on a graph of reactor power versus time. (adapted, with permission, from *Physics Today,* December 1986)

At Chernobyl, officials were impatient to continue the test. Instead of waiting for the xenon-135 to decay, they compensated for its neutron-absorbing presence by removing more control rods than regulations allowed. By 1:19 A.M. on April 26, this procedure had raised the reactor's power to 200 MW. At about the time they withdrew the control rods, the operators turned on two additional cooling-water pumps, as the test procedure dictated.

In a graphite-moderated reactor such as the one at Chernobyl, the primary effect of water is to absorb neutrons and therefore inhibit the chain reaction. This is the exact opposite of water's role in a U.S. light-water reactor, where its moderating properties make water essential to the reaction. So Chernobyl was in a situation where loss of water would increase the rate of fission.

The Chernobyl operators found there was too much water flowing in the reactor, and at 1:22 A.M. they reduced the flow rate. They did not immediately reinsert the control rods, however, and 30 seconds later a computer warned that the reactor was in a dangerous condition requiring immediate shutdown. But the operators proceeded with the test. They diverted steam from the turbine-generator, lowering the load on the reactor and causing more water to boil in the reactor core. The decreased water level led to less neutron absorption and more rapid fission. The chain reaction went supercritical from prompt neutrons alone, and the power soared to 500 times its design value in a mere 5 seconds. Cooling water flashed instantly to steam, blowing the heavy concrete lid off the reactor. A second explosion followed, probably due to hydrogen generated as water reacted with the fuel-rod cladding. The graphite moderator caught fire, and radioactive fission products were carried aloft with the smoke. The ensuing damage to the reactor is clearly evident in figure 9.8.

What caused the Chernobyl accident? As at Three Mile Island, many factors conspired. Human error and folly played a larger role than at TMI, but the design of the reactor contributed its own share. The bureaucratic pressure to complete the test had no immediate counterpart at TMI; on the other hand, the TMI accident happened amidst a long tradition of neglect for safety by the industry and the Nuclear Regulatory Commission. In both accidents operators became confused, made mistakes, and failed to heed warnings. Although the details of the two accidents are very different, both ultimately involved complex and potentially unstable systems operated by fallible human beings.

Two people died in the explosions at Chernobyl. Another 29—mostly power-plant workers and firefighters—died of acute radiation poisoning. Radioactive emissions continued for 10 days, spreading measurable fallout over much of Europe. Thousands of square miles of Ukranian agricultural land were contaminated, and 135,000 people living within 20 miles of the reactor were evacuated. Soviet authorities decided not to resettle the nearly 100 towns in the evacuation zone. People will continue to die of Chernobyl-induced cancers for years; the most likely estimate of worldwide cancer deaths over 50 years is about 17,000. The final cleanup bill will be staggering, possibly exceeding the cost of the entire Soviet nuclear power program.

How serious was the Chernobyl accident? Essentially all of the reactor's gaseous fission products escaped to the environment, along with about 3 percent of the solids. Nevertheless, many nuclear

Figure 9.8 A technician in a helicopter checks radiation levels above the damaged Chernobyl reactor. (Novosti Press Agency, Moscow)

experts see it as a worst case, and as something that could not happen in a U.S. light-water reactor. They take comfort in an ultimate death toll of "only" 17,000, and compare that 50-year, worldwide figure with the estimated 10,000 deaths caused *each year* in the United States alone by emissions from coal-burning power plants. Nuclear critics fear that our planet cannot tolerate many more Chernobyl-scale accidents, and they point to the many "near misses" among reactor mishaps as evidence that serious accidents remain inevitable. Nuclear advocates in the United States use Chernobyl to emphasize the supposed safety advantages of U.S. reactors, while critics stress that unexpected problems can arise with any reactor design. Proponents argue that even the rupture of the containment structure at Chernobyl caused relatively few deaths; opponents argue that a similar accident at a reactor near a major city— for example, at the Indian Point nuclear plant, 26 miles north of New York—could take a far greater toll. Nuclear advocates draw attention to serious non-nuclear accidents—such as the 1984 chemical leak at Bhopal, India, which killed 2,500 people and injured 200,000—in comparison with which, they suggest, nuclear power is relatively benign despite occasional severe accidents. Critics judge the nuclear industry on its own demerits, not in comparison with others, and find it unsafe.

Can Reactors Be Safer?

Light-water reactors, such as the one at Three Mile Island, evolved as scaled-up versions of the submarine reactors developed in the 1950s. Graphite-moderated reactors, such as that at Chernobyl, are relatives of the graphite "piles" first built in the 1940s to produce plutonium for nuclear weapons. Modern light-water and graphite reactors incorporate sophisticated safety features not found in their ancestors, and more safety features have been added in response to the accidents at TMI and Chernobyl. But fundamentally, today's power reactors remain rooted in designs that originated decades before safety became a paramount concern. Safety systems for modern reactors are in this sense an afterthought, not an essential design feature.

Can a really safe reactor be made? Spurred by rising concern over the global climatic effects of fossil fuels, the nuclear industry is moving toward a new generation of power reactors that could mitigate critics' safety concerns. Simply redesigning and downscaling the light-water reactor might go a long way toward providing intrinsic safety. In 1989 the Westinghouse Electric Corporation, a major reactor manufacturer, received $50 million from the U.S. Department of Energy to help develop such a reactor. Shown in figure 9.9, the Westinghouse design is a simplified pressurized-water reactor with a steel containment structure designed so natural air flow helps cool it in the event of an accident. The emergency core-cooling system includes a large reservoir mounted above the reactor, eliminating the need for ECCS pumps. More gradual bends in pipes mean fewer welds (which can leak or burst), and lower fuel enrichment means lower core temperature even in the event of a coolant loss. General Electric, which manufactures boiling-water reactors, is working on simplifications of that design, including a cooling system that requires no pumps. In principle these advanced light-water reactors could still melt down, but the likelihood of such an occurrence would be greatly reduced.

Another design has been proposed by General Atomics, a manufacturer of gas-cooled, graphite-moderated reactors. The General Atomics reactor's fuel consists of tiny uranium spheres surrounded by ceramic materials capable of withstanding very high temperatures (figure 9.10). Each composite pellet is just under 1 millimeter in diameter (about 1/25 inch) and is, in effect, a miniature containment system, trapping fission products generated in its uranium core. In the proposed reactor, pellets would be loaded into fuel rods mounted in hexagonal graphite blocks. The arrangement of the fuel would ensure that even in the event of a total coolant loss, the temperature of the fuel would remain below the failure point of the ceramic fuel coatings; therefore a meltdown should be intrinsically impossible. (An experimental German reactor of similar design was run in 1989 with all its control rods withdrawn and without coolant; no fuel failure or other damage occurred, despite conditions that would have destroyed a conventional reactor.) A final safety feature of the General Atomics design would be installation of the reactor vessel in an underground containment silo in order to minimize the possibility of airborne radioactivity.

All these advanced reactor designs share a common philosophy:

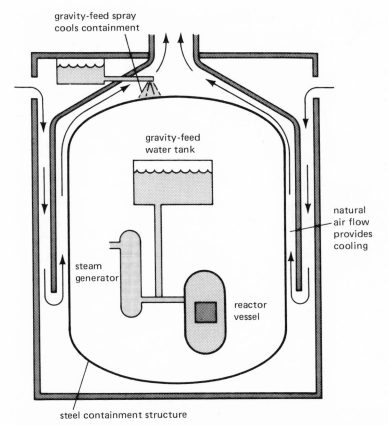

gravity-feed spray
cools containment

gravity-feed
water tank

natural
air flow
provides
cooling

steam
generator

reactor
vessel

steel containment structure

Figure 9.9 Diagram of a simplified pressurized-water reactor designed for inherent safety. Safety features include gravity-feed water tank for core cooling, additional gravity-feed sprays outside the containment, and a natural air-circulation pattern to provide additional cooling. (Westinghouse Nuclear Reactor Division)

Figure 9.10 Microphotograph of fuel pellet for General Atomics' Modular High Temperature Gas Reactor. The pellet, less than a millimeter across, has been broken open to show the enriched-uranium core and the surrounding layers of high-temperature ceramic materials. (General Atomics)

reliance on simple, inherent design features rather than complex mechanical systems to prevent dangerous accidents. Systems like gravity-fed emergency core cooling or fuel bundles that cannot exceed damaging temperatures are not only less likely to fail than the many-layered safety systems of contemporary reactors; they are also more immune to human tampering—an important consideration, in view of the fact that emergency core-cooling systems were turned off by human action at both Three Mile Island and Chernobyl.

On the other hand, nuclear critics are alarmed that the new reactors would eliminate the multi-layered safety systems of today's nuclear plants, that a few operators in a single control room might be handling as many as nine reactors, tht licensing procedures for the new reactors would be streamlined to the point where much public input would be eliminated, and that off-site emergency planning might be essentially terminated in light of the new reactors' purported safety. Indeed, a 1990 study commissioned by the Union of Concerned Scientists found safety claims for the new reactors to be overoptimistic, and raised the possibility that the new designs might actually prove more vulnerable to certain accidents.

Summary

Is nuclear power safe? No: The huge inventory of radioactive materials within a nuclear reactor carries with it the possibility that an accident could have serious health consequences for thousands of people. The heat generated in that radioactive material is a serious hazard that could lead to dispersal of radiation to the environment. Three Mile Island, Chernobyl, and dozens of less publicized incidents underscore the inevitability of nuclear accidents. And government agencies intended to protect the public have shown contemptible disregard for safety in the face of industry pressure to bring nuclear plants on line and keep them operating.

Is nuclear power safe? Yes: Multiple safety systems ensure that radiation will be contained even when malfunctions and operator errors occur. Major accidents are possible, but are so unlikely that the risk is negligible. You are far more likely to die in an automobile accident, a fall, or a fire than in a nuclear accident. And no other industry can match the safety record of the U.S. commercial nuclear power enterprise. Furthermore, the participation of citizens in the licensing process makes it difficult to bring a nuclear plant into operation without addressing countless safety and environmental concerns.

Can nuclear power be made safer? Probably: New reactor designs promise greater inherent safety. The nuclear industry has learned a lesson from Three Mile Island and Chernobyl. On the other hand, not all the safety recommendations that arose from TMI have been implemented, even though more than 10 years have passed since the accident. And even "inherently safe" reactor designs may be subject to unexpected failures not predicted by computer analyses.

The question isn't whether nuclear power is safe or not. It will never be perfectly safe. The question is whether it is safe enough.

Note

1. See J. P. McBride et al., "Radiological Impact of Airborne Effluents of Coal and Nuclear Plants," *Science* 202 (December 8, 1978), p. 1045.

Further Reading

Michelle Adato et al., *Safety Second: The NRC and America's Nuclear Power Plants* (Indiana University Press, 1987). The Union of Concerned Scientists, a leading anti-nuclear group, produced this authoritative criticism of the Nuclear Regulatory Commission's failings in regard to nuclear plant safety.

Bernard L. Cohen, *Before It's Too Late: A Scientist's Case For Nuclear Power* (Plenum, 1983). A leading pro-nuclear scientist presents a strong case for nuclear power. Chapter 3 deals specifically with reactor accidents.

Alan Cottrell, *How Safe Is Nuclear Energy?* (Heinemann, 1981). A survey of nuclear power safety by a former Chief Scientific Adviser to the British government.

Christopher Flavin, *Reassessing Nuclear Power: The Fallout from Chernobyl* (Worldwatch Institute, 1987). This brief report probes the Chernobyl accident's effects on public opinion. The author takes a decidedly anti-nuclear stand.

John W. Gofman and Arthur Tamplin, *Poisoned Power: The Case Against Nuclear Power Plants* (Rodale, 1971). Predating Three Mile Island and even the formation of the NRC, this older book remains one of the more influential anti-nuclear works for the general public.

Michael W. Golay and Neil E. Todreas, "Advanced Light-Water Reactors," *Scientific American* 262 (April 1990), p. 82. Nuclear engineers from MIT survey new developments in reactor technology.

Fred Hoyle, *Energy or Extinction? The Case for Nuclear Power* (Heinemann, 1977). In this pre-TMI book, a well-known British scientist argues strongly for nuclear power.

Richard Hubbard and Gregory Minor, eds., *The Risks of Nuclear Power Reactors* (Union of Concerned Scientists, 1977). This reassessment of the government's Rasmussen Report (WASH-1400) suggests that the government underestimated the risks of nuclear power by a factor of more than 100.

Michio Kaku and Jennifer Trainer, *Nuclear Power: Both Sides* (Norton, 1982). A collection of essays on opposite sides of nuclear issues. Chapter 3 deals with reactor safety.

John G. Kemeny et al., *Report of the President's Commission on the Accident at Three Mile Island* (Government Printing Office, 1979). This multi-volume work details all aspects of the TMI accident. The casual reader will find more than enough in the executive summary, entitled *The Need for Change: The Legacy of TMI*.

Harold W. Lewis, "The Safety of Fission Reactors," *Scientific American* 242 (March 1980), p. 33. A thorough look at reactor safety systems, and at the fault-tree analysis used to assess the probability of reactor accidents. Includes a discussion of the TMI accident.

David R. Marples, *Chernobyl and Nuclear Power in the USSR* (St. Martin's, 1986). A detailed look at the Chernobyl accident in the context of the Soviet nuclear power program.

Zhores A. Medvedev, *The Legacy of Chernobyl* (Norton, 1990). An exiled Soviet biochemist details the bureaucratic and technological failings that led to the Chernobyl accident.

Ralph Nader and John Abbotts, *The Menace of Nuclear Energy* (Norton, 1977). Consumer advocate Ralph Nader makes a strong case against nuclear power.

U.S. Senate, *Nuclear Accident and Recovery at Three Mile Island* (Government Printing Office, 1980). This well-illustrated report of a special Senate investigation makes dramatic reading.

Harvey Wasserman and Norman Solomon, *Killing Our Own: The Disaster of America's Experience with Atomic Radiation* (Dell, 1982). Among the more stridently anti-nuclear books, this volume presents an alarmist view of the effects of nuclear incidents. Part IV deals specifically with nuclear power, including the TMI accident.

Glossary

containment structure   Structure surrounding a nuclear reactor, designed to contain radioactivity in the event of an accident. Most containment structures are made of heavy concrete, and are designed to withstand some pressure buildup resulting from overheating.

emergency core-cooling system (ECCS)   Safety system designed to supply water to a reactor core in the event of a loss of normal coolant.

loss-of-coolant accident (LOCA)   A situation in which coolant is lost from the vicinity of a reactor core. Heat from the decay of radioactive fission products then builds up in the core, and may result in core damage, release of radioactive materials, and melting of fuel.

meltdown   Melting of a nuclear reactor's core after loss of coolant.

Nuclear Regulatory Commission (NRC)   United States government agency responsible for licensing and regulation of nuclear power plants and other nuclear facilities and materials.

What About Nuclear Waste?

10

You may or may not deem nuclear power plants acceptably safe, but in any event, what about nuclear waste? A nuclear reactor generates highly radioactive fission products. During normal operation—and, one hopes, even during accidents—these materials remain within the reactor. But eventually, spent fuel must be removed with its fission products still locked inside, and the reactor refueled. Then what happens to the spent fuel?

For U.S. commercial nuclear reactors, the answer is this: Highly radioactive spent fuel is stored at individual reactor sites, in pools of water that provide radiation shielding and cooling (figure 10.1). But few would regard that solution as permanent. Whereas the lifetime of a nuclear power plant is measured in decades, some radioisotopes in spent fuel have half-lives of 24,000 years or more.

In the 1960s and the 1970s, when nearly all of today's nuclear power plants were ordered, no one anticipated long-term storage of nuclear waste at the plant sites. But through the 1970s and the 1980s the United States was unable to arrive at a technologically and politically acceptable long-term storage plan, and it now appears that resolution of the United States' nuclear waste dilemma is unlikely before the year 2010. Power plants face premature shutdown once their on-site waste-storage facilities are full. As a result, many plants seek Nuclear Regulatory Commission approval to increase the storage density of spent fuel, or to open additional storage areas. These on-site storage options are costly to the utility companies, and opponents of nuclear power find in NRC on-site

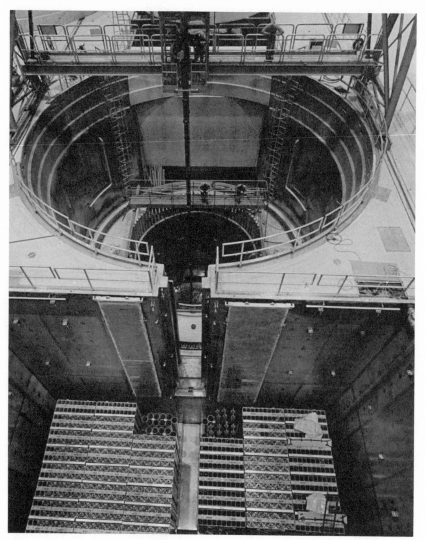

Figure 10.1 Storage of spent fuel at a boiling-water reactor. Deep in the circular pit at rear is the reactor vessel with its cover removed. A narrow channel connects the reactor compartment with the spent-fuel pool in the foreground. The spent-fuel pool is normally flooded with water to provide radiation shielding. (GE Nuclear Energy)

Nuclear News: Waste-Dump Delay

In 1982, the U.S. Congress passed the Nuclear Waste Policy Act, directing the Department of Energy (DOE) to develop a long-term facility for the storage of high-level nuclear waste that would begin operation by 1998. By 1986 three possible sites had been identified for further study, and in 1987 Congress directed the DOE to focus on a single site in Nevada. This site, at Yucca Mountain just west of the DOE's nuclear weapons test area, would receive waste from the more than 100 commercial nuclear reactors in the United States. Two years of intensive study followed, examining Yucca Mountain for geological stability and the possibility of groundwater contamination.

Then in 1989, amid criticisms of its technical work, the DOE announced that it was abandoning its plans for Yucca Mountain and would take a fresh look at the site. A storage facility at Yucca Mountain is now far from certain, and even if the site is developed it could not begin receiving waste before the year 2010.

For years utilities have been paying 1/10¢ for every kilowatt-hour of nuclear-generated electricity into a special fund to finance permanent storage of nuclear waste. With significant funds spent on the abandoned Yucca Mountain plans, utilities wonder when they will see a return on their waste-storage investment. With the DOE obligated to begin accepting nuclear waste in 1998, the agency will ask Congress to approve an interim storage facility.

The nuclear power industry began in the 1950s. Even if a permanent waste-storage facility opens in 2010, this means more than half a century of nuclear power operation without a satisfactory means of waste disposal. Public impatience with that situation could have a strongly negative impact on the nuclear industry's plans for a new generation of safer nuclear reactors.

News sources: "U.S. Will Start Over on Planning for Nevada Nuclear Waste Dump," *New York Times*, November 29, 1989; "Finding a Burial Place for Nuclear Wastes Grows More Difficult," *New York Times*, December 5, 1989; "Just One More Flop: Politics and Nuclear Waste Disposal," *New York Times*, December 8, 1989.

waste hearings new opportunities for curtailing the operation of nuclear plants.

The Nuclear Fuel Cycle

Assessing nuclear waste issues requires an understanding of the entire **nuclear fuel cycle**—the process whereby natural uranium becomes nuclear fuel, undergoes fission in a reactor, and emerges as nuclear waste. Figure 10.2 is a simplified diagram of the so-called **once-through** fuel cycle.

The first step in the fuel cycle is mining. Although uranium is widely distributed throughout Earth's crust, it is economically mineable only in certain ores where it occurs in concentrations of 0.1 percent or more. There is abundant uranium in the Western and Southwestern United States. Other countries with significant uranium resources include Australia, Argentina, Brazil, Canada, Na-

Figure 10.2 Once-through nuclear fuel cycle, taking natural uranium through fuel preparation, fission in a reactor, and storage as high-level nuclear waste.

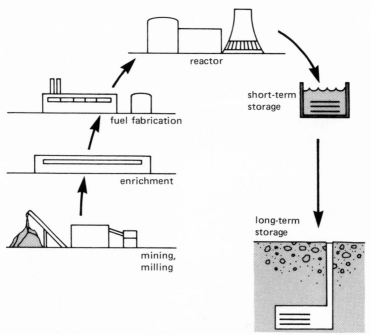

mibia, Niger, South Africa, and possibly the Soviet Union and China.

After mining, ore is pulverized and processed to remove the uranium. Left behind are vast piles of crushed rock called *tailings*. These piles contain radioisotopes formed in the uranium decay chain (recall figure 3.13), including radium, which decays continuously to radon gas. Because they exude radon, tailings piles should be covered to prevent excessive radioactivity in the surrounding air. Since that was not done in the early days of uranium mining, radon levels were high in the vicinity of tailings piles, and slightly radioactive dust was dispersed by the wind. Some tailings were even used in construction, resulting in buildings with excessive radiation levels.

Removed from its ore, uranium takes the form of "yellowcake," a chemical compound of uranium and oxygen (U_3O_8). Although yellowcake is mostly uranium, the fissile isotope U-235 occurs in yellowcake at its low natural abundance of 0.7 percent;

Nuclear News: Radioactive Houses

In the early 1950s, uranium mills in southwestern Colorado began offering their tailings to builders for use as fill under new building foundations. Builders liked the smooth, compact base the tailings provided, and some 4,000 buildings in the area incorporated uranium tailings. The practice continued for 15 years, then was halted in the 1960s amidst concern over high radiation levels in buildings constructed on uranium tailings. Radon gas arising from radium left in the tailings seeped into the buildings, in some cases giving radon concentrations close to the level that would trigger evacuation in a uranium mine, and hundreds of times the recommended safe level. Hundreds of buildings required evacuation, reconstruction, or corrective action. Hardest hit was Grand Junction, Colorado, where houses, offices, public buildings, stores, factories, and one wing of a high school were all affected. The federal government eventually paid 75 percent of the cleanup expenses—one of the many hidden costs of nuclear power and weapons.

News sources: "Colorado Studies Radiation Peril," *New York Times,* February 17, 1970; "Radioactive Ore Waste Stirs Fears," *New York Times,* May 16, 1971; "Radioactive Homes," *New York Times,* November 10, 1974.

---

essentially all the rest is nonfissile U-238. Heavy-water reactors, as we saw in chapter 8, can use natural uranium directly, so for these reactors the yellowcake need only undergo chemical and physical processing to become nuclear fuel. For the light-water reactors used in the United States, however, uranium must be enriched to about 3 percent U-235 in order to sustain fission.

Uranium Enrichment

Enrichment is a difficult process, because it involves separating uranium isotopes that are chemically similar and that have very nearly the same mass. That mass difference—235 versus 238—is the only distinguishing characteristic that an enrichment process can use to separate the two isotopes. Enrichment technology is politically sensitive, since a nation possessing it can enrich uranium to the high levels needed for nuclear weapons.

Uranium enrichment in the United States is currently done by **gaseous diffusion**, a scheme that relies on the fact that lighter molecules in a gas move faster. To get it into gaseous form, uranium is combined chemically with fluorine to form uranium hexafluoride gas (UF_6). The gas is introduced into a chamber containing a membrane through which UF_6 molecules can pass, but not easily. Those few molecules containing U-235 move slightly faster than their more numerous U-238 counterparts, and therefore hit the membrane more often. So molecules with U-235 are slightly more likely to get through the membrane, giving a mixture on the far side that is enriched in U-235. Figure 10.3 illustrates the enrichment process. A single chamber gives only a minuscule increase in U-235 content, so in practice many chambers must be connected one after another—about 500 for 3 percent enrichment to light-water reactor fuel, and several thousand to make weapons-grade uranium. Gaseous-diffusion enrichment plants are huge and expensive, and they consume enormous amounts of electrical energy.

Gaseous diffusion became the United States' preferred process for uranium enrichment as a result of the urgent need to produce bomb-grade uranium during World War II. Since then, other promising enrichment schemes have developed, some of them more accessible to smaller nations. Centrifuges—chambers that spin at

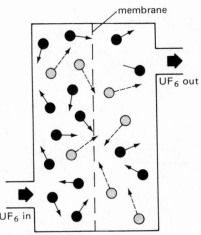

Figure 10.3 Enrichment of uranium by gaseous diffusion. Uranium hexafluoride gas enters the chamber at lower left. Molecules containing U-235 (light shading, with longer arrows indicating higher speed) are lighter than their U-238 counterparts (dark shading), so they penetrate the membrane more readily. The result is slightly enriched uranium to the right of the membrane, ready for processing in additional chambers to any desired level of enrichment.

high speed and force heavier material to the chamber walls—are popular in Europe and represent a potential route to nuclear weapons capability. Iraq and Pakistan are believed to have or to be developing centrifuge enrichment facilities for their nuclear weapons programs. South Africa has developed a technique whereby uranium hexafluoride is enriched when a stream of the gas impinges on a specially shaped nozzle. A promising method under development in the United States and elsewhere is laser enrichment, which uses precision lasers to remove electrons selectively from atoms of U-235 but not U-238; the electrically charged U-235 is then separated by means of electric forces.

After enrichment, uranium is processed chemically to make uranium dioxide (UO_2), then formed into pellets about 1/2 inch long and 3/8 inch in diameter. Thanks to the nuclear difference (chapter 2), each pellet contains the energy equivalent of 150 gallons of gasoline or a ton of coal. The pellets are loaded into special zirconium-alloy tubes to form fuel rods, and groups of rods are assembled into fuel bundles for delivery to nuclear power plants (figures 10.5–10.7).

Figure 10.4 This uranium-enrichment complex, at Oak Ridge, Tennessee, provided fuel for U.S. nuclear weapons and reactors from World War II through 1985. Gaseous-diffusion cells occupy the huge building at right center, with more than 40 acres under one roof. Today, uranium enrichment continues at a similar facility in Paducah, Kentucky. (Martin Marietta)

In the Reactor

What happens to uranium once it is in the reactor core? Uranium-235 undergoes fission, of course, producing energy that ultimately leaves the power plant as electricity and waste heat. Fission also results in the highly radioactive fission products that were discussed in chapter 5. Something else happens in the reactor fuel, too: Some of the uranium-238 absorbs neutrons, soon becoming plutonium-239. Pu-239 is fissile, so it too can undergo fission to produce energy and additional fission products. Or a plutonium nucleus may absorb a neutron without fissioning, giving rise to a still heavier nucleus. After a while, the original U-235 and U-238 have become a more complicated mix of uranium isotopes, fission products, plutonium, and heavier transuranic isotopes. (Chapter 5 introduced the term *transuranic* for plutonium and other nuclei beyond uranium in atomic number.) The details of that isotopic mix have implications

Figure 10.5 Uranium fuel pellets. (GE Nuclear Energy)

for the disposal of nuclear waste, the proliferation of nuclear weapons, and even the economics of nuclear power.

In a typical light-water reactor, a given fuel bundle remains in the core for about 3 years. Figure 10.8 shows the mix of substances at the end of that time. Every 1,000 pounds of 3.3-percent-enriched uranium contains 967 pounds of U-238 and 33 pounds of U-235. Three years later the U-235 is down to 8 pounds, the other 25 pounds having fissioned. Some of the U-238 is gone, too, 24 pounds of it having transformed into plutonium and other heavy isotopes. Much of the plutonium-239 thus formed has fissioned, contributing with the U-235 to make a total of 35 pounds of fission products. In fuel near the end of its useful life, in fact, most of the energy comes from fissioning plutonium rather than uranium.

Before proceeding to the nuclear waste dump, let us look more closely at plutonium. The fissile isotope Pu-239 is only one neutron

Figure 10.6 Cutaway view showing how fuel pellets are assembled into a zirconium-clad fuel rod. (GE Nuclear Energy)

away from U-238, and it builds up quickly. Heavier plutonium isotopes require additional neutron capture, so they increase more slowly. Fuel that has been in the reactor for a short time thus has a significant amount of Pu-239 and not much else in the way of plutonium isotopes. But Pu-239 is fissile, and as it builds up it begins to undergo significant fission. After 3 years the rate at which Pu-239 is produced and the rate at which it is destroyed by fission are nearly equal, and Pu-239 buildup essentially stops. But other plutonium isotopes continue to increase, giving "old" reactor fuel a higher proportion of these heavier plutonium isotopes in relation to Pu-239.

So what? As we will see in subsequent chapters, the presence of plutonium-240 makes the construction of plutonium weapons more difficult, and renders those weapons somewhat less predict-

Figure 10.7 A bundle of fuel rods being lowered into a reactor core during refueling. (GE Nuclear Energy)

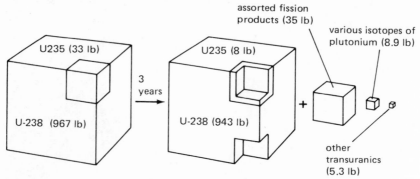

Figure 10.8 After 3 years in a reactor, an initial 1,000 pounds of 3.3-per-cent-enriched uranium includes fission products along with plutonium and other transuranic isotopes. (adapted from "The Disposal of Radioactive Wastes from Fission Reactors," by Bernard Cohen; copyright 1977 Scientific American, Inc.; all rights reserved)

able and less effective than bombs made with pure Pu-239. As a result, it is often argued that spent reactor fuel is "safe" from a weapons–proliferation standpoint, and that the connection between nuclear power and nuclear weapons is therefore tenuous. It is true that a weapon contaminated with Pu-240 would have a smaller explosive yield, but it would still be far more powerful than any conventional explosive. Fuel that has spent a long time in a reactor is certainly not the best choice for weapons—but it would do the trick, bringing a nation or a terrorist group into the nuclear weapons "club." This is why chapter 8 stressed the weapons-proliferation potential of continuously refueled reactors; with those designs it is easy to remove fuel after any desired time in the reactor. "Cooking" nuclear fuel for a shorter time is an inefficient way to make electricity, but it results in much better bomb material.

Out of the Reactor: Spent Fuel

Fresh from the reactor, spent fuel is nasty stuff. It is so intensely radioactive that a few minutes' exposure in the vicinity of a spent fuel bundle would be fatal. Energy released in radioactive decay makes the fuel physically hot, so it requires active cooling. The heat and the radioactivity impose severe constraints on facilities for transportation and storage of spent fuel.

Nuclear News: Clothes Dryers, Plutonium, and Bombs

How much plutonium do nuclear reactors make? In a typical light-water reactor, roughly two plutonium-239 nuclei are formed for every three uranium-235 fission events. That answer isn't too enlightening. But it translates into this: For every hour's operation, a 1,000-megawatt nuclear power plant produces about 10 grams—1/3 ounce—of plutonium. In other words, the power plant forms 10 grams of plutonium for every 1,000 megawatt-hours of electrical energy it produces.

To make those numbers really meaningful, suppose that half of the United States' roughly 100 million families own electric clothes dryers, and that they run those dryers an average of 1 hour each day. How many bombs' worth of plutonium would *not* be made each year if they all used clotheslines instead?

A clothes dryer uses energy at the rate of about 5 kilowatts. So 50 million families using their clothes dryers for 1 hour each day consume 250 million kilowatt-hours of electricity, or 250,000 megawatt-hours (1 MWh=1,000 kWh). About 20 percent of the United States' electricity is from nuclear reactors, so the families' clothes dryers use 50,000 megawatt-hours of nuclear-generated electricity each day. At 10 grams of plutonium per 1,000 mega-watt-hours, that means the clothes dryers are responsible for 500 grams, or 0.5 kilograms, of plutonium each day. In a year, that makes about 180 kilograms, or 400 pounds, of plutonium.

How many bombs will that make? In chapter 5 we found that as little as 5 pounds of plutonium could constitute a critical mass. Plutonium from light-water reactors is contaminated with Pu-240, so 10–20 pounds might be needed to make a bomb. Still, that is 20–40 bombs per year just from using clothes dryers instead of clotheslines! Of course, it isn't easy to get that plutonium out of spent reactor fuel, and it is not trivial to make bombs from it. But the potential is there, and it is closely coupled to our everyday energy-consumption habits.

The most intense radiation, and therefore the most intense heating, come from short-lived fission products. Figure 10.9 shows that radiation and therefore heat drop rapidly as these fission products decay. Holding spent fuel at the reactor site for several years therefore reduces the problems associated with transportation and long-term storage. But even years-old spent fuel is far from benign. Longer-lived fission products—especially strontium-90 (29-year half-life) and cesium-137 (30-year half-life) retain significant radioactivity for several hundred years. Beyond that time, as figure 10.9 shows, the fission products have essentially decayed. But the waste remains radioactive for tens of thousands of years, thanks to the longer-lived transuranic isotopes and their decay products.

How dangerous is spent fuel? The world's nuclear power re-

Figure 10.9 Radioactivity in the waste from a typical light-water reactor, versus the number of years the waste has been out of the reactor. Note the scales on the graph; each division corresponds to a factor of 10 in time or radioactivity. The radioactivity drops by nearly a factor of 10 in the first 10 years. The leveling of the curve at 1,000 years is due to the long-lived transuranic isotopes, such as plutonium. (adapted from Anthony Nero, *A Guidebook to Nuclear Reactors* [University of California Press, 1979], p. 172)

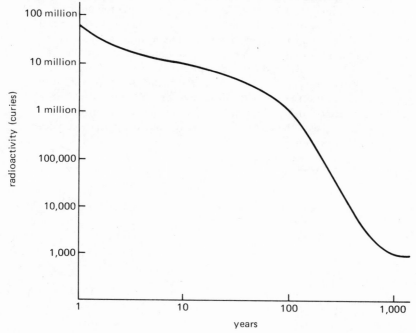

actors produce enough waste each year to kill the planet's human population many times over, if the wastes were to be ingested or inhaled. On the other hand, radioactive decay takes its inexorable toll on the wastes' lethality; nuclear advocates point out that 1,000-year-old nuclear waste may be less toxic than some cleaning products you probably store under your kitchen sink. (This presumes removal of plutonium from the waste, something which we will consider shortly.) Critics counter that nuclear wastes are an entirely new class of hazardous materials that have never before existed on Earth, and that 1,000-year-old nuclear waste decays far more slowly than the rapid chemical breakdown of most household poisons. In any event, nuclear advocates and critics agree that spent fuel is dangerous, and that it requires long-term isolation from human beings—indeed, from the entire biosphere.

How much nuclear waste is produced? In volume, not much. A 1,000-megawatt nuclear power plant produces only a few cubic yards of high-level nuclear waste in a year—little enough that it would fit under your dining room table. By the year 2000, all the waste generated since the beginning of the commercial nuclear power industry would cover a single football field to a depth of about 10 feet. Contrast that with the sprawling mountains of landfill that handle our more mundane domestic and industrial waste, or the huge volumes of gaseous waste that spew every second from coal-burning power plants. The small volume of radioactive waste is the reason on-site storage has been a viable temporary option, and is the reason we can consider a single waste-storage facility for the entire United States, occupying at most a few square miles. On the other hand, its virulent radioactivity and its long life make nuclear garbage unlike any other waste.

Disposing of Nuclear Waste

Can we find a means of disposal to ensure that radioactive spent fuel remains safely isolated for thousands of years? The problem is daunting, in part because we have no examples of human institutions that have lasted that long. How can we know enough about future conditions, both social and physical, to be sure that our waste doesn't put impossible burdens on our descendants?

Many who have worked on nuclear waste issues claim that satisfactory means of disposal are at hand now, and that the reason they have not been implemented is a lack of political will. Others lack confidence in long-term predictions of geological, climatic, and sociological factors involved in waste isolation. Some argue for further study; others claim that no solution will be forthcoming. Meanwhile, nuclear waste accumulates. Even if nuclear power plants ceased operating immediately, we would still need to dispose of several decades' worth of spent fuel now stored at reactor sites.

The options considered for nuclear waste disposal are numerous. They include the following:

- Launching waste into space, either in distant orbits about the Sun or the Earth or into the Sun. The cost of this option would be prohibitive, and the dangers of a launch accident—about 6 percent of rocket launches fail—make this an unacceptable option at present.

- Burial in thick sediment layers below the ocean bottom. Containers that would fall freely through the ocean waters and would bury themselves in sediments might make this an inexpensive method. On the other hand, little is known about the effects of deep ocean currents, and dispersal of waste could concentrate radioactivity in the food chain. Disposing of radioactive waste *on* rather than below the seabed is currently prohibited by international treaty, and increasing interest in mining and other seabed development makes many nations reluctant to pursue any kind of seabed waste disposal.

- Allowing the hot wastes to melt their way into or through the 2-mile-thick ice sheets of Greenland or Antarctica. But the long-term evolution of ice sheets is uncertain, and heating by the radioactive waste could induce unexpected changes in the behavior of the ice.

- Disposal in very deep holes, about 4 miles below the Earth's surface. The isolation provided by the extreme depth might be augmented by concentrating waste so that its heat would melt the surrounding rock—when the rock later cooled, the radioactive waste would be locked tightly into the rock structure. But drilling holes of the required depth is extraordinarily expensive. Fail-safe technology would be required to ensure that waste would reach the bottom of the hole. And little is known about the interaction of hot waste with the Earth's crust 4 miles down.

- Disposal at modest depths (about 1/2 mile) in geologically stable rock formations free from contact with groundwater. This method would ensure more human control over the emplacement and storage of the waste, and might allow for retrieval in the event of leakage, development of more appropriate disposal means, or discovery of new uses for the waste material.

- Transmutation of waste to shorter-lived isotopes that decay rapidly. This could destroy most of the long-lived transuranic elements in reactor waste, and might be accomplished by exposure to neutrons in appropriately designed reactors. A longer-term prospect is the "fusion torch," a hypothetical device that would use nuclear fusion to generate such high temperatures that nuclear waste would decompose into nonradioactive hydrogen.

Although several of these possibilities remain under study, the United States has opted for disposal in rock formations at modest depth. A nuclear waste repository would be a system of tunnels connected to the surface by shafts for the transport of personnel, equipment, and nuclear waste. Suitable rock formations must show long-term geological stability, and must be free of water that might eventually carry radioactive material to the surface. Figure 10.10 shows the proposed storage facility at Yucca Mountain, Nevada, while figure 10.11 shows the interior of an experimental waste-storage area at the nearby Nevada Nuclear Test Site.

Salt domes—huge underground deposits of naturally occurring rock salt—were originally chosen as especially appropriate for nuclear waste disposal. Since salt dissolves in water, its very presence indicates that no significant amount of water has been present for a very long time. Salt has the additional benefit of sealing itself in the event that cracks develop. On the other hand, studies show that even the minuscule amounts of water in a salt dome will migrate toward hot waste containers, surrounding them with highly corrosive salt water. Expansion of the salt due to heat from radioactive decay might open new avenues for water, even in deposits that had been dry for millions of years. And salt domes have many other potential uses, including mining of the salt itself and storage of oil reserves. What other uses might arise in 1,000 years, and how will

Figure 10.10 Artist's conception of the proposed repository for high-level nuclear waste at Yucca Mountain, Nevada. The underground storage area is shown in cutaway view at upper left; the remainder of the facility comprises receiving and preparation areas. (U.S. Department of Energy)

our descendants know that we have appropriated a particular salt dome for radioactive waste storage? Despite these potential problems, salt storage remains a serious option in the United States, and is currently employed in Germany.

Other rock structures under consideration in the United States include granite and the volcanic rocks tuff and basalt. Political pressures in the more densely populated eastern states have essentially eliminated the possibility of storing high-level radioactive waste in eastern granite. Neither of the other rock storage media has yet proved itself. In 1988 the federal government postponed indefinitely the opening of a storage facility for military nuclear waste, more than 2,000 feet below the New Mexico desert, that had been built at a cost of nearly $1 billion. This postponement came amid reports of excessive water leakage into the facility and of possible design and construction inadequacies. Similar concerns led to the 1989 reevaluation of plans for the first storage facility for

Figure 10.11 This experimental storage facility at the Nevada Nuclear Test Site houses commercial nuclear waste in cylindrical chambers beneath the floor of a tunnel bored in granitic rock. The facility is designed to test the rock's response to heat from the radioactive waste. (U.S. Department of Energy)

high-level waste from commercial nuclear power reactors, to be sited in volcanic tuff underlying Yucca Mountain in Nevada. These concerns centered on a rising water table that might leach radioactivity into groundwater, on the proximity of a supposedly dormant volcano whose eruption could breach the storage facility, and even on long-term climatic changes that could bring more rain and a higher population to the now arid and desolate site.

Despite these setbacks, many industry and government scientists are confident that underground waste repositories will ultimately prove safe and effective. They point to studies of the natural reactor at Oklo in West Africa, which show negligible migration of fission products and transuranics over 2 billion years. Even if waste containers fail—as is expected after 1,000 years or so—and if water seeps into the repository, the Oklo studies suggest it could be hundreds of thousands of years before waste would make its way to the surface. By that time, it would be essentially harmless. Critics respond with skepticism that realistic predictions can be made about events thousands of years hence, and point to a host of unknown factors—from intentional sabotage to earthquakes—that could compromise a waste-storage facility.

Transporting Waste and Preparing It for Disposal

It is one thing to store dangerously radioactive waste in a secure facility far below the Earth's surface; it is quite another thing to move the waste from commercial power reactors scattered around the United States to the storage repository. Many nuclear critics regard transportation as a weak link in the nuclear fuel cycle.

Spent fuel leaves a power plant in the same form in which it arrives: as bundles of fuel rods. But the presence of fission products makes it vastly more radioactive than fresh uranium fuel. Once a suitable repository has been established, spent fuel that has been out of the reactor for several years is to be removed from power plants and transported elsewhere for processing and eventual storage. Under current plans, spent fuel from all U.S. power plants would go first to a so-called *monitored retrievable storage facility* to be processed, packaged, and stored temporarily while awaiting transportation to the permanent underground repository.

Nuclear critics worry about using the nation's highways and railroads to transport nuclear waste. They point to numerous accidents, involving a variety of radioactive materials. Nuclear advocates counter with a record of over 3 million shipments of radioactive materials yearly in the United States, with no documented loss of life due to radiation. Only a minute fraction of these shipments involve the high radioactivity typical of spent reactor fuel. This highly radioactive waste is transported in special casks that provide for radiation shielding, cooling, and containment in the event of an accident. Among other criteria, these casks must survive falling 30 feet onto a hard surface, exposure to a temperature of 1,475°F for 30 minutes, and immersion in water for 8 hours without leaking any radioactive material. More dramatic tests have confirmed the integrity of some shipping-cask designs (figure

Figure 10.12 A truck carrying low-level radioactive waste from the Yankee Atomic Electric plant at Rowe, Massachusetts, crosses the yellow line as it rounds a bend on a mountain road. (Vermont Public Interest Research Group)

Figure 10.13 In a test, a locomotive moving at 82 mph slams into a flatbed truck carrying a cask designed for transport of highly radioactive waste. (U.S. Department of Energy)

10.13). Critics fault these seemingly impressive test results by noting that the casks used were obsolete designs that might prove more crashworthy than their contemporary counterparts. Critics also claim that the widespread publicity failed to note that cask leakage occurred in several tests, and that cask temperature ratings might be exceeded in fires involving commonly transported fuels and chemicals. Finally, critics see transportation as especially vulnerable to theft and sabotage, perhaps by nations or terrorist groups hoping to build nuclear weapons or to employ radioactive materials for blackmail. Is transportation of radioactive waste a serious danger to society? That nuclear question bears on our overall assessment of the safety of nuclear power.

Once it reaches a processing facility, spent fuel will be prepared for final disposal. The simplest preparation is to encase fuel bundles in heavy stainless steel cannisters for emplacement in the underground repository. These containers can be expected to last up to 1,000 years, after which the repository rock must provide containment for the wastes. More efficient use of repository space can be achieved by compacting the waste. A proposed compaction scheme involves incorporating ground-up nuclear waste into molten glass. The solidified glass rods would be encased in stainless steel for emplacement in the repository. Even after the steel corrodes, the glass should permit only a very slow release of radioactivity. Glassification is already in use in a number of European countries.

Reprocessing

As figure 10.8 shows, spent fuel still contains significant amounts of uranium-235 and plutonium-239. Removing these substances would greatly reduce the amount of long-lived radioactivity in the waste; also, it would allow further use of these fissile materials as nuclear fuels—either for power generation or for weapons. **Reprocessing** is the separation of fissile uranium and plutonium from spent fuel.

Reprocessing involves chopping up the spent fuel rods and dissolving the fuel pellets in acid. Chemical separation then removes the uranium and plutonium. The remaining fission products are ready for glassification and subsequent disposal as nuclear waste. The uranium, containing both U-235 and U-238, is put through gaseous diffusion or another enrichment process. The plutonium is mixed with enriched uranium to make "mixed oxide"—a mixture of uranium and plutonium oxides that can substitute for uranium as a reactor fuel. Alternatively, plutonium can be diverted to the production of nuclear weapons. A nuclear fuel cycle involving reprocessing is illustrated in figure 10.14; compare this with the simpler once-through cycle of figure 10.2.

Is reprocessing a good thing? Several European and Asian countries, heavily committed to nuclear power and without major uranium resources of their own, think so. They have developed sophisticated reprocessing plants in which robot systems take the highly radioactive spent fuel through a complex sequence of chemical and physical processing stages. These plants are even available for export, often to less advanced nations seeking to exercise their nuclear options. An international market in reprocessed plutonium is developing, with intercontinental shipment of plutonium now a reality.

Most nuclear critics—and even some proponents of nuclear power—consider plutonium reprocessing undesirable. Highly fissile plutonium would be widely available in a world economy encouraging reprocessing. Through theft, terrorism, or legal purchase, a

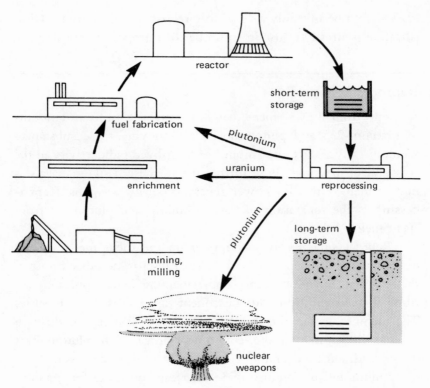

Figure 10.14 Nuclear fuel cycle with reprocessing. Uranium-235 and plu-
tonium-239 are removed from spent fuel and recycled into new reactor fuel.
Alternatively, plutonium can be used for weapons production. Compare
with the once-through cycle of figure 10.2.

nation or group seeking nuclear weapons capability might easily
obtain the necessary plutonium. And although plutonium from
spent reactor fuel is not ideal bomb material, it is, as was noted
earlier, sufficient for a crude fission bomb. And a nation with
reprocessing facilities could easily produce bomb-grade plutonium
from specialized reactors.

In 1977, concern over weapons proliferation led President
Jimmy Carter to ban reprocessing of commercial reactor fuel in the
United States. At the same time, Carter deferred work on the
Clinch River Breeder Reactor, further signaling the United States'
turn away from a plutonium-based commercial nuclear power industry. In 1981 President Ronald Reagan lifted the reprocessing
ban, but since then the nuclear industry has shown little interest in

pursuing reprocessing. With its major reserves of uranium, the United States has no urgent need to reprocess spent fuel. (Reprocessing has occurred in the United States, though, at specialized plants that separate weapons-grade plutonium made in special plutonium-production reactors.)

Would a global reprocessing ban help limit the spread of nuclear weapons? That question will be explored further in chapter 17; for now, note that five of the six known nuclear weapons nations used plutonium in their first nuclear explosives. (Only China chose the enriched-uranium route. Plutonium and uranium weapons evolved simultaneously in the United States.) But none of that plutonium came from civilian power reactors. On the other hand, that pattern might not hold in the future, as reprocessing plants and reprocessed plutonium increasingly become commercial commodities. How to handle the international development of nuclear power and its plutonium by-product is among the most sensitive nuclear issues.

Other Nuclear Waste

So far we have concentrated exclusively on nuclear waste from spent reactor fuel; because of its intense radioactivity, spent fuel is designated **high-level waste** (HLW). But there are many other sources of nuclear waste. In the United States, the nuclear weapons industry has produced considerably more high-level waste than all the nation's power reactors. And a great many activities produce **low-level waste** (LLW), whose radiation emission is low enough and brief enough to permit less stringent handling and disposal requirements. Low-level waste is generated in the nuclear fuel cycle during refinement, enrichment, and fabrication of nuclear fuel. Contaminated clothing and tools, filters used to clean primary-loop cooling water, piping, and hardware are among the low-level wastes generated in nuclear power plants. Radioisotope tracing techniques used in medicine, research, and industry provide additional low-level waste; in states lacking nuclear power plants or weapons facilities, hospitals and universities are generally the largest producers of low-level radioactive waste. A third officially designated waste category is **transuranic waste** (TRU), consisting of mildly radioactive but very long-lived heavy radioisotopes from the production of weapons or the reprocessing of nuclear fuel.

High-level waste from the weapons program and spent fuel from commercial reactors are destined for the same permanent underground repositories. In the absence of any such repositories, on-site storage of weapons wastes is currently practiced under far less safe conditions than prevail at commercial nuclear power plants. Low-level waste from all sources is handled very differently, typically by burial in shallow trenches at designated dump sites (figure 10.15). Still another source of mixed nuclear waste is the nuclear power plants themselves, as they reach the ends of their roughly 40-year life spans. Should they be decontaminated, disassembled, and transported to appropriate storage sites? Or entombed and guarded as monuments to a technology whose short-term benefits masked the long-term problem of radioactive waste storage? The **decommissioning** of old reactors remains a largely unsolved problem.

Figure 10.15 Burial trench for low-level radioactive waste at the Hanford nuclear reservation. (Westinghouse Hanford Company)

Nuclear News: Not in My Backyard!

Ultimately, we are all responsible for the generation of nuclear waste. But few of us relish the prospect of a waste dump in our backyard or even in our state. That is one reason why storage facilities for high-level waste are most likely to be built in sparsely populated areas of the western United States. Low-level waste, much less radioactive but occupying much more volume, calls for more localized disposal.

For years nearly all the United States' low-level waste was shipped to sites in South Carolina, Nevada, and Washington. But a 1980 federal law, passed at the urging of those three states, mandated that each state be responsible for its own low-level waste. Since then, states have been scrambling to find technically and politically suitable sites within their own borders. That is not easy: Citizen opposition to nuclear-waste dumps has been vigorous. Identification of potential waste sites has united neighbors of diverse political backgrounds into active waste-dump opponents who often practice civil disobedience in efforts to prevent exploration of the dump sites.

Are these citizen actions justified? They are driven in part by the general public fear—partly founded in ignorance—of all things nuclear and particularly of radiation. And the difference between high- and low-level waste may not always be clear. On the other hand, citizens learn almost daily of new radiation hazards and of unsafe conditions at nuclear facilities, especially in the U.S. nuclear weapons industry. Caution seems in order. And yet, where is the waste to go?

News source: "Counties Battle Radioactive Waste Sites," *New York Times,* February 26, 1990.

Summary

Nuclear power plants and a host of other nuclear endeavors produce radioactive waste. Spent reactor fuel contains shorter-lived but highly radioactive fission products and longer-lived transuranic isotopes, including plutonium. Disposal of these materials with safety ensured for thousands of years poses technical and political challenges that have prevented the United States from opening a waste repository over a half-century since the first nuclear reactor went critical. Although many would argue that technical solutions are at hand, others point to continuing scientific uncertainty and lack of confidence in government agencies as evidence that the waste-disposal problem may not be solved in the near future. Reprocessing of spent fuel reduces the burden of long-lived radioactive materials and reduces the need for fresh uranium; at the same time, reprocessed plutonium and the technology to prepare it may facilitate the spread of nuclear weapons. Nuclear waste remains among the thorniest problems of the nuclear age.

Further Reading

Niel A. Chapman and Ian G. McKinley, *The Geological Disposal of Nuclear Waste* (Wiley, 1987). A detailed technical description of nuclear waste and its underground disposal, from a British perspective.

Bernard L. Cohen, "The Disposal of Radioactive Wastes from Fission Reactors," *Scientific American* 236 (June 1977), p. 21. A well-written article detailing many of the issues associated with underground nuclear waste disposal. Excellent graphs and diagrams accompany the text. The author is an independent scientist who is among the United States' most eloquent and persistent advocates of nuclear power.

Michio Kaku and Jennifer Trainer, *Nuclear Power: Both Sides* (Norton, 1982), chapter 4. An introduction to nuclear waste disposal, followed by two diametrically opposed essays on the subject.

Raymond L. Murray, *Nuclear Energy,* third edition (Pergamon, 1988), chapter 22. A straightforward but technical account of nuclear issues.

Marvin Resnikoff, *The Next Nuclear Gamble: Transportation and Storage of Nuclear Waste* (Council on Economic Priorities, 1983). A thoroughly detailed and carefully researched criticism directed primarily at transportation and short-term storage of nuclear wastes.

U.S. Department of Energy, *Answers to Your Questions on High-Level Nuclear*

Waste (DOE Office of Civilian Radioactive Waste Management, 1987). A government pamphlet (DOE/RW-0152), authoritative but clearly designed to assuage citizens' fears of nuclear waste.

Charles A. Walker, Leroy C. Gould, and Edward J. Woodhouse, *Too Hot to Handle? Social and Policy Issues in the Management of Radioactive Wastes* (Yale University Press, 1983). A good discussion of the social and political issues surrounding nuclear waste disposal, including a section on public perception of the risks from nuclear power and other technologies.

Glossary

decommissioning   The process of disassembling or entombing a nuclear reactor at the end of its useful life, and of safeguarding the associated radioactive material.

enrichment   The process of increasing the proportion of fissile U-235 in uranium, to approximately 3 percent for light-water reactor fuel and to 90 percent or more for weapons.

fuel rod   A structure composed of uranium fuel pellets contained in a tube of stainless steel or zirconium alloy, forming the basic fuel structure in a nuclear reactor.

gaseous diffusion   A uranium enrichment process in which U-235-containing molecules pass more readily through a membrane than do slower-moving molecules containing U-238.

high-level waste (HLW)   Nuclear waste derived from spent fuel in power or weapons-production reactors, and typically characterized by a mix of intensely radioactive fission products and longer-lived but less radioactive transuranic isotopes.

low-level waste (LLW)   Any radioactive waste not classified as high-level or transuranic, and characterized by low radioactivity and relatively short half-life.

nuclear fuel cycle   The sequence by which uranium progresses from its natural state through fuel fabrication, fission in a reactor, processing, and eventual long-term storage of nuclear waste.

once-through fuel cycle   Simple fuel cycle in which spent reactor fuel is stored as radioactive waste, without any attempt to remove and reuse fissile U-235 or Pu-239.

reprocessing   The removal of fissile uranium-235 and plutonium-239 from spent reactor fuel, making these materials available for use as reactor or weapons fuel.

salt dome   An underground structure consisting of a large volume of natural rock salt, often considered among the best candidates for underground nuclear waste storage.

transuranic waste (TRU)   Radioactive waste consisting of transuranic isotopes (heavier than uranium), and characterized by modest radioactivity and half-lives that may range to tens of thousands of years and more.

Alternatives to Nuclear Fission

11

Is nuclear power safe? There is no simple answer to that question, except for the certainty that no technology can be 100 percent safe. The question then becomes "Is nuclear power *acceptably* safe?" The answer, ultimately, is a personal judgment. Technical know-how can help you with that judgment, but vigorous disagreement even among nuclear experts indicates that technical considerations alone cannot decide the issue.

But one factor that *can* help you in your personal decision about nuclear power is how it stacks up against its alternatives. You may not like nuclear power, and you may not think it particularly safe. But if you want electricity and if you find the alternatives *less* safe, then you may have to judge nuclear power acceptable. Or you may conclude that existing alternatives are preferable, or that new energy sources should be developed. Finally, contemplation of nuclear power and its alternatives might lead you to a basic reconsideration of your appetite for energy.

In this chapter we will consider alternatives to nuclear power, weighing such factors as safety, economics, environmental effects, and availability to meet our short-term and long-term energy needs. But first we need to look at basic energy sources and the flow of energy about Planet Earth.

Earth's Energy Balance

Only a few basic sources of energy are available to us inhabitants of the Earth's surface. They include sunlight (whose ultimate origin is nuclear fusion), heat energy from within the Earth (some left over from the planet's formation and some arising from natural radioactive decay), mechanical energy from the motion of the Earth and the Moon, and energy trapped in the nuclei of the atoms that make up the Earth. Those four are not what you would probably think of when asked to list prospective energy sources. The things that might come to mind—coal, oil, gas, falling water, nuclear fission, solar energy, wood, wind, harnessing of the tides, geothermal energy, burning garbage, methane from cow manure, nuclear fusion—are all specific examples of the four basic sources: sunlight, terrestrial heat, planetary motion, and nuclear energy.

In the natural course of things, those four are far from equal. More than 99.98 percent of the energy supplied to the Earth's surface comes from the Sun. Nearly all the rest—less than 0.02 percent of the total—comes from the Earth's interior, along with an even tinier amount of tidal energy, whose ultimate origin is the motion of the Earth and the Moon. We humans have hardly altered those proportions, largely because human energy consumption remains a minuscule fraction—about 1/10,000—of the total global energy flow. Of the human portion of that energy flow in the late twentieth century, nearly 90 percent comes from fossil fuels, which contain solar energy stored by plants that lived millions of years ago.[1] Most of the remaining energy we use is also ultimately solar, extracted primarily from falling water that was earlier evaporated from oceans warmed by sunlight. Only a small fraction of global human energy consumption—at most a few percent—is not ultimately solar in origin. Today, the bulk of that nonsolar energy is from nuclear fission.

What happens to all the energy reaching the Earth's surface? Figure 11.1 depicts the global **energy balance**, showing the energy flows that drive weather and ocean circulation, sustain life, and power human activities. But ultimately, as the figure shows, essentially all the energy becomes heat that is radiated into space. If the

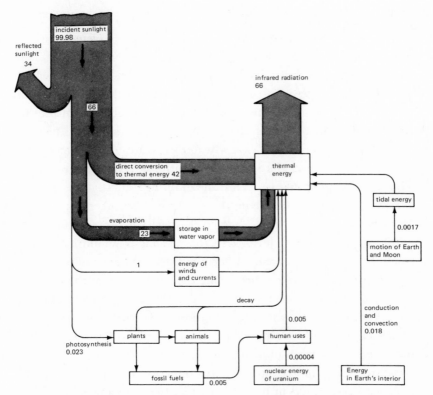

Figure 11.1 Earth's energy balance. Virtually all the energy reaching the Earth's surface comes from the Sun, with additional tiny amounts from heat in the Earth's interior, from the motion of the Earth and the Moon via the tides, and from nuclear energy released in fission reactors. This diagram shows the flows of energy that drive the Earth's weather, sustain living things, and power technological civilization. The numbers are percentages of the total energy flow. Note that the Earth ultimately radiates to space essentially the same amount of energy it receives; if this balance did not hold, the planet's temperature would change. (adapted, with permission, from Romer, *Energy Facts and Figures* [Spring Street, 1985])

rate at which energy leaves Earth weren't equal to the rate at which it comes in, then energy would build up or be depleted, and the planet's temperature would change.

Energetically, Earth is like a bathtub that is partly full of water, with the drain open and the faucet on. If water flows into the tub as fast as it goes down the drain, then the water level in the tub remains the same. But if the inflow increases slightly, then the water level will rise until excess water pressure forces more water down the drain and the inflow and outflow again become equal. The same thing would happen if the drain became slightly clogged: The water level would rise until excess pressure forced the outflow again to equal the inflow. Conversely, a slight decrease in inflow would drop the water level. (All these changes would have to be small; large changes would either overflow or drain the tub.)

The Earth's temperature is like the water level in the bathtub. Changes in either the inflow or the outflow of energy alter the temperature. Changes in inflow could have natural causes, such as variations in the Sun's power or changes in the Earth's orbit, or they might arise from human activities, such as blockage of sunlight by atmospheric pollution or the introduction of additional energy that had been stored in nuclear or fossil fuels. Changes in outflow could arise from natural changes in vegetation or ice cover that affect the radiation of energy into space, or from such human activities as the introduction of carbon dioxide and other gases that block the escape of energy through the atmosphere. This latter effect is at the root of contemporary concern about global climate change. As we explore energy alternatives, however, it is important to recognize that the Earth's energy balance can be upset in other ways as well.

In considering the advantages and disadvantages of various alternatives to nuclear power, we will often focus on short-term technical, environmental, or economic comparisons that have little to do with the Earth's energy balance. But in the long run, every energy source must be weighed against its potential for disrupting the energy flows shown in figure 11.1, and we will have to apply that demanding test to any sources that show long-term potential for meeting human energy needs.

The Fossil-Fuel Alternative

We hardly think of fossil fuels as "alternative energy sources," since they are so much in the mainstream of our energy economy. But fossil-fuel power plants are an alternative to nuclear plants, and they are the alternative that today's electric utilities are most likely to entertain.

The formation of fossil fuels represents a very small deviation from perfect global energy balance. Solar energy captured by plants is normally returned to the global energy flow in a short time, as plants and the animals that eat them die and decay. But a minute fraction of dead material is transformed, over hundreds of millions of years, into deposits of coal, oil, and natural gas, whose chemical structure locks in the energy that the plants originally captured from the Sun.

The fossil fuels we use today were formed over hundreds of millions of years. But we are burning them at such a rate that even the most optimistic estimates suggest that the known reserves will be exhausted in at most a few hundred years. Whatever the short-term prospect for fossil fuels, our long-term consumption of them is but a brief burst of frenzied burning, like a match that flares momentarily in the dark. Fossil fuels are not, in any long-term sense, a sustainable energy resource.

Fossil fuels are no long-term alternative, but today they supply nearly 90 percent of humanity's energy needs, and they will continue to be important for at least several decades. How do fossil fuels compare with nuclear fission over this shorter time scale?

Fossil fuels release energy when they burn, combining with oxygen in the air. These fuels consist primarily of carbon and hydrogen, so the principal products of combustion are carbon dioxide (CO_2) and water (H_2O). But smaller quantities of other substances, ranging from sulfur to uranium and radon, also occur in fossil fuels, and these may be released in combustion. Coal contains sulfur, and when burned it produces sulfur dioxide gas (SO_2), which acidifies rain; when the acid rain falls, it may damage lakes, forests, and even steel and concrete. Other contaminants end up as tiny solid particles that cloud the air and damage human lungs. Still other chemical substances, many of them harmful, are formed when burning is incomplete. The result is that fossil-fuel power plants (to say nothing of motor vehicles) are major sources of atmospheric pollution.

Air pollution does more than harm trees and lakes; it kills people, primarily through respiratory disease. It is as hard to pin a particular case of emphysema on the burning of coal as it is to blame a specific case of cancer on radiation, but epidemiological studies suggest that some 10,000 people die prematurely each year in the United States as a result of emissions from coal-burning power plants. That yearly U.S. figure is comparable to the number who will die *worldwide in 50 years* from the Chernobyl nuclear accident. Coal is hardly benign, and it almost certainly kills far more people than nuclear power. On the other hand, coal cannot be made into bombs that threaten hundreds of millions of deaths. So which is worse, coal or nuclear fission?

Air pollution at today's levels is not a necessary consequence of fossil-fuel consumption. Processes exist for removing pollutants, either before the fuel is burned or as it goes up the smokestack. Devices called *electrostatic precipitators* use static electricity to remove particulate pollution from coal-burning power plants; installed to meet air-quality standards, precipitators have dramatically reduced the visible pollution from power plants (figure 11.2). Flue gases can also be "scrubbed" to remove sulfur dioxide. A more radical approach is to treat coal chemically to turn it into a clean-burning gas. These and many other technologies for pollution reduction are available, but they are costly and they consume a significant percentage of a power plant's electrical output. Cost is a principal reason why coal-burning utilities in the Midwest have resisted strengthening of the U.S. Clean Air Act. Ultimately, economics

Figure 11.2 In foreground is the Coal Quality Development Center in Homer, Pennsylvania, a joint project of the U.S. Department of Energy and the Electric Power Research Institute aimed at developing cleaner-burning coal. In background, the Homer City coal-fired generating station shows little visible emission from its tall stacks, although background haze may be related to sulfur dioxide. The visible clouds are condensation from the cooling towers. (U.S. Department of Energy)

and politics, not technology, impede further reduction of atmospheric pollution.

Air pollution is not the only harmful effect of fossil energy sources. Extracting and transporting oil take an egregious environmental toll, especially on oceans and coastlines. Storms and fires kill workers on offshore oil-drilling platforms. Coal-mine accidents in the United States alone have killed nearly 100,000 miners since 1905, and many more have died of "black lung" disease. Strip mining of coal can devastate the landscape (figure 11.3). Accidents involving coal trains alone may average one death per year for each 1,000-MW coal-burning power plant.

Of course accidental deaths and environmental damage result from the mining and transportation of uranium, too. But here the nuclear difference is a big advantage. Because nuclear fuel packs a million times more energy than its fossil counterparts, nuclear power requires far fewer mines, far fewer miners, and much less transportation. (Recall figures 2.1 and 2.2.)

Suppose we really took coal accidents and oil spills seriously,

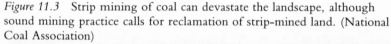

Figure 11.3 Strip mining of coal can devastate the landscape, although sound mining practice calls for reclamation of strip-mined land. (National Coal Association)

installed state-of-the-art pollution controls on fossil-fuel power plants, and got the mining industry to reclaim strip-mined land. Even then fossil fuels could pose a grave environmental danger. Burning these fuels inherently produces carbon dioxide. CO_2 is not a minor contaminant, nor (in the usual sense of the word) is it a pollutant. Instead, it is the natural product of the burning process. And CO_2 is like a wad of hair that clogs the bathtub drain in our analogy for Earth's energy balance: It impedes the flow of energy from Earth into space. In the decades since we began significant use of fossil fuels, the CO_2 content of our atmosphere has been climbing steadily. If no compensatory effects occur, increased atmospheric CO_2 must inevitably result in a higher global temperature. Some scientists believe such a **global warming** has already begun, and point to abnormally high temperatures during the 1980s as evidence. Others think we have yet to detect global warming, but few doubt that it will inevitably occur. The consequences could dwarf any nuclear disaster short of all-out nuclear war. Even a temperature increase of a few degrees could raise sea level enough to endanger coastal cities, where a significant portion of the populace now lives. Changing precipitation patterns might turn productive land into deserts, possibly exacerbating the problem by destroying plants that now absorb carbon dioxide. Social and political upheaval would surely accompany such drastic environmental changes.

Although all fossil fuels contribute to global warming, coal is significantly worse because it is predominantly carbon. Natural gas, richer in hydrogen, produces less carbon dioxide per unit of energy released. Oil is intermediate between gas and coal. Unfortunately, oil and gas are in the shortest supply. Furthermore, there are huge coal reserves in China, making much greater increases in atmospheric CO_2 likely as that most populous nation industrializes. Another international nuclear question: Is it better to supply developing nations with nuclear-power technology, which could lead to the production of weapons, or to allow those countries that have fossil-fuel reserves to pour carbon dioxide into our common atmosphere? Or are there, perhaps, alternatives more palatable than either of these?

Which is the safer way to make electricity: with fossil fuel, or with nuclear fission? Table 11.1 compares some undesirable effects of nuclear and coal-burning power plants. Clearly neither energy source is without risk. But which is worse? Do we prefer the steady and rather high level of death and damage from the coal plant to the seemingly more benign nuclear plant, with its uncertain possibility of a major accident and its still-undisposable wastes? And how do the uncertainties of global warming figure into our assessment? Or the dangers of nuclear war from diversion of nuclear materials to weapons production?

Fossil fuel is often held up as the only real alternative to nuclear fission. This has forced people to argue strongly for one or the other, even when neither may be particularly attractive. But that either/or choice is profoundly limiting. There *are* alternatives to coal and nuclear power plants. Some are not particularly viable at this time; others may never be. On the other hand, some are surprisingly effective and economical, and these could play a much larger role in our energy supply.

Centralized or Decentralized?

What do we mean by "alternatives to nuclear fission"? Today, fission is used almost exclusively to generate electricity, so we mean alternative ways of generating electricity. Given a proposal for a 1,000-MW nuclear power plant, we can most easily visualize as an alternative some other facility that would also generate 1,000 MW. This is one reason why fossil-fuel and nuclear power plants are often the only candidates when a new power plant is considered: Each can be built in large, centralized facilities.

But it doesn't have to be this way. Ten 100-MW power plants produce as much electricity as a single 1,000-MW plant; so do a hundred 10-MW plants. And today there are more energy options available for the smaller plants. That alone doesn't make the smaller options better; it is certainly important to consider economies of scale along with the environmental and safety consequences of having many smaller plants utilizing different energy sources. But it does mean that smaller plants afford a wider variety of energy options, and that is important when an individual or society is uncomfortable with both the nuclear and the fossil option.

Table 11.1
Coal versus nuclear fission.

1,000-MW coal-burning plant	1,000-MW nuclear plant
15 tons CO_2 every minute; 1 ton SO_2 (sulfur dioxide) every 5 minutes (causes acid rain); nitrogen oxides, equivalent to exhaust of 200,000 cars; miscellaneous chemical carcinogens; uranium, thorium, radon, and other natural radioactive substances	Small quantities of radioactive gases: radon, xenon, tritium, iodine
30 pounds of ashes per second	2 cubic yards of high-level radioactive waste per year
30 deaths to general public each year from respiratory disease associated with air pollution	1 death every 2 years to general public from radiation-induced cancer associated with normal emissions, nuclear waste, and reactor accidents ("official" estimate); "anti-nuclear" estimate 2–10 deaths per year
1.5 deaths each year from coal-mining accidents	0.2 deaths each year from uranium-mining accidents
4 deaths each year among coal miners from "black lung" disease	0.1 deaths each year among uranium miners from radon-induced lung cancer
20,000 cases each year of nonfatal respiratory disease from air pollution	1 case per year of nonfatal radiation-induced thyroid disease

Source of health and safety data: William Ramsay, *Unpaid Costs of Electrical Energy* (Resources for the Future and Johns Hopkins University Press, 1979).

Some would argue that **decentralization** of electric-power generation is an important end in itself. Carried to its extreme, a decentralized electrical economy would have a mix of small generating facilities, ranging from the scale of an individual home to that of an industry or a small community. Such a system could be less vulnerable to major disruption and would naturally involve "gentler" energy sources, such as wind, water, and sunlight. Proximity to one's energy source would also encourage more active conservation efforts, and advocates of extreme decentralization envision a society with considerably lower per-capita energy use than our current 100 "energy slaves." On the other hand, a fully decentralized electrical system would require radical economic changes in our society, and would not be without its own environmental consequences. Is it better, for instance, to dam every available stream for small-scale hydroelectric production, or to concentrate the equivalent of hundreds of small hydro plants into a single nuclear or coal plant, or a large hydroelectric dam?

For those who do not advocate decentralization for its own sake, the immediate importance of small-scale electric-power generation lies in the diversification of energy sources. Sources that could not be considered for large power plants might work in a mix of smaller plants, thus broadening the range of viable energy alternatives.

Hydroelectric Energy

Hydropower is a prime example of a sustainable energy resource. As long as the Sun continues evaporating water from the oceans, rain and snow will continue to fall on the continents. This water makes its way into rivers and flows toward the oceans, and the energy of its motion can be tapped to generate electricity. Hydropower is one of our more benign sources of electricity. It produces no atmospheric pollution and no waste heat. It is quiet and requires no fuel. The flow of water can easily be adjusted to match the output of electrical power to the demand, something that is not easily done with large thermal power plants. Hydroelectric plants come in all sizes, extracting energy from small streams and mighty rivers (figures 11.4, 11.5).

Nuclear News: Decentralization

Advocacy of decentralized electric-power generation is sometimes viewed as a politically radical position. Actually, the utility industry is quietly pursuing decentralization because it makes economic sense. In an age of fuel-price fluctuations and uncertain growth in energy demand, it is good business to add generating capacity in small increments rather than in 1,000-MW power plants that can take a decade or more to bring on line. Most power plants now being ordered are smaller gas-fired power units, often incorporating cogeneration to put waste heat to work. But more radical decentralization may be coming. The Edison Electric Institute, the utility industry's trade association, is sponsoring tests of miniaturized power plants for individual homes. The gas-fired units, only 5 feet long by 2½ feet square, would use cogeneration to supply both electricity and heat. The 1990 cost of $8,000 per house could drop to $3,000 by the mid-1990s. Widespread use of this decentralized technology could radically alter our patterns of energy use.

News sources: "Smaller-Scale Technologies Dominate New Generation After 1995," in U.S. Department of Energy, *Annual Energy Outlook* (DOE/EIA-0383, 1989); "Mini Power Plants May Light Homes of the Next Decade," *Wall Street Journal,* January 26, 1989; "Better Ways to Make Electricity," *New York Times,* April 11, 1990.

But even hydropower is not without its problems. Dams block free-flowing rivers, flooding large areas and radically altering the local ecology. Even at the end of the twentieth century, Native Americans are being displaced from their lands to make way for hydroelectric development in the Canadian wilderness. Increased evaporation from the dammed water raises the level of salt and pollutants in the water, making it less fit for agricultural irrigation or human consumption. When river water reaches the slow-moving lake water behind a hydroelectric dam, it drops its load of mud and sediments, gradually filling the lake with silt. In 50–200 years most hydroelectric sites will become useless because of this silting.

But at least it's safe, right? No: A hydroelectric dam can break, sending a wall of water toward downstream communities. And dam failures do occur, typically at the rate of several per decade in

Figure 11.4 This small hydroelectric generating station on Vermont's Lamoille River produces 5 MW of hydroelectric power. (Central Vermont Public Service Corporation)

the United States. In the middle years of this century, U.S. dam failures caused an average of 42 deaths per year. The rate of dam failures and the resulting death rate may be significantly higher than the rate of life-threatening accidents at nuclear power plants. And failure of a very large dam could cause as many fatalities as a major meltdown at a nuclear plant.

Hydroelectric power is the dominant source of electricity in a number of countries. But in the United States hydropower supplies only about 8 percent of our electricity, well under half of nuclear fission's share. About 30 percent of the United States' hydroelectric capacity has been developed. Even complete exploitation of our hydroelectric resources would therefore bring hydro's share of U.S. electricity production to only about 25 percent. But complete exploitation is unlikely on environmental grounds, since we would surely want to leave some rivers flowing freely. The numbers make it clear that hydroelectric power alone could not replace nuclear and fossil fuels as the United States' predominant source of electricity at our current consumption rate.

Figure 11.5 The Grand Coulee Dam in Washington State produces nearly 6,500 MW, the equivalent of six large nuclear or coal-burning plants. (Bureau of Reclamation)

Wind Energy

The Sun warms parts of the Earth to different temperatures, resulting in differences in atmospheric pressure that drive the flows of air we call *wind*. Harnessing the wind for transportation dates back thousands of years, and the use of land-based windmills to pump water or to power industry is at least centuries old. Nearly all modern windmills turn generators to produce electric power. Wind, like falling water, is a sustainable energy source that originates with nuclear fusion in the Sun's core.

The energy available from wind is not insignificant. In a 20-mile-per-hour wind, the wind in an area of 1 square yard carries energy at the rate of about 400 watts. A well-designed windmill can extract somewhat under half this power. Power increases by a factor of 8 for each doubling of wind speed. But wind speeds fluctuate, giving much lower average wind power. Even so, the United States' wind energy resources are many times its current energy use.

On the other hand, this substantial energy resource is spread thinly. Even in windy locations, the *average* power one can expect to extract from a square yard of wind is only about 50 watts. Thus very large windmills are needed to produce significant power. For an average power of 1 MW, a windmill would require blades over 500 feet in diameter, and 1,000 such machines would be needed to replace a single 1,000-MW nuclear or coal-fired power plant. Furthermore, the variability of the wind means that windmills can't directly replace a steady power source, such as a coal or nuclear plant, unless some means of energy storage is provided. At present, the technology for storing energy lags well behind the technology for generating electricity.

The energy crisis of the 1970s led to a revival of interest in small-scale wind generators as well as to serious proposals for large-scale production of electric power from the wind. Wind machines with peak power outputs of several megawatts were built, and "energy farms" consisting of hundreds and even thousands of windmills were constructed. Figure 11.6 and 11.7 show some modern windmills.

How has wind power fared economically? As recently as 1980,

Figure 11.6 A modern windmill for generating electric power. Designed for individual homes, this unit produces 2.1 kW in a 28-mph wind. (Enertech)

Figure 11.7 Part of the large wind-energy farm at Altamont Pass, California. More than 6,000 windmills occupy the site, generating a total average power of over 100 megawatts. (Pacific Gas & Electric Company)

wind-generated electricity averaged 25¢ per kilowatt-hour, well above the typical retail price of about 10¢/kWh. But improved windmill design and utility companies' experience with wind energy systems have now lowered that cost to a more competitive 7–8¢/kWh, and installation costs per unit of generating capacity today are comparable to those of fossil-fuel power plants and well below those of nuclear plants. Today's economic and regulatory climate makes it hard for wind to penetrate the commercial electricity market, but continued cost reductions could see wind supplying a significant amount of electricity by the year 2000.

Are windmills safe? Are they environmentally benign? Are they a serious alternative to nuclear fission? Windmills emit no pollution and do not share hydroelectric dams' potential for disastrous failure. Because of the need to service equipment on high towers, wind power does present greater occupational hazards than other energy sources. Although windmill towers take up land area, the surrounding land is still available for farming and grazing. The principal environmental effect is aesthetic: Do we want our landscapes dotted with large numbers of windmills? One estimate suggests that half of the United States' electricity could be supplied by 6 million wind generators mounted on 300,000 towers scattered

Nuclear News: Killer Windmills

The world's largest wind-powered electricity-generating facility is the Altamont Pass Energy Farm in California, shown in figure 11.7. Its more than 6,000 windmills have a total average power output over 100 MW. But Altamont Pass lies on the migration route for large birds of prey that winter in Mexico. Since 1986, nearly 100 eagles, hawks, owls, and falcons have been killed by flying into the whirling 40-foot blades of the Altamont Pass windmills.

At a cost of $120,000, the California Energy Commission has hired biologists to study the problem. Protecting the birds may require exacerbation of the windmills' "aesthetic pollution"—they may be painted in bright colors, or noisemakers may be added to frighten oncoming birds.

News source: "California Windmills Killing Big Birds of Prey," *New York Times*, December 3, 1989.

---

throughout the American Great Plains. Each tower would be the height of a 70-story building, and there would be nearly one tower for every square mile. Is that radical change in our landscape preferable to a few hundred localized nuclear power plants?

Wind energy may well make an increasingly important contribution to our electrical energy supply. However, it is unlikely to be developed on such a scale that wind alone could replace nuclear and fossil-fuel power plants. In conjunction with other alternatives, though, wind certainly appears to be a potentially realistic sustainable energy source.

Biomass Energy

Roughly 1/5,000 of the solar energy that reaches Earth is absorbed by plants and stored chemically through the process of photosynthesis. Plants and animals, ourselves included, live on the solar energy trapped by photosynthetic plants. Our cells release that energy in the chemical processes of metabolism.

Energy stored by living things can also be released quickly, through combustion. Wood and animal dung are burned as the primary energy sources in many developing countries. The energy crisis of the 1970s prompted renewed interest in the so-called **biomass** energy sources, even in industrialized nations. Wood-fired electric-power plants up to 50 MW in size have been built in the United States (figure 11.8). In other applications, bacterial digestion converts biomass into convenient liquid and solid fuels, including methane from animal wastes and alcohol from corn and other grains.

If biomass fuels are consumed at the same rate at which they are grown, then there is no net contribution to atmospheric carbon dioxide (since growing plants remove CO_2 from the air). But large-scale harvesting of forests for fuel would be doubly harmful, since the CO_2-utilizing plants would be gone and their combustion would contribute additional CO_2 to the atmosphere. Biomass could constitute a sustainable energy source, but sustainability isn't automatic.

What about pollution? Burning wood is notorious for produc-

Figure 11.8 The 50-MW McNeil Generating Station in Burlington, Vermont, is the world's largest wood-fired electric-power plant. Note the large pile of wood chips behind the plant. The structure between the main building and the smokestack is an electrostatic precipitator for removing particulate air pollution. (Burlington Electric Department)

ing particulate matter and various organic chemicals, some of them carcinogenic. On the other hand, wood has less sulfur than either coal or oil. Conversion of biomass to methane and alcohol results in very clean-burning fuels; this is one reason why the automobile industry is increasingly interested in "gasohol," a blend of biomass alcohol and gasoline from fossil petroleum.

Is there enough biomass? Estimates suggest that the United States could fill virtually all its current energy needs if biomass were grown for energy on all U.S. cropland. That, of course, is impossible, since we need to eat. And it is very uncertain whether a large-scale switch to biomass would be economically feasible. There is also an ethical question about diverting land from food crops to energy production in a world where much of the population is malnourished. On the other hand, energy from waste material alone—agricultural waste, slash from lumbering operations, and municipal garbage—might satisfy 5–10 percent of the U.S. energy demand. Biomass could make a significant contribution to our energy supply, but alone it will not replace our dominant energy sources.

Tidal Energy

Figure 11.1 shows that a tiny fraction of the energy input to the Earth comes from the motions of the Earth and the Moon. This energy drives the tides, sloshing the waters of the ocean back and forth in response to the differing pull of the Moon's gravity on opposite sides of the Earth. Tidal energy is normally dissipated as heat when the rising tides churn against the continents, but it is possible to harness tidal energy to generate electricity.

A tidal power plant is much like a hydroelectric plant, using the energy of moving water to spin turbines attached to generators. Tidal generators are mounted in a dike that blocks the entrance to a bay or a river mouth where there is large tidal variation. As the tide rises, water flows through the dike to turn the generators, filling the bay. When the tide ebbs, water returns to the ocean, again turning the generators. This cyclic process means the tidal plant cannot produce electricity continuously.

The total amount of tidal power is small, and there are few

suitable sites for tidal power plants. Development of all the tidal sites in the United States and Canada might replace only ten or fifteen 1,000-MW power plants. Furthermore, tidal plants disrupt the ecology of the coastal waters, which serve as breeding grounds for marine life. Although tidal power might be significant in a few localized regions, it is no replacement for our current means of large-scale electric-power generation.

Geothermal Energy

Heat flows from the hot interior of the Earth to the surface. The average rate of energy flow is very low, except at "hot spots" where this **geothermal energy** accumulates. These hot spots are responsible for volcanoes, geysers, and hot springs. They are also sites where we can "mine" geothermal energy for heat, industrial processes, and electric-power production.

The geothermal areas most suitable for electric-power generation contain reservoirs of naturally occurring steam that can drive turbines directly. Such a steam field runs the world's largest geothermal power plant, a 1,350-MW facility in California (figure 11.9). Other hot spots have volumes of heated water; hot, dry rock; or molten magma. Technologies to exploit these other geothermal resources are less advanced.

Geothermal energy is not without its problems. As currently exploited, it is certainly not sustainable; at the Geysers plant in California, heat is withdrawn at 80 times the rate at which it is replenished from below, giving the plant a useful life of at most a few decades. Extraction of heat also brings up water often high in salt and other potential contaminants; concomitant release of hydrogen sulfide and other subterranean gases may pollute the surrounding air. Ground collapses and even earthquakes could be triggered by the extraction of geothermal energy. And development of geothermal resources can damage areas of ecological or scenic value.

Geothermal energy may be an energy alternative—albeit an imperfect one—for Hawaii and California. But could it really make a dent in the U.S. energy supply? Even the most optimistic estimates suggest that a full national commitment to geothermal de-

Figure 11.9 Pipes carry steam from geothermal wells to a generator building at the Geysers geothermal power plant in California. (Pacific Gas & Electric Company)

velopment would bring geothermal energy's share to less than 10 percent of our capacity to generate electric power. Geothermal energy alone is not the alternative to nuclear fission.

Nuclear Fusion

Nuclear fusion powers the sun, driving hydroelectric, wind, solar, and biomass energy systems; earlier it created the fossil fuels that now supply 90 percent of our electricity. What about harnessing nuclear fusion on Earth? Can our technology mimic the Sun? A little more space will be devoted here to fusion than to other energy alternatives, in part because of fusion's long-term promise and in part because fusion is, in its own right, a *nuclear* technology.

Fusion involves the joining of light nuclei, a process made difficult by the mutual electrical repulsion of the positively charged nuclei. We looked at the basics of fusion in chapter 5, and identified

Nuclear News: Endangered Rain Forest

The Hawaiian Islands are the product of one of the world's most active geothermal hot spots, currently situated below the southeast coast of the big island of Hawaii. Now, as increasing development raises Hawaii's demand for electricity, plans are afoot to construct a 500-MW geothermal power plant in the midst of the United States' last significant tropical rain forest. Controversy over the plant has pitted environmentalists and native peoples against advocates of economic growth.

How is the issue to be decided? Is the desire to run air conditioners in new tourist hotels reason to destroy a rain forest? Are there other options? Would a nuclear plant be better? How about an OTEC plant (to be described below)? These energy choices cannot be made without a clear understanding of the alternatives.

News source: "Energy Project Imperils a Rain Forest," *New York Times,* January 26, 1990.

the technological challenges: heating fusing material to 100 million degrees[2] and containing that hot stuff.

In the Sun, fusion involves conversion of ordinary hydrogen ($^1_1$H) to helium. But fusion is easier with the heavy hydrogen isotope deuterium ($^2_1$H), and even easier with a mixture of deuterium and radioactive tritium ($^3_1$H). Terrestrial fusion efforts have therefore concentrated on deuterium and deuterium-tritium fuels. D-D (deuterium-deuterium) fusion joins two deuterium nuclei to form helium-3 and a neutron[3]:

$$^2_1H + {}^2_1H \rightarrow {}^3_2He + {}^1_0n + \text{energy}.$$

D-T (deuterium-tritium) fusion yields helium-4 and a neutron:

$$^2_1H + {}^3_1H \rightarrow {}^4_2He + {}^1_0n + \text{energy}.$$

Both reactions produce lots of energy; for a given weight of fuel, these reactions are nearly 4 times as energetic as nuclear fission.

It isn't necessary to contain the fusing hydrogen isotopes forever—just long enough to extract at least as much energy from the fusing material as it took to heat and confine it. The more densely packed the fusion fuel, the shorter the necessary confinement time. The two principal approaches to fusion attack the confinement problem in diametrically opposite ways.

Magnetic-confinement fusion, under development since the 1950s, makes use of the fact that electrically charged particles are "trapped" by magnetic fields (figure 11.10). At fusion temperatures atoms are stripped of their electrons, giving rise to a charged-particle gas called a **plasma**. In the most promising devices, the fusion plasma is heated by electric currents, high-energy radio waves, or beams of particles. It is confined in a chamber shaped like a donut so that its magnetic field wraps around, with no end where particles could escape. The **tokamak**, a Russian invention now under study throughout the world, is currently the most likely candidate for a magnetic fusion reactor.

Inertial-confinement schemes compress fusion fuel to such very high density that significant fusion energy is produced before the material has time to disperse; it is confined by its own inertia. Devices now under study bombard tiny pellets of deuterium-tritium fuel with converging beams from high-powered lasers. So much power is required that, for the billionth of a second they are actually firing, the lasers consume more power than all the world's electric-power plants produce. (This is a good place to recall the difference between energy and power: While on, fusion lasers consume energy at an enormous rate, so the power level is high; but

---

Figure 11.10 Charged particles are "trapped" in a magnetic field, spiraling around the magnetic lines of force. Bending the lines into circles makes a magnetic "bottle" that confines the particles.

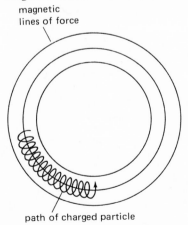

magnetic
lines of force

path of charged particle

Figure 11.11 The toroidal (donut) shape of the Tokamak Fusion Test Reactor at Princeton University is most effective for confining fusing material with magnetic fields. (Princeton Plasma Physics Laboratory)

they are on for only a billionth of a second, so the energy used is relatively small.) The conditions within the pellet mimic those inside a star or an exploding hydrogen bomb, and a flash of fusion energy results. A practical power plant might fuse several fuel pellets each second to produce a continuous supply of energy.

Unlike the other energy alternatives we are considering, controlled fusion is not yet a proven technology. (Uncontrolled fusion—in the form of thermonuclear weapons—has been with us since the 1950s.) After decades of research, fusion experiments are just nearing "scientific breakeven," the point at which the energy produced from fusion exceeds that needed to heat the fusing material. A practical fusion power plant is much further off; many design and engineering problems remain unsolved. Today's best estimates suggest that fusion reactors might begin contributing to our energy supply sometime after the year 2010. It is not clear that fusion will compete economically with still-abundant fossil fuels or uranium, even then.

Figure 11.12 The target chamber of the Nova laser fusion facility at the Lawrence Livermore National Laboratory. Ten intense laser beams converge onto a pellet just a few millimeters in diameter. (Lawrence Livermore National Laboratory)

Figure 11.13 For a brief instant, the fusing pellet shines like a miniature star. (Lawrence Livermore National Laboratory)

Nuclear News: Progress toward Fusion

Large-scale fusion research began in the 1950s, amidst claims that 20 years' development would make commercial fusion reactors a reality. In 1990 that goal is still at least 20 years off, but considerable progress has been made. Baffling problems with magnetic confinement have been solved, and the most recent generation of fusion devices is approaching scientific break-even. Princeton University's next-generation machine, the Compact Ignition Tokamak, could ignite a self-sustaining fusion reaction by the end of the century. Laser-fusion devices, relative newcomers in the field, have made even more rapid gains. Figure 11.14 traces progress toward a fusion reactor.

The pursuit of fusion is an international effort, with significant contributions from the United States, the Soviet Union, Japan, and the European Community. The United States, long a fusion leader, is now falling behind. Budget cuts have pushed the U.S. milestone of scientific breakeven toward the mid-1990s, making it likely that the Joint European Torus in England will be the first fusion device to achieve breakeven. Japan is taking the lead in inertial-confinement fusion; U.S. research on the subject is hampered by secrecy regulations, since inertial-fusion technologies also have applications in "Star Wars" defense systems. In fact, Japanese journals regularly publish laser-fusion articles that would be classified in the United States. Despite fusion's immense promise, U.S. funding for controlled-fusion research has declined in recent years. In contrast, funding for the destructive use of fusion—thermonuclear weapons—continues at some 20 times the level for peaceful fusion research.

News source: "U.S. Losing Ground in Worldwide Race for 'Hot' Fusion," *New York Times,* June 20, 1989.

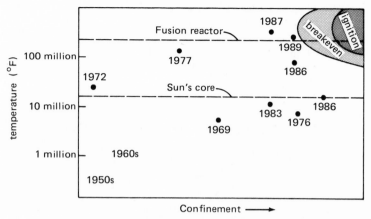

Figure 11.14 Controlled thermonuclear fusion requires both high temperature and confinement of the fusion fuel. Dates marked show when fusion
experiments throughout the world have achieved particular combinations of
temperature and confinement. The goals are breakeven (a net production of
energy from fusion) and ignition (initiation of a self-sustaining fusion reaction). A prototype fusion power plant might follow, perhaps by the year
2005. (Princeton Plasma Physics Laboratory, adapted)

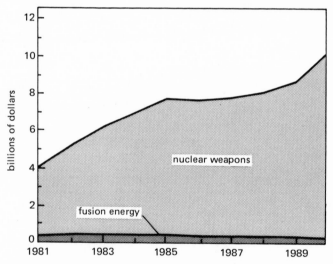

Figure 11.15 U.S. Department of Energy budgets for controlled fusion and
for nuclear weapons. (data sources: U.S. House of Representatives, Subcommittee on Energy and Power and U.S. Department of Energy, Budget History Table)

Would fusion be safe? It would be far safer than fission, and far less polluting than fossil fuels. The end product of both D–D and D–T fusion is harmless, nonradioactive helium gas. D–T fusion does use radioactive tritium as a fuel, but tritium can be "bred" right in the reactor, eliminating the need to transport and store large quantities of radioactive fuel. Furthermore, its 12-year half-life makes tritium far safer to handle than the longer-lived radioisotopes from fission reactors. Some radioactive materials would also result from neutrons absorbed in the walls of the fusion reactor. But the amount of radioactivity involved is orders of magnitude less than what is produced in a fission reactor, and the material would remain significantly radioactive only for decades rather than millennia.

Could a fusion reactor explode, or run out of control like Chernobyl? No: Fusion does not involve a chain reaction, with its potential for explosive neutron multiplication. Fuel would be supplied continuously to a fusion reactor, whose reaction chamber would contain only a few seconds' worth of fuel—in sharp contrast to the several years' fuel supply in the core of a fission reactor. It is so difficult to maintain the high temperature needed for fusion that any deviation from normal operation would immediately cool the fusion fuel and halt the reaction. And once fusion stops, there is essentially no residual radioactivity and therefore no need for additional cooling. Meltdown of a fusion reactor is impossible.

How about weapons proliferation? Thermonuclear weapons—"hydrogen bombs"—do involve fusion reactions. But fusion weapons must incorporate fission explosives to generate the temperature needed to ignite fusion, so fissile materials remain crucial. Furthermore, fusion weapons are technologically far more difficult to construct than fission weapons. There has been some thought of using the neutrons produced in fusion reactors to "breed" fissile plutonium for fueling fission reactors, especially if fusion should prove economically marginal on its own merits. A single fusion breeder could supply plutonium for twenty fission reactors. That plutonium might also be used for weapons. On the other hand, the full technological and economic success of fusion would eliminate the need for fission fuel and would therefore make fusion breeders unattractive.

Is there enough fuel to make fusion a practical alternative to fission and fossil fuels? Yes! One out of every 6,500 hydrogen atoms is deuterium, and the oceans are full of hydrogen-containing water. The deuterium in a gallon of ordinary water makes that water the energy equivalent of 360 gallons of gasoline. At humanity's current energy-consumption rate, the deuterium in the oceans could meet our entire energy demand for 300 billion years—nearly 100 times longer than the Sun will continue to shine. If any energy source is "unlimited," surely it is nuclear fusion.

Fusion fuel is so readily available that successful fusion technology might supply all our energy, not just electricity. Since fusion reactors would be safe enough to be located near industrial or population centers, their waste heat might be used for domestic heating or industrial processes. Fusion-generated electricity could decompose water into hydrogen and oxygen, with the hydrogen used in place of oil and natural gas. We would then have an easily transportable fuel whose only combustion product would be water.

Fusion sounds like a nearly ideal energy source. If fusion power plants become available in the next few decades, they may reduce or eliminate our dependence on fossil fuels and nuclear fission. Widespread use of fusion might avert the global climatic crisis threatened by the buildup of carbon dioxide from our continued burning of fossil fuels.

But in the long run, fusion itself could bring on the ultimate climatic crisis. The energy released in fusion would not otherwise be available on Earth; it would represent a new input to the global energy flow. Like all the rest of the global energy depicted in figure 11.1, fusion energy would ultimately become heat that Earth would have to radiate into space. As long as humanity kept its energy consumption a tiny fraction of the global energy flow there would be no major problem. But history shows that human energy consumption grows rapidly when it is not limited by shortages of fuel. Fusion fuel would be unlimited, so our species might expand its energy consumption to the point where the output of our fusion reactors became significant relative to the global input of solar energy. At that point Earth's temperature would inevitably rise. This long-term criticism of fusion holds for any energy source that could add to Earth's energy flow even a few percent of what the Sun provides. Only solar energy itself escapes this criticism.

Solar Energy

Solar energy is the favorite of many alternative-energy advocates, while traditionalists dismiss it as unlikely to make a major contribution to our energy supply. Who is right?

There is *plenty* of solar energy available. Figure 11.1 shows that solar energy reaches Earth at 10,000 times humanity's energy-consumption rate. Solar energy is also sustainable, since the Sun will shine for another 5 billion years.[4] And because it provides 99.98 percent of the energy reaching the Earth's surface, solar energy is the only source that could provide for greatly increased human energy consumption without upsetting the global energy balance.

Solar energy's main disadvantage is that it is relatively diffuse, with the enormous solar energy flow spread over the entire daytime half of the globe. But even though it is spread out, solar energy is concentrated enough to be impressive. Direct sunlight falling on a square yard carries energy at the rate of about 1 kilowatt. Thus your 100 energy slaves' worth of energy consumption—10 kilowatts—could be supplied by 10 square yards of direct sunlight. You would have to multiply that by a factor of about 5 to compensate for night and clouds. Still, 50 square yards—about one-third the floor area of a typical American home—is not much space to devote to your entire energy supply.

How might solar energy be harnessed? Many solar technologies are very simple. The right building design alone can allow solar energy to supplement or even supplant other sources of light and heat. Active solar collector systems can replace gas and electric water heaters, which are the second-largest users of energy (after space heating) in typical homes. Inexpensive solar ovens made from a few pieces of reflective material could curb the deforestation now occurring as burgeoning populations of developing nations seek wood for cooking.

But can solar energy compete with high-technology nuclear reactors in generating electricity? There are, in fact, several ways to produce electricity from sunlight. One approach treats the sun as the heat source for a thermal power plant. Mirrors focus sunlight to boil water or other fluids and drive a turbine as in a conventional fossil-fuel or nuclear plant. Energy-storage systems may allow power generation to continue for several hours in the absence of sunlight. **Solar-thermal power plants** have been built in the American southwest, with several hundred megawatts' worth installed in California alone in the 1980s. By 1990 these plants were producing electricity at a wholesale cost of 7–8¢ per kilowatt-hour, just a few cents above the cost of electricity from conventional power plants. Solar-thermal power plants represent a solar alternative that may well be competitive with conventional energy sources in a very few years.

A second approach to solar-thermal power generation uses the tropical oceans as a vast solar collector. Sunlight warms surface waters to temperatures over 80°F, while water several thousand feet down remains at 40°F. A thermal power plant can be operated by using a fluid that boils between these temperatures. The fluid would boil on contact with the warm surface water, turn a turbine, and condense on contact with cool water pumped from below. The

Figure 11.16   Aerial view of a solar-thermal power complex built in the late 1980s by Luz Engineering in California's Mojave Desert. The peak power output is 90 MW, and the facility produces electricity at a cost of 7–8¢ per kilowatt-hour. The land area used is about 500 acres, comparable to the site areas of larger nuclear power plants. (Luz Engineering)

Figure 11.17 Close-up of sun-tracking parabolic concentrators of the complex shown in figure 11.16. Synthetic oil flows through the thin pipes, is heated by concentrated sunlight, and transfers its energy to boil water that drives a steam turbine. (Luz Engineering)

efficiency of such a plant would be very low, for the thermodynamic reasons we considered in chapter 7. But since it would require no fuel, inefficiency would be no great disadvantage. Many engineering and ecological problems will have to be solved before such **ocean thermal energy conversion** (OTEC) schemes become practical, but a concerted effort could see OTEC plants producing as much electricity by the year 2010 as the United States now gets from nuclear fission.

A most promising solar application is direct conversion of sunlight to electricity by means of **photovoltaic cells.** These are thin wafers of specially treated silicon. When sunlight strikes the silicon, it dislodges electrons from their atoms and sets them into motion, producing electric current. Photovoltaic cells have the advantage of no moving parts, and a photovoltaic power plant could be as simple as an array of cells electrically connected together.

The physical basis of photovoltaic devices lies in the same technology of silicon semiconductors that drives the ongoing revolution in microelectronics. Silicon, the essential material for photovoltaic cells and computer chips, is the second most abundant

Figure 11.18 Energy past and energy future? These photovoltaic cells are part of a 1-MW solar photovoltaic power plant operated by the Sacramento Municipal Utility District in California, and are the first stage of a planned 100-MW system. In the background are the reactor containment building and the cooling towers of the Rancho Seco nuclear plant, since closed by a voter referendum. (U.S. Department of Energy)

element (after oxygen) in the Earth's crust; ordinary sand is largely silicon. The problem lies in forming high-purity silicon crystals in enough quantity to cover large areas.

How large an area must be covered? Typical solar cells convert about 15 percent of the incident sunlight to electricity. Coupling this cell efficiency with a conservative average of 200 watts of available solar power per square meter means that 30 watts of solar power could be generated for each square meter of cells. The United States consumes electrical energy at an average rate of 300,000 megawatts, which would require 10 billion square meters of solar cells. That sounds like a big area, but it is less than 3 percent of the state of New Mexico. Claims that there isn't enough solar energy, or that its widespread use would gobble up too much land, are plainly wrong.

Solar photovoltaic cells have no moving parts, are made from an abundant resource, and could supply all our electrical energy while occupying a relatively small area. So what is the hitch? Many would argue that photovoltaic cells are too expensive to compete with traditional energy sources. That certainly *was* true: Before the energy crisis of the 1970s, solar cells cost several hundred dollars for an average power output of 1 watt—versus a dollar or so for fossil-fuel power. But by 1980 the cost of photovoltaic cells had fallen below $100 per average watt.[5] And through the 1980s, as energy traditionalists continued to dismiss photovoltaics by quoting 1970 costs, the price of solar cells dropped by a factor of 10, to just under $10 per average watt in 1990. Meanwhile the cost of nuclear plants in the United States rose dramatically; for example, New Hampshire's 1,150-MW Seabrook plant came on line in 1990 at a cost $6.5 billion. That's $5.65 per watt if we assume continuous full-power operation, and almost $9 per watt if Seabrook matches the average 63 percent of rated power output for nuclear plants in the United States. On the other hand, photovoltaic power plants require more than just photocells; mounting structures and auxiliary equipment might double the cost per watt. Nevertheless, the quiet revolution in photovoltaics has made this benign energy source increasingly competitive with nuclear fission (figure 11.19).

If solar photovoltaic cells are so good, why aren't they already in widespread use? Shouldn't market forces naturally result in the adoption of this economical and nonpolluting energy technology? No, and for several reasons. First, there is no large-scale photovoltaic industry today. In 1990, the world's total production of photovoltaic cells was just over 100 megawatts' worth—1/10 the power output of a single large nuclear plant. Second, environmental and other costs of competing energy sources are not usually included in economic calculations. Utilities don't take the cost of global warming into account in planning new coal-burning plants, nor do they figure the cost of U.S. military operations in the Persian Gulf when considering oil-burning facilities. And since there is no operational facility for the storage of nuclear waste in the United States, the waste-disposal costs of nuclear plants are at best estimates. Finally, market forces are not the only factor in determining our mix of energy sources. As figure 11.20 shows, government subsidies for energy research and development favor fossil fuels and

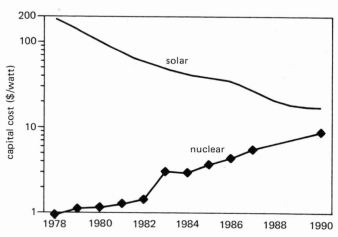

Figure 11.19 Comparative costs of constructing photovoltaic and nuclear power plants. Photovoltaic costs assume an average of 250 watts of solar energy per square meter and suppose that solar cells account for half the total cost of a solar power plant. The steeper cost drop in 1987 represents the introduction of lower-cost "amorphous" silicon solar cells. Fission costs are those of actual power plants that became operational in the years indicated, adjusted for the fact that nuclear plants today average about 63 percent of their rated capacity. Nuclear costs do not include fuel, waste disposal, decommissioning, or other operating expenses. Note the uneven scale; nuclear costs have risen by a factor of 10, and solar costs have fallen by a similar factor, during the 1980s. (data sources: solar costs from Yoshihiro Hamakawa, "Recent Advances in Solar Photovoltaic Technology," *Business Japan,* September 1989, figure 2; nuclear costs from Christopher Flavin, *Reassessing Nuclear Power: The Fallout from Chernobyl* [Worldwatch Institute, 1987], figure 3, with Seabrook added in 1989)

nuclear fission over solar energy. A reordering of government priorities could help make solar photovoltaics a safe, economical, and widespread alternative to nuclear fission.

Energy Conservation

Do Americans really need the energy that alternatives to fission could provide? We received a partial answer in figure 6.2, which showed that several of our global neighbors use energy twice as efficiently as we do. More of the answer came out in chapter 7, with the recognition that two-thirds of the energy released in generating electric power gets dumped to the environment as waste

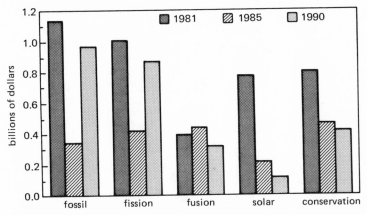

Figure 11.20 U.S. Department of Energy budgets for energy research. Despite growing concern about nuclear safety and global warming, well-funded fission and fossil-fuel efforts increasingly overpower solar and conservation. (data source: U.S. House of Representatives, Subcommittee on Energy and Power)

heat. The fact is that our high-energy industrial society was built on the promise of cheap energy. Conservation came into vogue with the huge energy-cost increases of the 1970s, but the United States and many of its industrialized neighbors remain profligate wasters of energy.

How much better could we do? Serious, mainstream analyses suggest that we Americans might cut our energy consumption in half without significantly changing our lifestyles. Figure 11.21 shows one reason for this: The energy-consuming devices of our everyday lives are far from their maximum possible energy efficiencies.

Wouldn't energy conservation be costly? No. In fact, conservation measures often eliminate the need for energy production costing many times more. Investing $8 million in low-loss windows, for example, could save $300 million in oil. Improving the insulation of electric water heaters costs 30¢ for every kilowatt reduction in the energy-consumption rate; building nuclear plants to produce that power would cost several dollars per kilowatt. In other cases, changes in outmoded tax laws and regulations would enhance the economic viability of energy conservation. Dramatic reduction in our energy consumption—dismissal of at least 50 of

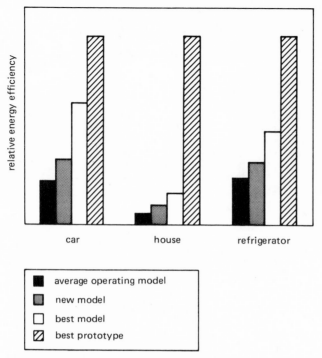

Figure 11.21 Potential for energy efficiency in common energy-consuming technologies. Graph shows relative energy efficiency for the average existing devices, for the average new models, for the best commercially available models, and for the best experimental prototypes. For cars, today's U.S. average is 18 miles per gallon; the best prototypes get 77 miles per gallon. (data source: *Scientific American,* September 1989, p. 141)

our 100 "energy slaves"—is within our grasp; political and social will are required to accomplish it.

Summary

Energy alternatives and conservation demand a book of their own. Here we have touched briefly on short-term and long-term alternatives to nuclear fission. Your author does not find fission among the worst of the alternatives, or among the best. In the coming years of global climatic change, fission may well be preferable to continued burning of fossil fuels. But fission and fossil fuels are not the only options. An expanded solar industry could soon displace the need for more conventional sources. The successful development of fusion would give us a safer nuclear alternative. And conservation measures now available could postpone the need to choose among energy alternatives. Finally, the prospect of a growing and increasingly industrialized human population suggests that the only long-term energy alternatives that are compatible with a stable global climate are those that make use of solar energy, which is already part of the global energy flow.

Notes

1. A small minority of scientists believe that at least some of the substances we consider fossil fuels are actually primordial constituents of the Earth. If that is the case, their energy is not solar in origin.

2. At least, that is the conventional wisdom. "Cold fusion" schemes seek fusion at room temperature. At present they appear improbable; see Nuclear News: Cold Fusion in chapter 5.

3. About half the time, D-D fusion results in tritium ($^3_1$H) rather than helium-3. The tritium is then consumed by the D-T reaction.

4. We will need more than a replacement energy source after that time. In its death throes, the Sun will expand to the orbit of Mars, engulfing Earth.

5. The price per average watt assumes an average solar-energy availability of 250 watts per square meter, one-fourth that of direct sunlight. Solar costs are often given per peak watt, which would be one-fourth the cost given here.

Further Reading

Gordon Aubrecht, *Energy* (Merrill, 1989). A general introduction to energy concepts, including a careful treatment of energy alternatives.

Michael Brower, *Cool Energy: The Renewable Solution to Global Warming* (Union of Concerned Scientists, 1990). Up-to-date information on the prospects for renewable energy sources by a group known for its opposition to nuclear power and its concern for the global environment. Although the book puts renewable energy in its best light, the data presented are authoritative and current.

John H. Gibbons, Peter D. Blair, and Holly L. Gwin, "Strategies for Energy Use," *Scientific American* 261 (September 1989), p. 136. An up-to-date analysis of energy conservation strategies by researchers at the U.S. Congressional Office of Technology Assessment.

Jack J. Kraushaar and Robert A. Ristinen, *Energy and Problems of a Technical Society,* revised edition (Wiley, 1988). A good, accessible, detailed discussion of all aspects of energy use in technological societies. But this 1988 edition still quotes 1978 figures for the cost of photovoltaic cells.

Amory Lovins and John H. Price, *Non-Nuclear Futures: The Case for an Ethical Energy Strategy* (Ballinger, 1975). An articulate criticism of the nuclear alternative by a leading advocate of efficient energy use.

Peter L. Murphy, editor, *How to Make a Star: The Promise of Laser Fusion.* Publication LLL-TB-85, Lawrence Livermore National Laboratory, 1986. A thorough but simplified account of laser fusion, with a discussion of energy-use trends and comparisons with fission and fossil fuels.

William Ramsay, *Unpaid Costs of Electrical Energy: Health and Environmental Impacts from Coal and Nuclear Power* (Resources for the Future and Johns Hopkins University Press, 1979). Quantitative comparisons of all aspects of coal and nuclear power, from mining accidents to pollution and waste disposal.

Scientific American 263 (September 1990). This issue, devoted to the theme "Energy for Planet Earth," includes articles on nuclear fission and its many alternatives.

Glossary

biomass   Material of biological origin used as an energy source. Biomass energy is ultimately solar, trapped by plants through the process of photosynthesis.

D-D fusion   Nuclear fusion involving the reaction of two deuterium nuclei to produce helium-3 and a neutron. Another D-D reaction produces tritium and a proton, with the tritium consumed in subsequent D-T fusion.

D-T fusion   Nuclear fusion involving the reaction of a deuterium nucleus and a tritium nucleus to produce helium-4 and a neutron. D-T fusion is easier to achieve than D-D fusion.

decentralization   The spreading of electric-power-generation capability over a large number of smaller sources, as opposed to concentration in fewer large power plants.

energy balance   A situation in which a system (e.g., the Earth) is maintained at constant temperature because the energy entering the system is equal to the energy leaving the system.

geothermal energy   Energy from the Earth's interior. Geothermal energy arises from the decay of natural radioactive elements within the Earth, and some may be left over from the planet's formation billions of years ago.

global warming   Heating of the Earth due to upsetting of the global energy balance. Today, global warming appears to be resulting from the emission of CO_2 and other gases that trap energy that would otherwise be radiated into space.

hydropower   Power generated from moving or falling water, as at dams and waterfalls. Hydropower is ultimately solar in origin.

inertial confinement A fusion scheme in which deuterium or deuterium-tritium fuel is confined briefly by its own inertia after being compressed by high-powered laser beams.

magnetic confinement A fusion scheme in which deuterium or deuterium-tritium fuel at 100 million degrees is contained by magnetic fields.

ocean thermal energy conversion (OTEC) Use of the temperature difference between warm ocean surface water and cool deep water to drive a turbine-generator, producing electric power.

photovoltaic cell A device, made of specially treated silicon or other material, that converts sunlight directly into electricity.

plasma A gas of electrically charged particles. At the high temperatures needed for nuclear fusion, matter is necessarily in the plasma state.

solar-thermal power plant A power plant in which the heat from concentrated sunlight is used to boil water, with the steam driving a turbine-generator.

tokamak A promising fusion device whose donut shape helps keep particles from escaping the magnetic confinement.

Nuclear Weapons

III

History and Technology

12

Our nation can be destroyed in half an hour, and there is nothing we can do to prevent it. That is a basic truth of the nuclear age. The potential for national and even global devastation lies fundamentally in the nuclear difference, that millionfold increase in the energy associated with nuclear reactions. The energy stored in the tens of thousands of nuclear weapons in the world's arsenals exceeds by many orders of magnitude the destructive capability that humankind has invoked in all its wars.

Quantifying Destruction

Nothing in our experience compares even remotely to the destructive effects of nuclear weapons. Relatively few human beings—residents of Hiroshima and Nagasaki—have ever witnessed those effects. Most of them are dead. Even those who survived the effects of the small, crude nuclear weapons used on those two cities cannot imagine warfare with today's thousands of multiple-warhead missiles, each of which carries the destructive power of up to 300 Hiroshima bombs.

How are we to quantify these weapons, whose effects lie so far beyond experience or even imagination? The scientists who worked on the first nuclear bombs were well aware of the million-fold nuclear difference; they knew that a few pounds of nuclear explosive could yield the equivalent of thousands of tons of chem-

Figure 12.1   A 12.5-kiloton bomb, small by today's standards, damaged or destroyed over 90 percent of Hiroshima's buildings. 100,000 people died immediately; by the end of 1945 the toll had reached 140,000, and by 1950 200,000 had died from the bombing. (Department of Defense)

ical explosive. Thus was born the **kiloton** (kt), a unit of nuclear explosive energy (also called *yield*) equal to that of 1,000 tons of the chemical explosive TNT. By the 1950s, technology had outpaced the kiloton, and the **megaton** (Mt)—1,000 kilotons, or the equivalent of 1 million tons TNT—became the appropriate unit for the largest weapons.

How can we make these abstract numbers meaningful? Only by comparisons with realized destruction. The medium-size city of Hiroshima was essentially destroyed by a single 12-kiloton bomb; figure 12.1 thus provides a yardstick for imagining nuclear destruction. The largest nuclear weapon ever tested yielded nearly 60 megatons. Today's arsenals range from multi-megaton weapons down to battlefield nuclear explosives with sub-Hiroshima yields. The strategic nuclear weapons that the superpowers have aimed at each other's homelands range typically from a few hundred kilotons to several megatons. The United States' MX missile, for example, carries ten 300-kiloton warheads—making each missile the equivalent of nearly 250 Hiroshimas.[1]

How are we to visualize entire arsenals of nuclear weapons? In all of World War II, the explosive energy released in all weapons used by all sides was the equivalent of 6 million tons of TNT.[2] Most of that destructive power was produced by chemical explosives, from the gunpowder driving individual bullets to bombs weighing thousands of pounds. This 6 megatons also includes the two nuclear bombs dropped on Japan. If we visualize 6 megatons as being "one World War II" of destruction, then we have an appropriate unit for describing the destructive potential of today's nuclear arsenals. In those terms, a single MX missile, with its ten 300-kt (0.3-Mt) warheads, carries half a World War II of destructive potential. And the newest Trident submarine—armed with 24 missiles, each with up to fifteen 500-kt warheads—carries nearly 30 World War IIs.

Today's Nuclear Arsenals

A year after World War II ended, the world's nuclear arsenal consisted of nine Nagasaki-type bombs, possessed by the United States; their combined explosive yield of 200 kt was the equivalent of 1/30 World War II. By 1950, just after the Soviet Union exploded its first nuclear weapon, the global arsenal held the nuclear equivalent of 12 World War IIs. The world's arsenal peaked in 1973 at 25,000 megatons, or more than 4,000 World War IIs. Today it is about two-thirds of that value—still almost 3,000 World War IIs.

What, exactly, do these weapons do for us? What could they do *to* us? How might we intend to use them? How many do we really need? Can we protect ourselves from them? Will nuclear weapons spread to other nations? What is the connection between nuclear power and nuclear weapons? What can be done to control these instruments of unparalleled destruction? We will explore these nuclear questions in subsequent chapters. First, let us look briefly at the history of nuclear weapons and at the technology that makes them work.

A Brief History

Hiroshima's destruction occurred less than 7 years after Lise Meitner and Otto Frisch first recognized the occurrence of nuclear fission in uranium bombarded by neutrons. The world scientific community immediately saw the potential for a weapon of unprecedented destructiveness, although its technical feasibility remained

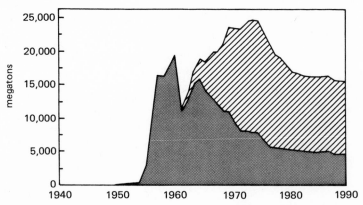

Figure 12.2 Evolution of the American (solid) and Soviet (shaded) nuclear arsenals, as measured in megatons. The decrease is associated with the deployment of smaller but more accurate warheads. For reasons discussed in note 1, megatonnage is a crude measure of destructive capability. That capability has not decreased significantly, and it remains essentially equal for the two nuclear superpowers. Including the other nuclear-weapon nations would add about 500 Mt to the totals shown here. (data source: *Bulletin of the Atomic Scientists,* December 1989, p. 52)

far from certain. That the actual fission experiments had first occurred in Germany alarmed many, including refugee scientists who had fled Hitler for safer havens in England and the United States. Frightened by the prospect of a nuclear-armed Third Reich, scientists lobbied the American and British governments to begin all-out efforts to develop fission weapons. The most famous of these efforts was a 1939 letter to President Franklin Roosevelt, carrying Albert Einstein's signature. The letter was actually the idea of three Hungarian refugee physicists—Leo Szilard, Eugene Wigner, and Edward Teller—and the American businessman Alexander Sachs, an associate of Roosevelt. The Hungarians informed Einstein of the fission work and prepared draft letters; Einstein's signature lent his prestige and probably contributed, along with $E=mc^2$, to the mistaken view of Einstein as a major progenitor of the bomb.[3]

Despite the urgings of famous scientists, including also the Danish physicist Niels Bohr, the Allied governments and their military establishments were slow to act. Germany launched a more vigorous nuclear weapons program, and the Japanese military began nuclear explorations. In the war-torn Soviet Union, a nuclear weapons program begun in 1939 made only slow progress during the war years. Meanwhile, basic research on nuclear fission continued at a number of laboratories in the United States and England. It soon became apparent that only the rare isotope uranium-235 could sustain a fission chain reaction, but that the greater effectiveness of slow neutrons in causing U-235 fission might nevertheless permit a slow-neutron chain reaction in uranium with U-235 at its low natural abundance. This possibility led to the search for effective moderators that would make a chain reaction possible in natural uranium. The American efforts soon focused primarily on graphite, while, fatefully, the Germans chose heavy water. In December 1942, the Italian-born physicist Enrico Fermi and his team at the University of Chicago achieved the first self-sustaining chain reaction in their uranium-and-graphite "pile."

A moderated reaction employing slow neutrons could never release its energy fast enough for the catastrophic explosiveness of a weapon. It became obvious that nuclear fission bombs must be fast-neutron devices, highly enriched in U-235 so the less effective fast neutrons could sustain the chain reaction. By mid-1941 a British group had arrived at a rough design for a simple U-235 fission bomb. The proposed weapon was to use 25 pounds of U-235 and would have an explosive yield estimated at 1,800 pounds (1.8 kt). Although different in some details, the British design anticipated the larger Hiroshima bomb.

Because a uranium bomb must contain a high proportion of U-235, it seemed that uranium enrichment—separation of the uranium isotopes—was essential to the bomb project, although it was not necessary for a reactor. Experimental enrichment schemes produced small quantities of enriched uranium in the laboratory, but large-scale production of enough bomb-grade uranium seemed a daunting task.

Another realization dawned on the nuclear scientists in the early 1940s: that the element with atomic number 94, not yet named and not occurring naturally, might be fissile. In early 1941, University of California chemist Glenn Seaborg first identified element 94 and began producing microscopic samples. Within months Seaborg and his colleagues had strong evidence that element 94 was fissile. In 1942 Seaborg named his new element *plutonium*, after the Sun's outermost planet. Now there were two approaches to a nuclear fission weapon: (1) to separate enough fissile uranium-235 from natural uranium, and (2) to produce fissile plutonium in a moderated reactor powered by natural uranium.

By 1942 the U.S. nuclear weapons program was finally moving, with efforts underway at the University of Chicago, at Berkeley, at Columbia University, and elsewhere. The following year saw the establishment of major government centers for nuclear weapons work: a huge uranium-enrichment complex at Oak Ridge, Tennessee; plutonium-production reactors at Hanford, Washington; and the famous weapons laboratory at Los Alamos, New Mexico. Army General Leslie Groves oversaw the entire operation, now code-named the Manhattan Project. Groves chose the Berkeley physicist J. Robert Oppenheimer for the project's scientific leader. Oppenheimer brought many of the world's leading scientists to the hastily assembled, isolated laboratory in the mountains of northern New Mexico.

The Los Alamos scientists pursued two bombs: a simpler U-235 device and a more sophisticated plutonium weapon. By the summer of 1944 intelligence efforts had determined that a German nuclear bomb was highly unlikely, but the Manhattan Project pushed on. Early in 1945 sizable quantities of bomb-grade uranium from Oak Ridge and plutonium from Hanford began arriving at Los Alamos. On May 8, 1945, Germany's unconditional surrender ended the war in Europe. At Los Alamos, momentum continued building toward the first nuclear test, and nuclear scientists met with government officials to advise them on the possible use of the bomb against Japan.

By July 1945, the uranium and plutonium bomb designs were complete. The nuclear scientists were confident that the uranium device would work if any nuclear weapon would, so they chose to test the plutonium design first. In the early morning of July 16, at the so-called Trinity site in southern New Mexico, the first nuclear weapon exploded with a yield equivalent to nearly 20,000 tons of TNT. For many who witnessed it, the Trinity explosion ushered in a new and different world. Oppenheimer is quoted as follows: "We waited until the blast had passed, walked out of the shelter and then it was extremely solemn. We knew the world would not be the same . . . I remembered the line from the Hindu scripture, the Bhagavad-Gita: . . . 'Now I am become Death, the destroyer of worlds.' "[4]

Figure 12.3   J. Robert Oppenheimer and General Leslie Groves at the Trinity test site. Metal rods are all that remain of the 100-foot tower on which the first nuclear device exploded. (Los Alamos National Laboratory)

Figure 12.4 The fireball of the Trinity test, shown just 0.025 second after detonation. (Los Alamos National Laboratory)

The successful test spurred a vigorous debate among government, military, and scientific leaders. Should a demonstration explosion be staged to persuade the Japanese to surrender? Or should the bomb be used against the enemy? Should they be warned? Should the target be military, or civilian? Would dropping the bomb save the lives of American soldiers, poised for a bloody invasion of Japan? And what about the Soviet Union? Might a quick, nuclear end to the war forestall Soviet involvement? The decision was made, and the bomb was dropped—not once, but twice. The uranium bomb "Little Boy" exploded over Hiroshima on August 6, and the plutonium "Fat Man" devastated Nagasaki 3 days later.

Many nuclear scientists shared Oppenheimer's view that a world with nuclear weapons would not be the same, and they fervently believed their creation would put an end to war and lead to an era of openness and global cooperation. Their view did not prevail. In the climate of worsening relations with the Soviet Union, military and political opinion embraced the notion that nuclear superiority would make the United States invulnerable. The

Nuclear News: The Hiroshima Debate Continues

Every August 6 the world marks Hiroshima Day, expressing the fervent hope that the last days of World War II will remain Earth's only experience of nuclear war. Yet the debate continues: Was it really necessary to drop the bomb? The conventional argument is that the bombing saved countless American lives that would have been lost in the planned invasion of Japan. But there was much more behind the first use of nuclear weapons. Bomb or no bomb, the Pacific war was clearly ending, and American leaders looked beyond to postwar relations with an expansionist Soviet Union that stood poised to enter the war against Japan. In that context, it seemed desirable to defeat Japan without Soviet help, and to impress the Soviets with U.S. nuclear might. Domestic politics played a role, too: The United States had just invested $2 billion in the Manhattan Project. How could the growing nuclear establishment expect additional funding if it did not show concrete wartime results? A related reason was sheer technological and political momentum. In his prize-winning history *The Making of the Atomic Bomb*, Richard Rhodes notes that "once Trinity proved that the atomic bomb worked, men discovered reasons to use it."[5]

But were those reasons legitimate? Morally defensible? The debate continues as each year Hiroshima Day pricks the conscience of humanity.

News sources: "A-Bombing of Japan Was Unnecessary," *New York Times*, October 29, 1988; "Did We Have to Drop the Bomb?," *New York Times*, August 3, 1989; "I Was Reborn at Hiroshima," *New York Times*, August 6, 1989; "Good Translation Might Have Prevented Hiroshima," *New York Times*, August 21, 1989.

idea of a fusion-based "superbomb," first suggested by Enrico Fermi and Edward Teller early in World War II, gained new support after the Soviet Union's 1949 fission test. By the mid-1950s, both the United States and the Soviet Union had fusion weapons with yields in the 10-megaton range—1,000 times the yield of the Hiroshima bomb. The nuclear arms race was underway, and since then nuclear weapons have proliferated, in number and in variety, within and beyond the American and Soviet arsenals.

How They Work

How do nuclear weapons achieve their devastating explosive yields? How easy is it to make a nuclear bomb? How are "atomic" and "hydrogen" bombs different? These nuclear questions are best not left to scientists, since they have fundamental implications for global security. Your survival hinges on the nuclear weapons policies of your government and on those of other governments, and to understand and influence those policies it helps to know how nuclear weapons work, how they are made, and what they can do.

Fundamentally, of course, nuclear weapons exploit the nuclear difference—that millionfold energy advantage of nuclear over chemical reactions. A nuclear weapon simply entails a rapid nuclear reaction. The nuclear difference ensures that even a small quantity of nuclear material can yield enormous explosive energy. The fission and fusion reactions we considered in chapters 5, 8, and 11 in regard to power generation are also suitable for weapons. In the remainder of this chapter we will take a closer look at those weapons.

Fission Weapons

In principle, making a fission weapon is simple: Just assemble a supercritical mass of fissile material, and a chain reaction will grow rapidly as each generation of neutrons begets more and more additional fission. You could do that by bringing together two subcritical chunks of fissile material. The trick is to assemble them *quickly;* otherwise, energy released in the initial fission events will drive the chunks apart and halt the reaction.

An obvious and simple approach to a fission weapon is the

gun design, in which a uranium-235 "bullet" is fired into a U-235 "target" (figure 12.5). Surrounding the fissioning mass with a heavy "tamper" holds the exploding mass together for a fleeting instant, allowing more complete fission before the weapon blows itself apart. That's basically all there is to it.

Figure 12.6 shows "Little Boy," the gun-type fission weapon that destroyed Hiroshima. Despite the fact that it destroyed a city and killed 200,000 people, Little Boy was not particularly efficient. The huge, 9,700-pound bomb contained just over 100 pounds of uranium, enriched to 70 percent U-235. Of this material, less than 2 pounds actually underwent fission. The gun design's simplicity made it a good choice for the first combat nuclear weapon, but because of its inefficient use of scarce fissile material it fell out of favor among weapons designers. The world may not have seen its last gun-type fission weapon, however; its extreme simplicity makes the gun design a logical choice for first-time nuclear nations or terrorist groups. The only significant barrier to the construction of a gun-type fission bomb is the acquisition of enough highly enriched uranium. The actual amount can be somewhat less than Little Boy's 100 pounds, and in any event is roughly equivalent to a grapefruit-size uranium sphere.

What starts the chain reaction in a fission bomb? We addressed

---

Figure 12.5 Diagram of a gun-type fission weapon. Both the bullet and the target are subcritical masses of uranium-235. Igniting the gunpowder fires the bullet into the target, creating a critical mass that undergoes a fission explosion. The heavy steel tamper holds the exploding mass together briefly, ensuring more complete fission.

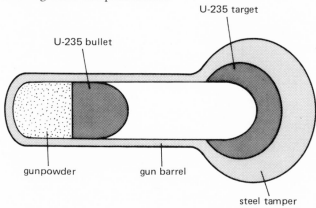

U-235 target

U-235 bullet

gunpowder   gun barrel

steel tamper

Figure 12.6 "Little Boy," the gun-type fission bomb that destroyed Hiroshima. The bomb weighed 9,700 pounds. Of that, only about 100 pounds was fissile uranium-235, and of this less than 2 pounds actually underwent fission. (Los Alamos National Laboratory)

this question abstractly in chapter 5, noting that stray neutrons from cosmic-ray interactions or from relatively rare spontaneous fission events guarantee a chain reaction in a supercritical mass. But a bomb requires precise timing, down to a fraction of a millionth of a second. Will there be enough neutrons just as the subcritical pieces come together? Or will there be too many, initiating the chain reaction before the pieces are fully joined? That would "preignite" the bomb, blowing it apart before much fission could occur and greatly reducing its explosive yield. Preignition could also be hazardous to a first-time bomb maker experimenting with fissile material, for it blurs the transition to criticality and could result in unanticipated detonation.

For uranium bombs, the available neutrons are sufficient to start the chain reaction in a gun-type design, without being so prolific as to cause preignition. But this is not true in the case of plutonium, for a subtle reason which was alluded to earlier. In chapter 10 we saw how plutonium bred in fission reactors is a mix of mostly fissile Pu-239 with smaller quantities of Pu-240; the longer nuclear fuel sits in a reactor, the greater the portion of Pu-240. Pu-240 happens to undergo spontaneous fission at a significant rate, giving rise to neutrons that make preignition a real problem in plutonium fission weapons. This is why light-water reactor fuel, typically 3 years in a reactor, is a poor source of plutonium for weapons. It is also why continuously refueled reactors, such as the Canadian CANDU and the Soviet RBMK, have greater potential for making weapons-grade plutonium. The difference, though, has sometimes been exaggerated with the statement that light-water reactors are "safe" because they can't make bomb-grade plutonium. Preignition would reduce the yield of a bomb made from reactor-grade plutonium, but the device would still be a decidedly *nuclear* weapon.

Even with weapons-grade plutonium, the preignition problem requires more rapid assembly of the critical mass than is allowed by the gun-type design. The alternative, originated in 1943 and used in nearly every nuclear weapon today, involves surrounding a subcritical plutonium sphere with conventional (that is, chemical) high explosives. When fired, the explosives compress the plutonium to higher density, making it supercritical and initiating the nuclear explosion. The inward compression of the plutonium core gives this means of nuclear ignition its name: the **implosion** method.

The difficulty with implosion is to compress the plutonium uniformly from all directions. That problem nearly discouraged the Los Alamos scientists until they found a solution: Shape the high explosives so they act as lenses, focusing their blast into a spherical shock wave that converges to compress the plutonium. Making a nuclear implosion bomb is not so much a nuclear challenge as it is a challenge requiring expertise with conventional high explosives. Figure 12.7 illustrates the workings of an implosion bomb.

Several improvements greatly increase the yield and the reliability of fission explosives. An "initiator" supplies extra neutrons at the precise moment of detonation. Surrounding the plutonium core are materials that reflect neutrons back into the fissioning mass, increasing the efficiency of fission. Surrounding materials may also act as a tamper, holding the plutonium together briefly to ensure more complete fission.

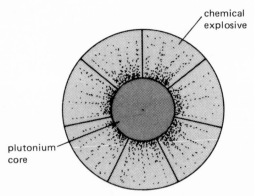

chemical
explosive

plutonium
core

Figure 12.7 In the implosion method, high explosives compress a plutonium sphere to supercritical density. In practice, a tamper would surround the plutonium.

Another technique employed in today's fission explosives was developed in the early 1950s in anticipation of fusion weapons. Recall that the deuterium–tritium fusion reaction considered in chapter 11 produces helium, energy, and a neutron. Getting energy from fusion is not easy; we saw that in chapter 11 when we looked at the feasibility of controlled fusion for electric-power generation, and in the next section we will see how the inventors of fusion weapons cleverly surmounted that difficulty. But the neutrons from fusion can be useful even when the energy isn't, as they help sustain *fission* reactions. In a "fusion-boosted" fission weapon, a small quantity of deuterium-tritium mixture at the center of the imploding plutonium reaches such a high temperature that it undergoes fusion. The resulting neutrons contribute to much more complete fission of the surrounding plutonium, greatly increasing the weapon's explosive yield. Together with the other enhancements discussed above, fusion boosting helps modern fission explosives utilize about 30 percent of their fissile material, in contrast to the Hiroshima bomb's roughly 1 percent efficiency. Fusion-boosted weapons, however, require a source of tritium, and that requirement brought the U.S. nuclear weapons industry to a crisis situation in the late 1980s.

Why all the fuss about fission weapons, tritium-boosted or otherwise? Aren't today's nuclear weapons "hydrogen bombs," and doesn't that make them fusion weapons? Generally, yes. The larger

Nuclear News: Tritium

Today's nuclear warheads use tritium to boost fission yields with the extra neutrons from deuterium-tritium fusion. But tritium's 12-year half-life means that the tritium in nuclear weapons must be replaced every few years. The United States and other advanced nuclear-weapons nations therefore require steady supplies of tritium.

Tritium is easily produced in nuclear reactors by bombarding lithium with neutrons. For years, most of the tritium for U.S. weapons came from special reactors at the government's Savannah River nuclear facility in South Carolina. But the late 1980s saw the entire U.S. nuclear weapons industry falter, after decades of neglect for environmental and safety concerns. Tritium-production reactors were shut down pending safety checks, and U.S. tritium supplies began their inexorable decline.

The tritium crisis prompted responses from across the political spectrum. The government pushed for a new tritium reactor and for reopening of the Savannah River facilities even before full safety checks had been completed. In one of his last acts as president, Ronald Reagan issued an executive order directing the Nuclear Regulatory Commission to draw plans for seizing civilian nuclear power plants to produce tritium in the event of a "national security emergency." Scientists from government laboratories suggested alternate means of tritium production. But proponents of nuclear disarmament saw the tritium crisis in a very different light. They advocated no further tritium production, and saw tritium's 12-year half-life setting a natural time scale for disarmament—especially in light of dramatic political changes in the Soviet Union and difficulties with Soviet tritium production.

What is the right way to handle the tritium crisis? Pour billions of dollars into refurbishing the nuclear weapons industry? Turn plowshares into swords by pressing the civilian nuclear power industry into tritium production? Or seize the opportunity to encourage nuclear disarmament? Those are nuclear choices both the U.S. and Soviet governments must make in the coming years.

News sources: "Reactor Shutdown Could Impede Nuclear Deterrent, Officials Say," *New York Times,* October 9, 1988; "Atoms for Peace and War: Is There a Clear Distinction?" *New York Times,* November 27, 1988; "How a Vital Nuclear Material Came to Be in Short Supply," *New York Times,* December 31, 1988; "Report Details Cheaper Source for Bomb Fuel," *New York Times,* January 23, 1989; "U.S. to Reopen Reactor in 1990; Safety in Carolina Is Still at Issue," *New York Times,* September 8, 1989.

weapons in today's nuclear arsenals do employ fusion, but all fusion weapons require fission triggers to generate the enormous temperature necessary to ignite thermonuclear fusion. And, as we will see in the next section, most of today's nuclear weapons are actually complex hybrids whose explosive yield usually involves both fission and fusion.

Fusion Weapons

In September 1941, Enrico Fermi and Edward Teller were strolling down a New York street when Fermi suddenly asked if the fission bomb they were beginning to conceive might be used to ignite thermonuclear fusion. Teller became obsessed with the notion of a thermonuclear bomb, an obsession that decades later held him among the United States' foremost advocates of continuing nuclear weapons developments. A modest program of theoretical fusion research continued at Los Alamos through the war years, and the work advanced slowly after the war. The Soviet Union's first fission explosion, in late 1949, intensified debate on the moral and practical issues surrounding the development of fusion "superbombs." Early in 1950 the U.S. government authorized a full-scale fusion effort.

Igniting thermonuclear fusion proved complicated. Blowing up a fission bomb amidst fusion fuel would just disperse the fusion material. The problem was to use the immense energy of the fission bomb to ignite fusion without the fission explosion's simultaneously destroying the fusion fuel. The solution came in 1951, when Teller and the mathematician Stanislaw Ulam hit on the idea of physically separating the fission and fusion components of the weapon. X-ray and gamma-ray energy, traveling from the fission explosion at the speed of light, could then act to ignite fusion before the slower-moving fission blast arrived. This Teller-Ulam invention is at the heart of today's thermonuclear weapons. Although the invention remains classified, the journalist Howard Morland pieced together most of the details from unclassified sources in the late 1970s. After a celebrated legal case in which the government sought to halt publication of Morland's findings, the "H-bomb secret" was published in *The Progressive* in 1979.[6]

The Teller-Ulam weapon—the "hydrogen bomb"—involves a

carefully orchestrated sequence of events, including both fission and fusion reactions. Figure 12.8 is a schematic diagram of the entire device. At one end is the fission trigger—a fusion-boosted fission bomb of the sort described in the previous section. The entire trigger is about the size of a soccer ball, and it alone probably has an explosive yield comparable to the Hiroshima bomb. When it explodes, the fission trigger produces intense x rays that travel ahead of its physical blast, vaporizing the plastic foam that occupies most of the weapon. At the end opposite the trigger is a complex structure, probably consisting of a uranium-238 tamper surrounding a cylinder of solid lithium deuteride (a blend of lithium and deuterium) that in turn surrounds a rod of plutonium or uranium-235. The vaporized plastic foam compresses this entire structure, initiating a fission chain reaction in the fissile rod. Neutrons from this reaction convert lithium to tritium in the lithium deuteride fusion fuel, and the high temperature from the compression and

Figure 12.8 Diagram of a thermonuclear weapon using the Teller–Ulam configuration. Energy from the fission trigger compresses the lithium deuteride fuel and a central plutonium rod. The compression heats the fusion fuel, while fission neutrons convert lithium to tritium. Deuterium-tritium fusion ensues, giving the weapon about half its explosive yield. Neutrons from the fusion reactions cause fission in the surrounding uranium-238, producing most of the remaining explosive energy and nearly all of the radioactive fallout. (adapted from Howard Morland, *The Secret That Exploded* [Random House, 1981])

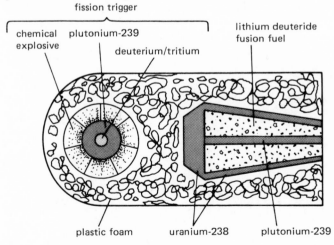

from the fission ignites deuterium-tritium fusion. The neutrons produced in this reaction convert more lithium to tritium, and D-T fusion quickly engulfs the mass of fusion fuel, giving the weapon about half its explosive yield. But the bang isn't over yet. High-energy neutrons from D-T fusion now induce fission in the surrounding uranium–238 tamper, producing the remaining explosive energy and nearly all the weapon's radioactive fallout. Fission in U–238? Shouldn't it read U–235? No. U–238 will undergo fission when struck by neutrons of high enough energy. It won't sustain a chain reaction, but that is not necessary when an external neutron source is available. In a thermonuclear weapon that source is the D-T fusion reaction. So a "hydrogen bomb" is really a **fission-fusion-fission** bomb: Fission initiates fusion, and fusion-produced neutrons cause additional fission.

In 1952 the United States tested its first thermonuclear weapon—a cumbersome device that weighed 62 tons and occupied an entire building. It yielded an impressive 10 megatons (nearly 1,000 Hiroshimas) and vaporized the Pacific island of Elugelab, leaving an underwater crater 2 miles wide and half a mile deep. By 1954 a practical design, deliverable by aircraft, yielded 15 megatons. The Soviet Union tested its first thermonuclear weapon in 1955,

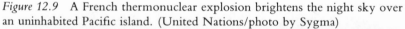

Figure 12.9 A French thermonuclear explosion brightens the night sky over an uninhabited Pacific island. (United Nations/photo by Sygma)

and within a few years both nations were deploying megaton-range thermonuclear weapons, first on bombers and later on missiles.

Advances in Nuclear Weaponry

The development of thermonuclear weapons brought a thousand-fold increase in explosive yield over simpler fission weapons. The requirements for a critical chain reaction put a limit—probably several hundred kilotons—on the yield of fission weapons. But there is no such thing as a critical mass for fusion, so fusion-based weapons can be made as big as is desired. In the early years of thermonuclear weaponry, both the United States and the Soviet Union experimented with very large weapons; in 1961 the Soviets exploded the largest ever, a 58-Mt device.

In the 1950s, interest in so-called tactical nuclear weapons had led to small, relatively low-yield fission weapons that could be placed in artillery shells for use on the battlefield. By 1955 a 12-kiloton device was available that could fit into an 8-inch-diameter cannon. Later innovations included a hand-carried "special atomic demolition munition" with a sub-kiloton yield. A host of small-scale fission weapons proliferated, many of them deployed on the front lines of a divided Europe during the 45 years of the Cold War.

The development of intercontinental missiles in the late 1950s brought pressure to decrease the size and weight of larger thermonuclear weapons as well. This led to modern weapons whose ratio of weight to explosive yield is up to 1,000 times greater than that of the nuclear bombs used in World War II. The development of multiple-warhead missiles in the 1970s continued the emphasis on smaller warheads, causing an actual drop in the total explosive yield of the U.S. arsenal. Increased accuracy and the ability to target individual warheads prevented a concomitant reduction in the capability for practical destruction, however.

Nuclear weapons can be designed for special purposes. "Dirty" bombs incorporate materials that become highly radioactive when they absorb neutrons; they would produce an abundance of lethal fallout. Replacing the uranium-238 tamper in a thermonuclear weapon with a nonfissionable material leads to a relatively "clean"

Figure 12.10 A model of the Nagasaki "Fat Man" bomb alongside the modern Mk-12A warhead used in Minuteman III missiles. "Fat Man" weighed nearly 11,000 pounds and yielded the equivalent of 22 kt of TNT. The Mk-12A weighs less than 800 pounds and yields 335 kt. Each Minuteman carries three Mk-12A warheads. (Los Alamos National Laboratory)

bomb whose radioactive fallout is limited to fission products from the relatively small fission trigger. The "neutron bomb," or enhanced-radiation weapon (ERW), is a thermonuclear device designed to minimize the blast and maximize the lethal effects of high-energy neutrons produced in fusion. ERWs would be used against tanks, killing or disabling their occupants in a burst of neutron radiation without producing much physical damage. After a protracted moral debate, the United States began deploying enhanced-radiation weapons in Europe in 1981. Another nuclear innovation is the variable-yield weapon, whose explosive violence is selected with the turn of a dial. Specialized "third-generation" nuclear weapons are under development at weapons laboratories. These include enhanced-radiation devices aimed at destroying electronic systems in attacking missiles or aircraft, and the nuclear-weapon-powered x-ray laser suggested for missile defense.

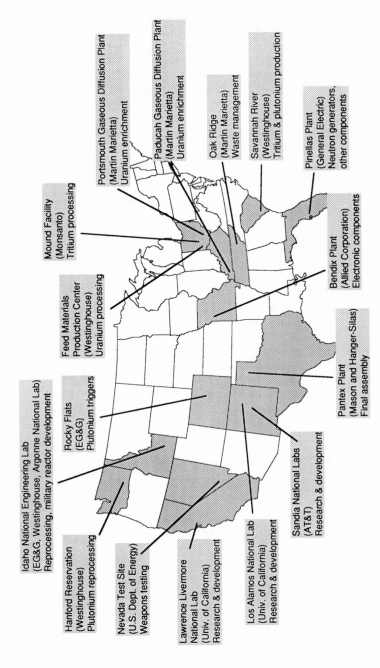

Portsmouth Gaseous Diffusion Plant
(Martin Marietta)
Uranium enrichment

Paducah Gaseous Diffusion Plant
(Martin Marietta)
Uranium enrichment

Oak Ridge
(Martin Marietta)
Waste management

Savannah River
(Westinghouse)
Tritium & plutonium production

Pinellas Plant
(General Electric)
Neutron generators,
other components

Mound Facility
(Monsanto)
Tritium processing

Bendix Plant
(Allied Corporation)
Electronic components

Feed Materials
Production Center
(Westinghouse)
Uranium processing

Pantex Plant
(Mason and Hanger-Silas)
Final assembly

Idaho National Engineering Lab
(EG&G, Westinghouse, Argonne National Lab)
Reprocessing, military reactor development

Rocky Flats
(EG&G)
Plutonium triggers

Sandia National Labs
(AT&T)
Research & development

Hanford Reservation
(Westinghouse)
Plutonium reprocessing

Nevada Test Site
(U.S. Dept. of Energy)
Weapons testing

Lawrence Livermore
National Lab
(Univ. of California)
Research & development

Los Alamos National Lab
(Univ. of California)
Research & development

Figure 12.11 Major facilities of the U.S. nuclear weapons industry are spread over thirteen states. Names in parentheses are contractors who operate the facilities for the Department of Energy; they include some of the United States' best known corporations. (data source: Cochran et al., *Nuclear Weapons Databook* [Ballinger, 1987], updated with 1990 contractors)

Nuclear News: The Weapons Business

Who makes nuclear weapons? Where? The United States' nuclear weapons industry is spread over thirteen states, and includes re-search laboratories, uranium-enrichment and fuel-processing plants, reactors for the production of nuclear materials, facilities that manufacture parts for weapons, a final assembly plant, and a test site (figure 12.11). Many of the facilities are operated by well-known U.S. corporations under government contract: General Electric, Westinghouse, AT&T, and even the University of California are among the contractors. The industry employs nearly 100,000 people and has a yearly budget approaching $10 billion. Many U.S. citizens will be surprised to learn that the government agency overseeing this vast enterprise is not the Department of Defense but the Department of Energy. The DOE's budget for nuclear weapons dwarfs its expenditures for research and development on energy supplies and conservation. Only when the weapons leave the final assembly plant are they turned over to the DOD.

For years the nation's nuclear weapons industry operated in secrecy, setting its own safety and environmental standards and dealing as it saw fit with its enormous output of nuclear waste. All that changed in 1984, when a federal judge ruled that nuclear weapons plants must comply with federal anti-pollution laws. A rash of past and present safety problems, accidents, and radiation leaks burst into the news media. As the media's interest peaked, nearly every issue of the *New York Times* carried new revelations about the weapons industry. A few examples follow, all from the *Times:*

September 18, 1987: "Explosion Risk at Nuclear Site is Reported High"
October 30, 1987: "U.S. Admits Safety Faults at Reactors for Bomb Fuel"
September 22, 1988: "The Bomb Maker Becomes a Bomb"
October 1, 1988: "Accidents at a U.S. Nuclear Plant Were Kept Secret Up to 31 Years"
October 6, 1988: "Chronic Failures at Atomic Plant Disclosed by U.S."
October 6, 1988: "Memo Says Error Almost Caused Catastrophe at U.S. Nuclear Plant"
October 7, 1988: "Inquiry Ordered at Nuclear Arms Site"
October 12, 1988: "U.S. Officials Concede Lapses in Security at Weapon Labs"
October 19, 1988: "Bitter Neighborhood Adjoins U.S. Uranium Plant in Ohio"
October 21, 1988: "Dispute on Wastes Poses Threat to Weapons Plant"

October 26, 1988: Operators Got Millions in Bonuses Despite Hazards at Atom Plants"
October 27, 1988: "Report Finds Perils at Atom Plant Greater Than Energy Dept. Said"
October 30, 1988: "Safety Snag Delays the Restarting of Carolina Arms Plant Reactor"
December 7, 1988: "U.S. Finds Health Threat From Nuclear Plant Contamination"
January 26, 1989: "Energy Dept. Says Savannah River's Radiation Safeguards Are Lax"
September 1, 1989: "Idaho Shuts Border to Nuclear Waste from Colorado Weapons Plant"
October 17, 1989: "Chemicals in Hanford A-Plant Create Risk of Explosion, Study Says"
November 19, 1989: "Tritium Released at a Weapons Lab"
February 8, 1990: "Energy Dept. Says It Still Plans to Repair Nuclear Arms Plant"
June 20, 1990: "Nuclear Arms Plant Program Stalls"
July 4, 1990: "Cost of Cleanup at Nuclear Sites is Raised by 50%"

The upshot of decades of neglect has been a virtual collapse of the entire U.S. nuclear weapons industry. The start of the 1990s found the Rocky Flats plant (near Denver) closed pending numerous safety and environmental improvements. Rocky Flats is the sole manufacturer of plutonium triggers for thermonuclear weapons. Its reopening was further threatened by the lack of a permanent repository for its plutonium waste. Earlier in the nuclear chain, the plutonium-producing "N" reactor at Hanford, Washington, had been closed in 1987, in part because of its similarity to the Chernobyl reactor that had exploded the previous year. Closed, too, was the uranium-processing plant at Fernald, Ohio, whose host state had successfully sued the plant for violation of its Clean Air Act. And, most critical for continued maintenance of nuclear weapons, all the tritium-production reactors at Savannah River, South Carolina, were closed for extensive safety checks and repairs.

Before its sudden demise, the U.S. nuclear weapons industry had been producing about five new warheads each day. With revolutionary changes in Europe and with improving relations between the nuclear superpowers, do we need to rebuild the nuclear industry of the Cold War years? Should the pursuit of nuclear might continue to be an important national goal? Or is the crisis in the nuclear weapons industry an opportunity to shape a world whose global neighbors face one another with more limited capability for mutual annihilation? These nuclear questions continue under active debate.

Summary

Since 1945 the world's nuclear arsenal has grown to the equivalent of 3,000 World War IIs. The offspring of the simple fission weapons that ended World War II have multiplied in number, variety, and explosive yield. Today's arsenals contain not only kiloton-range fission devices but also thermonuclear weapons employing a complex sequence of fission and fusion to deliver yields up to many megatons. The existence and the continuing production of these weapons poses grave moral and political questions.

Notes

1. Sheer megatonnage is only a rough way to classify the destructive effects of nuclear weapons. The area devastated by a nuclear weapon grows more slowly than in direct proportion to the weapon's explosive yield (see chapter 13). For that reason, the actual destructive effect of several smaller weapons is greater than that of a single weapon of equivalent yield. Nuclear strategists often use a more accurate measure called "equivalent megatonnage" in describing weapon's yields. The equivalent megatonnage is the actual megatonnage to the power 2/3.

2. See Arthur H. Westing, "Misspent Energy: Munition Expenditures Past and Future: The World Arsenal of Nuclear Weapons," *Bulletin of Peace Proposals* 16 (1985), p. 9.

3. These and other details of nuclear history are chronicled in Richard Rhodes' book *The Making of the Atomic Bomb* (Simon and Schuster, 1986).

4. J. Robert Oppenheimer, quoted in Len Giovannitti and Fred Freed, *The Decision to Drop the Bomb* (Coward McCann, 1965), p. 197.

5. Richard Rhodes, *The Making of the Atomic Bomb*, p. 696.

6. Howard Morland, "The H-Bomb Secret," *The Progressive* 43 (November 1979), p. 14.

Further Reading

Thomas Cochran et al., *Nuclear Weapons Databook* (Ballinger, 1984–1989). A four-volume compendium of data on nuclear weapons and delivery systems, and the nuclear weapons industry's research, production, and testing facilities, prepared by the Natural Resources Defense Council.

Leslie R. Groves, *Now It Can Be Told* (Harper & Row, 1962). The general in charge of the Manhattan Project recounts the project's history.

Alwyn McKay, *The Making of the Atomic Age* (Oxford University Press, 1984). A British nuclear scientist recounts the history of nuclear weaponry through the Cold War.

John McPhee, *The Curve of Binding Energy* (Farrar, Straus and Giroux, 1974). A popular writer explores the nuclear weapons business and gets bomb designer Ted Taylor to tell how a homemade nuclear bomb might be constructed.

Howard Morland, *The Secret that Exploded* (Random House, 1981). A journalist's account of his discovery and publication of the secret to making thermonuclear fusion weapons.

Richard Rhodes, *The Making of the Atomic Bomb* (Simon and Schuster, 1986). A Pulitzer Prize-winning history of nuclear weapons through the early 1950s. A gripping blend of history and technology.

Dietrich Schroeer, *Science, Technology, and the Nuclear Arms Race* (Wiley, 1984). A good account of nuclear weapons and related technologies, with some discussion of political issues. The technical level is somewhat higher than that of this book.

T. B. Taylor, "Third-Generation Nuclear Weapons," *Scientific American* 256 (April 1987), p. 30. A leading designer of nuclear weapons looks at the latest developments in his craft.

Kosta Tsipis, *Arsenal: Understanding Weapons in the Nuclear Age* (Simon and Schuster, 1983). A detailed technical account of nuclear weapons and delivery systems by the director of MIT's Program in Science and Technology for International Security.

Glossary

fission–fusion–fission weapon   A high-yield nuclear weapon in which fission initiates fusion, whence fusion-generated neutrons cause additional fission.

gun-type fission weapon   A nuclear fission weapon in which a subcritical U-235 "bullet" is fired into a U-235 target to form a critical mass, which then undergoes an explosive chain reaction.

implosion-type fission weapon   A fission weapon in which a subcritical sphere of fissile material (generally plutonium) is compressed by high explosives to critical density.

kiloton (kt)   A unit of explosive yield equivalent to that of 1,000 tons of the chemical explosive TNT.

megaton (Mt)   A unit of explosive yield equivalent to that of 1 million tons of the chemical explosive TNT, or 1,000 kilotons.

preignition   Premature detonation of a nuclear weapon due to excess neutrons, usually from spontaneous fission. Preignition blows apart the fissile material before it undergoes complete fission, greatly reducing the explosive yield.

tactical nuclear weapon   A nuclear weapon designed for use on the battle-field, typically characterized by small size and modest explosive yield.

thermonuclear weapon   A nuclear weapon involving thermonuclear fu-sion—the fusion of light nuclei at high temperature to produce heavier nuclei and energy.

The Effects of Nuclear Weapons

13

What can nuclear weapons do? How do they achieve their destructive purpose? What would a nuclear war be like? In this chapter we will explore these and related questions that reveal the most horrifying manifestations of the nuclear difference.

A Bomb Explodes: Short-Term Effects

In the preceding chapter we examined the carefully orchestrated sequence of events involved in the detonation of a modern nuclear weapon. Those events are over in a millionth of a second, but their effects on the surrounding environment continue for seconds, minutes, hours, days, and even weeks and more.

The most immediate effect of a nuclear explosion is an intense burst of nuclear radiation, primarily gamma rays and neutrons. This **direct radiation** is produced in the weapon's nuclear reactions themselves, and lasts well under a second. Lethal direct radiation extends nearly a mile from a 10-kiloton explosion. With most weapons, though, direct radiation is of little significance because other lethal effects generally encompass greater distances. An important exception is the enhanced-radiation weapon, or neutron bomb, which is designed to maximize direct radiation and minimize other destructive effects.

An exploding nuclear weapon instantly vaporizes itself. What was cold, solid material microseconds earlier becomes gas hotter than the Sun's 20-million-degree core. This hot gas begins to radiate away its energy in the form of x rays, which quickly heat the surrounding air. A **fireball** of superheated air forms and grows rapidly; 10 seconds after a 1-megaton explosion, the fireball is a mile in diameter. The fireball glows visibly from its own heat—so visibly that the early stages of a 1-megaton fireball are many times brighter than the Sun even at a distance of 50 miles.

Besides light, the glowing fireball radiates heat. The **thermal flash** lasts many seconds and accounts for more than one-third of the weapon's explosive energy. The intense heat can ignite fires and cause severe burns on exposed flesh as far as 20 miles from a large thermonuclear explosion. Two-thirds of injured Hiroshima survivors showed evidence of such flash burns.

As the rapidly expanding fireball pushes into the surrounding air, it creates a **blast wave** consisting of an abrupt jump in air pressure. The blast wave moves outward initially at thousands of miles per hour, but slows as it spreads. It carries about half the

Figure 13.1 The fireball of a megaton-range nuclear explosion, photographed from 50 miles away. (U.S. Department of Energy/National Atomic Museum)

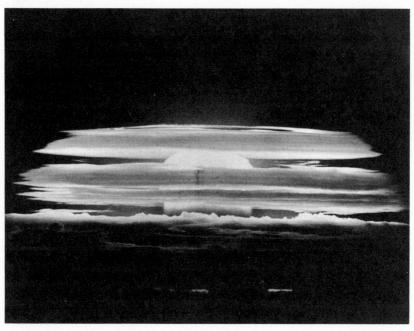

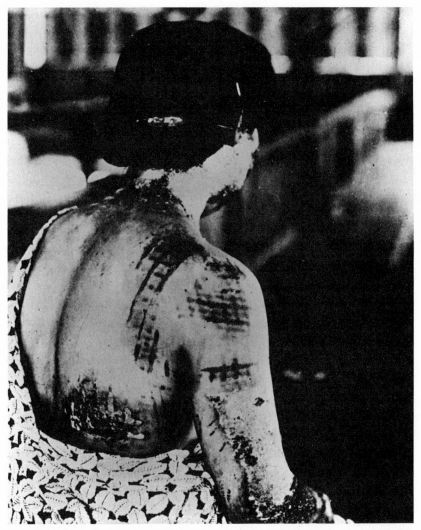

Figure 13.2 Flash burns on a Hiroshima victim's skin. Where dark clothing blocked the thermal flash, the skin was not burned. (Defense Nuclear Agency)

bomb's explosive energy, and is responsible for most of the physical destruction.

Normal air pressure is about 15 pounds per square inch (psi). That means every square inch of your body, or of your house, experiences a force of 15 pounds. You don't usually feel that force, because air pressure is normally exerted equally in all directions, so the 15 pounds pushing a square inch of your body one way is counterbalanced by 15 pounds pushing the other way. What you do feel is **overpressure**, caused by a greater air pressure on one side of an object. If you have ever tried to open a door against a strong wind, you have experienced overpressure. An overpressure of even 1/100 psi could make a door almost impossible to open. Why? Because a door has lots of square inches—about 3,000 or more. So 1/100 psi adds up to a lot of pounds.

The blast wave of a nuclear explosion may create overpressures of several psi many miles from the explosion site. Think about that! How many square inches are in the wall of a house? There are about 50,000 square inches in the front wall of a modest house— and that means 50,000 pounds or 25 tons of force even at 1 psi overpressure. Overpressures of 5 psi are enough to destroy most residential buildings, as is shown dramatically in figure 13.3. An overpressure of 10 psi collapses most factories and commercial buildings, and 20 psi will level even reinforced concrete structures. People, remarkably, are relatively immune to overpressure itself. But they aren't immune to collapsing buildings or to pieces of glass hurtling through the air at hundreds of miles per hour, or to having themselves hurled into concrete walls—all of which are direct consequences of a blast wave's overpressure. Blast effects therefore cause a great many fatalities (figure 13.4).

Blast effects depend in part on where a weapon is detonated. The most widespread damage to buildings occurs in an **air burst**, a detonation thousands of feet above the target. The blast wave from an air burst reflects off the ground, which reinforces it and adds to its destructive power. A **ground burst**, in contrast, digs a huge crater and pulverizes everything in the immediate vicinity, but its blast effects do not extend as far. Nuclear attacks on cities would probably employ air bursts, whereas ground bursts would be used on hardened military targets such as underground missile

Figure 13.3 Destruction of a wood-frame house during a nuclear test in the 1950s. First, the house is illuminated by visible radiation from the fireball; note the shadow cast by the house as it blocks the bomb's light. Almost immediately the house begins to smolder from the intense heat of the thermal flash. Two seconds later the blast wave arrives, bringing total destruction. (U.S. Department of Energy)

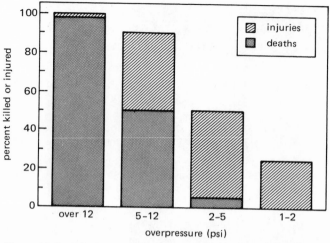

Figure 13.4 Injuries and fatalities caused by blast from a nuclear explosion. The four categories represent different ranges of blast-wave overpressure; for a 1-megaton weapon the 5–12 psi overpressure zone would extend over 4 miles from ground zero. (data source: U.S. Congress Office of Technology Assessment, *The Effects of Nuclear War,* p. 19)

silos. As we will soon see, the two types of blasts have different implications for radioactive fallout.

How far do a weapon's destructive effects extend? That distance—the **radius of destruction**—depends on the explosive yield. The volume containing a given level of destruction grows directly with yield. That means the radius of destruction grows approximately as the cube root of the yield (since volume is proportional to radius cubed). A tenfold increase in yield then gives just over a twofold increase in the radius of destruction. The area of destruction grows faster, but still not in direct proportion to the yield. That relatively slow increase in destruction with increasing yield is one reason why multiple smaller weapons are more effective agents of destruction than single larger weapons; twenty 50-kt warheads, for example, destroy nearly 3 times the area leveled by a numerically equivalent 1-Mt weapon. What constitutes the radius of destruction depends on how serious a level of destruction you want to consider. Roughly speaking, though, the distance at which overpressure has fallen to about 5 psi is a good definition of destructive radius. Many of the people within this distance would be killed, although some wouldn't. But some would be killed beyond the 5-psi distance, making the situation roughly equivalent to having everyone within the 5-psi circle killed and everyone outside surviving. Figure 13.5 shows how the destructive zone varies with explosive yield.

The blast wave is over in a minute or so, but the immediate destruction may not be. Fires started by the thermal flash or by blast effects still rage, and under some circumstances they may coalesce into a single gigantic blaze called a **firestorm**. Hot gases rise from the firestorm, replaced by air rushing inward along the surface at hundreds of miles per hour. Winds and fire compound the blast damage, and the fire consumes enough oxygen to suffocate any remaining survivors of the blast. The conditions necessary for a firestorm are unclear, but certainly include some minimum density of combustible material. During World War II, the bombing of Hamburg with incendiary chemicals resulted in a firestorm that claimed 45,000 lives. The nuclear bombing of Hiroshima resulted in a firestorm; that of Nagasaki did not. The question of firestorms is important not only to the residents of a target area; firestorms might also have significant long-term effects on the global climate.

Fallout

Both nuclear and conventional weapons produce destructive blast effects, although of vastly different magnitudes. But radioactive fallout is unique to nuclear weapons. Fallout consists primarily of fission products, although neutron capture and other nuclear reactions may contribute to the production of additional radioactive material. The term *fallout* generally applies to those isotopes whose half-lives exceed the time scale of the blast and other short-term effects. Although fallout contamination may linger for years and even decades, the dominant lethal effects last from days to weeks.

The amount of fallout produced in a nuclear explosion depends greatly on the type of weapon, its explosive yield, and where it is exploded. The neutron bomb, although it produces intense *direct* radiation, is primarily a fusion device and generates only slight fallout from its fission trigger. Small fission weapons like those used at Hiroshima and Nagasaki produce locally significant fallout. But the fission-fusion-fission design used in today's thermonuclear weapons introduces the new phenomenon of *global fallout*. Most of

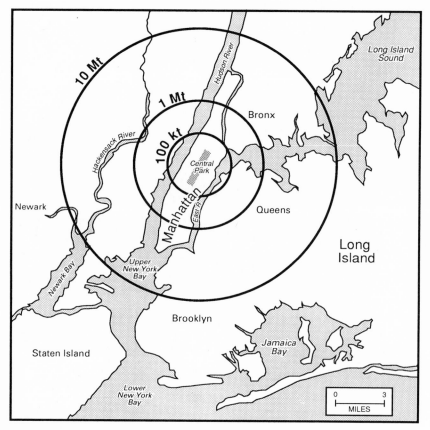

Figure 13.5 Destructive radii of 100-kt, 1-Mt, and 10-Mt weapons superimposed on a map of the New York City area. The destructive radius is defined as the distance within which blast overpressure exceeds 5 pounds per square inch, and it measures 2 miles, 4.4 miles, and 9.4 miles for the weapon yields shown. These values assume air-burst explosions at optimum altitudes over Central Park. In a modern-day attack, multiple smaller warheads would be used; the ten-warhead MX missile would destroy the same area as a single 5-Mt weapon.

this fallout comes from fission of the U-238 jacket that surrounds the fusion fuel (recall figure 12.8). The global effect of these huge weapons comes partly from the sheer quantity of radioactive material and partly from the fact that the radioactive cloud rises well into the stratosphere, where it may take months or even years to reach the ground. Even though we have had no nuclear war, we have had experience with fallout; figure 13.6 plots the Earth's contamination with strontium-90 due to the testing of nuclear weapons.

Fallout differs greatly, depending on whether a nuclear weapon is exploded at ground level or high in the atmosphere. In an air burst the fireball never touches the ground, and as a result radioactivity is carried rapidly into the stratosphere in the form of fine particles. Local fallout is reduced and global fallout enhanced. In a ground burst the explosion digs a huge crater and entrains tons of earth, rock, and other pulverized material into its rising cloud. Radioactive materials cling to these heavier particles, which drop back the ground in a relatively short time. Rain may wash down particularly large amounts of radioactive material, producing local "hot spots" of especially intense radioactivity. The exact distribution of fallout depends crucially on wind speed and direction; under

Figure 13.6 The buildup of strontium-90 on the Earth's surface in the early decades of the nuclear era. Sr-90 continued to rain down from the stratosphere for several years after the 1963 treaty banning above-ground weapons tests; only in the mid-1970s did radioactive decay significantly reduce the Sr-90 burden. (adapted from Glasstone and Dolan, *The Effects of Nuclear Weapons,* third edition, p. 449)

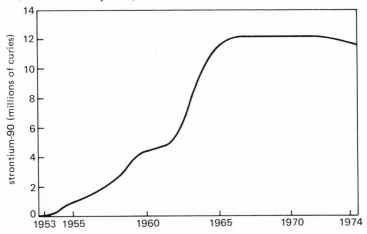

some conditions, lethal fallout may extend several hundred miles downwind of the explosion.

A 1-megaton ground burst on Detroit could bring lethal fallout to Cleveland and high radiation levels as far as Pittsburgh (figure 13.7). The levels shown in figure 13.7 are far in excess of the millirem quantities we have been discussing in connection with the normal operations of nuclear power plants. The zone in which a fatal dose of 900 rem would be received in a week would extend past Cleveland; the associated dose rate is over 5 rems per hour, giving a higher 1-hour dose than is considered acceptable for nuclear power plant workers in a year. Even in the disastrous Chernobyl accident, this level of radiation extended only about a mile from the damaged reactor, and radiation levels at the edge of the Chernobyl evacuation zone were some 500 times lower than what Cleveland would experience from a nuclear explosion at Detroit. (Considerably more fallout would result if a nuclear weapon scored a direct hit on a nuclear reactor, releasing *all* of the radioactive solids instead of the 3 percent released in the Chernobyl accident.)

Electromagnetic Pulse

A nuclear weapon exploded at very high altitude produces none of the blast or local fallout effects just described. But intense gamma rays from a nuclear explosion knock electrons out of atoms in the surrounding air, and when the explosion takes place in the rarefied air at high altitude this effect may extend hundreds of miles. As they gyrate in the Earth's magnetic field, these electrons generate an intense pulse of radio waves known as the **electromagnetic pulse** (EMP).

A single large weapon exploded some 200 miles over the central United States could blanket the entire country with an electromagnetic pulse intense enough to damage computers, communication systems, and other electronic devices. It could also affect satellites used for military communications, reconnaissance, and attack warning. The EMP phenomenon thus has profound implications for a military that is increasingly dependent on sophisticated electronics. Since the test-ban treaty of 1963 it has been impossible to study EMP effects directly, although elaborate devices have been

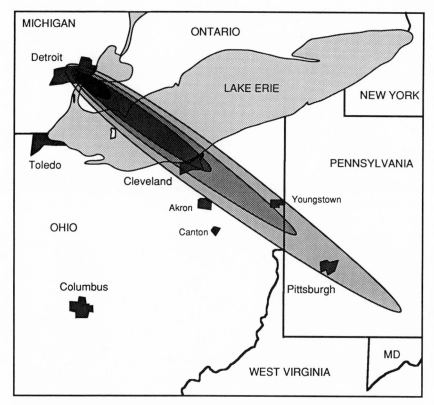

Figure 13.7 The fallout from a 1-megaton ground burst on Detroit extends well past Pittsburgh. Shaded contours, from darkest to lightest, correspond to 7-day radiation doses of 3,000 rems, 900 rems, 300 rems, and 90 rems; note that Cleveland residents would receive a fatal 900 rem. The fallout pattern assumes a 15-mph wind from the northwest. (adapted from U.S. Congress, Office of Technology Assessment, *The Effects of Nuclear War,* p. 25)

Nuclear News: Electromagnetic Pulse

On a July night in 1962, the sky over Hawaii lit briefly with daytime brilliance. The cause was a U.S. nuclear explosion 200 miles above Johnston Island, some 700 miles southwest of Honolulu. The blast was seen as far away as New Zealand, 3,000 miles distant. The purpose of the test, involving a 2-megaton device, was to determine the effects of high-altitude nuclear explosions on communications.

Trans-Pacific radio communications between the United States, Japan, and Australia were indeed disrupted, although for only an hour rather than the 16–30 hours some had expected. More severe electrical disturbances occurred on the Hawaiian island of Oahu, where burglar alarms went off and street lights failed.

The Partial Test Ban Treaty of 1963 put an end to experiments like the Johnston Island blast, but the military has continued to explore electromagnetic-pulse effects using non-nuclear sources. Even these tests are controversial, with mounting concern over environmental and health effects of the intense electromagnetic radiation used. By the late 1980s environmentalists had forced suspension of U.S. Navy EMP tests off North Carolina.

Do we know enough about EMP? Or is continued experimentation necessary? Does uncertainty over EMP make nuclear attack more or less likely? EMP is a serious effect, and the answers to these questions may influence our nuclear strategies.

News sources: "Hydrogen Blast Fired 200 Miles Above Pacific," *New York Times,* July 10, 1962; "Disruption of Communications in the Pacific Surprisingly Mild," *New York Times,* July 10, 1962; "Electromagnetic Pulse Tests Challenged," *New York Times,* June 7, 1988.

built to mimic the electronic effects of nuclear weapons. Increasingly, crucial electronic systems are "hardened" to minimize the effects of EMP. Nevertheless, the use of EMP in a war could wreak havoc with systems for communication and control of military forces.

Would the high-altitude detonation of a nuclear weapon to produce EMP be an act of war warranting nuclear retaliation? With its electronic warning systems in disarray, should the EMPed nation launch a nuclear strike on the chance that it would soon be attacked? How are nuclear decisions to be made in a climate of EMP-crippled communications? These are difficult questions, but the nuclear age demands that our military strategists have answers.

Nuclear War

So far we have looked at the effects of single nuclear explosions. But a nuclear war would involve hundreds to thousands of explosions, creating a situation for which we have simply no relevant experience. What would a nuclear war be like?

When you think of nuclear war, you probably envision an all-out holocaust in which adversaries unleash their arsenals in an attempt to inflict the most damage possible. Many people—including this author—believe that misfortune to be the likely outcome of almost any use of nuclear weapons between the superpowers. But nuclear strategists have explored many scenarios that fall short of the "all-out" nuclear exchange. What might these "limited" nuclear wars be like? Could they remain limited?

Limited Nuclear War

One form of limited nuclear war would be like a conventional battlefield conflict, but employing the low-yield tactical nuclear weapons that make up a large portion of the superpowers' nuclear arsenals. These weapons include the neutron bomb, designed specifically to stop advancing tanks by killing their crews with penetrating doses of neutrons; nuclear shells launched from artillery pieces; and small warheads carried on mobile short-range missiles (figure 13.8). Although these tactical weapons were designed primarily for use in Europe during the pre-1990 years of the Cold

Figure 13.8 "Davy Crockett," an early tactical nuclear weapon, could be launched from a jeep. (U.S. Department of Energy, National Atomic Museum)

War, U.S. military officers actually requested permission to use tactical nuclear weapons on a number of occasions in Vietnam and elsewhere. That permission was invariably denied in Washington.

Would a tactical nuclear battle remain limited? To Europeans that has been a hollow question; "limited," in the context of tactical nuclear weapons, usually meant "limited to Europe." Millions of European casualties would follow even the most "limited" nuclear battle. And many strategists believe that even a tactical nuclear battle would inevitably escalate to involve the large, strategic weapons aimed at the superpowers' homelands. As one side got the upper hand in battle, the losing side might be unable to resist the temptation to strike deeper and with ever-larger weapons. An all-out nuclear war could soon result.

Other scenarios for limited nuclear war envision selective attacks aimed at destroying an adversary's retaliatory capacity or a crucial industry. Even if such attacks marked the extent of the hostilities, they would still bring death and destruction in excess of anything the United States has experienced in all previous wars.

Suppose a nuclear adversary decided to cripple the United States' nuclear retaliatory forces (a virtual impossibility, given nuclear missile submarines, but nevertheless a scenario considered with deadly seriousness by nuclear planners). Figure 13.9 shows that many of the 48 contiguous states have at least one target—a nuclear bomber base, a submarine support base, missile silos—that would warrant destruction in such an attack. The attack, which would require only a tiny fraction of the roughly 10,000 strategic nuclear weapons in the Soviet arsenal, could kill millions of civilians. Those living near targeted bomber and submarine bases would suffer blast and local radiation effects. Intense fallout from ground-burst explosions on missile silos in the Midwest would extend all the way to the Atlantic coast. Fallout would also contaminate a significant fraction of the United States' cropland for periods up to a year, and would kill livestock. On the other hand, the U.S.

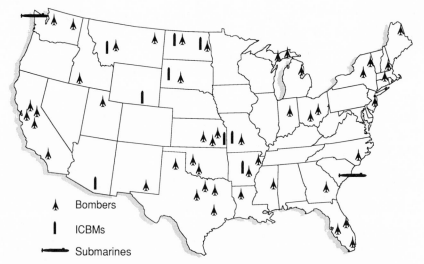

Figure 13.9 Strategic nuclear targets in the United States include bomber bases, missile sites, and submarine bases. Attacks on these targets alone could result in up to 7 million civilian deaths. (adapted from U.S. Congress, Office of Technology Assessment, *The Effects of Nuclear War,* p. 82)

industrial base would remain relatively unscathed, *if* no further hostilities occurred.

In contrast to attacking military targets, an adversary might seek to cripple the U.S. economy by destroying a vital industry. In a hypothetical attack considered by the congressional Office of Technology Assessment, ten Soviet SS-18 missiles, each with eight 1-Mt warheads, attack the United States' oil refineries. The result is destruction of two-thirds of the United States' oil-refining capability. And even with some evacuation of major cities in the hypothetical crisis leading to the attack, 5 million Americans are killed.

Each of these "limited nuclear attack" scenarios kills millions of Americans—many, many times the 1.2 million killed in all the wars in our nation's history. Do we want to entertain limited nuclear war as a realistic possibility? Do we believe nuclear war could be limited to "only" a few million casualties? Do we trust the professional strategic planners who prepare our possible nuclear responses to an adversary's threats? What level of nuclear preparedness do we need to deter attack? We will examine these difficult nuclear questions further when we explore nuclear strategies in chapter 15.

All-Out Nuclear War

Whether from escalation of a limited nuclear conflict or as an outright full-scale attack, an all-out nuclear war remains possible as long as the nuclear nations have thousands of weapons aimed at one another. What would be the consequences of all-out nuclear war?

Within individual target cities, the conditions described above for single explosions would prevail. (Most cities, though, would likely be targeted with multiple weapons.) Government estimates suggest that over half of the United States' population could be killed by the prompt effects of an all-out nuclear war. For those within the appropriate radii of destruction it would make little difference whether theirs was an isolated explosion or part of a war, but for the survivors in the less damaged areas the difference could be dramatic.

Consider the injured. Thermal flash burns extend well beyond the 5-psi radius of destruction. A single nuclear explosion might produce 10,000 cases of severe burns requiring specialized medical treatment; in an all-out war there might be several million such cases. Yet the United States has facilities to treat only a few thousand burn cases—virtually all of them in urban areas that would be leveled by nuclear blasts. Burn victims who might be saved had their injuries resulted from some isolated cause would succumb in the aftermath of nuclear war. The same goes for fractures, lacerations, missing limbs, crushed skulls, punctured lungs, and myriad other injuries suffered as a result of nuclear blast effects. Where would the doctors, the hospitals, the medicines, and the equipment needed for their treatment be? Most would lie in ruin, and those that remained would be inadequate to the overwhelming numbers of injured. Again, many would die whom modern medicine could normally save.

In an all-out war, lethal fallout would cover much of the United States (figure 13.10). Survivors could avoid fatal radiation exposure only in fallout shelters adequately stocked with food, water, and medical supplies. Even then, millions would be exposed to radiation

high enough to cause lowered disease resistance and greater incidence of subsequent fatal cancer. Lowered disease resistance could
lead to death from everyday infections in a population deprived of
adequate medical facilities. And the spread of diseases from contaminated water supplies, nonexistent sanitary facilities, lack of medicines, and the millions of dead could reach epidemic proportions.
Small wonder that the international group Physicians for Social
Responsibility has called nuclear war "the last epidemic."

Attempts to contain damage to cities, suburbs, and industries
would suffer analogously to the treatment of injured people. Fire-
fighting equipment, water supplies, electric power, heavy equipment, fuel supplies, and emergency communications would be
gone. Transportation into and out of stricken cities would be
blocked by debris. The scarcity of radiation-monitoring equipment
and of personnel trained to operate it would make it difficult to
know where emergency crews could safely work. Most of all, there
would be no healthy neighboring cities to call on for help; all would
be crippled in an all-out war.

What about the psychological effects on individuals and society? Extreme and cooperative efforts would be needed for survival,

Figure 13.10   After all-out nuclear war, unsheltered persons in the shaded
areas would receive weekly radiation doses in excess of 100 rems; 60 percent
would receive fatal doses in excess of 1,000 rems.

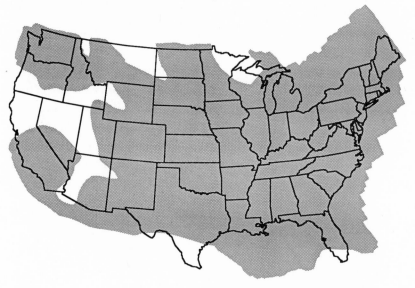

but would the shocked and weakened survivors be up to those efforts? How would individuals react to watching their loved ones die of radiation sickness or untreated injuries? Would a selfish attitude prevail, preventing the cooperation necessary to rebuild society? How would residents of undamaged rural areas react to the streams of urban refugees flooding their communities? What governmental structures could function in the postwar climate? How could people know what was happening throughout the country?

Some students of nuclear war see postwar society in a race against time. An all-out war would have destroyed much of the nation's productive capacity, and would have killed many of the experts who could help guide social and physical reconstruction. The war also would have destroyed stocks of food and other materials needed for survival.

On the other hand, the remaining supplies would have to support only the much smaller postwar population. The challenge to the survivors would be to establish production of food and other necessities before the supplies left from before the war were exhausted. Could the war-shocked survivors, their social and governmental structure shattered, meet that challenge? That is a very big nuclear question—so big that it is best left unanswered, since only an all-out nuclear war could decide it definitively.

Global Effects

The effects of an all-out nuclear war would extend far beyond the borders of the belligerent nations. A government study suggests that more than 20 million people outside the United States would eventually die of radiation-induced cancer if the United States were attacked with 3,325 nuclear weapons yielding a total of 6,500 megatons. In a war with a nuclear adversary, another 20 million outside the adversary's territory could also die. And these radiation deaths might be dwarfed by the results of major changes in the global climate—effects from which no part of the world would be immune.

Climatic Effects

A large-scale nuclear war would pump huge quantities of chemicals and dust into the upper atmosphere. Humanity was well into the nuclear age before scientists took a good look at the possible consequences of this. What they found was not at all reassuring.

The upper atmosphere includes a layer enhanced in ozone gas, an unusual form of oxygen that vigorously absorbs the Sun's ultraviolet radiation. In the absence of this ozone layer, more ultraviolet radiation would reach the Earth's surface, with a variety of harmful effects. Concern about possible depletion of the ozone layer has grown in recent years, as theory and observations suggest that a variety of human pollutants are consuming ozone. A nuclear war would produce huge quantities of ozone-consuming chemicals, and recent models predict postwar ozone depletion of 50 percent or more lasting for several years. Plants and marine life could be severely damaged by the increased ultraviolet radiation, and light-skinned humans would receive blistering sunburns after only 30 minutes' exposure outdoors. More ultraviolet radiation would also lead to a greater incidence of fatal skin cancers and to general weakening of the human immune system.

Even more alarming is the possible effect of the dust that would be injected into the atmosphere from burning cities ignited in a nuclear conflict. A 1983 study by Richard Turco, Carl Sagan, and others shocked the world with the suggestion that even a modest nuclear attack—as few as 100 warheads—could trigger drastic global cooling as airborne dust blocked incoming sunlight. In its most extreme form, this **"nuclear winter"** hypothesis raised the possibility of extinction of the human species. (This is not the first dust-induced extinction pondered by science. A view popular with some scientists holds that the dinosaurs went extinct as a result of climatic change brought about by atmospheric dust from the collision of an asteroid or a comet with Earth; indeed, that hypothesis helped prompt the nuclear-winter research.)

The original nuclear-winter study used a relatively unsophisticated computer model, and it spurred vigorous controversy among atmospheric scientists. More realistic calculations have shown that the effects originally predicted are exaggerated, prompting some to use the term "nuclear fall" in place of "nuclear winter." Still, the possible climatic effects of nuclear war are substantial. They include temperature drops of 18–36° F lasting from several weeks to a month or more. If this occurred in the summer, the effect on food production could be devastating—especially in a world already reeling from the immediate effects of nuclear war. Whether that degree of cooling is enough to constitute "nuclear winter" is still a matter of debate. The debate is further compounded by substantial uncertainties in the amount of material that would actually burn in a nuclear attack, in the length of time soot would remain in the atmosphere, and in the light-blocking effect of the soot. Scientists may continue the debate for years, but they have already given citizens more than enough reason to hope their governments' nuclear choices diminish the likelihood of nuclear war.

Summary

Nuclear weapons have devastating effects. Destructive blast effects extend miles from the detonation point of a typical nuclear weapon, and lethal fallout may blanket communities hundreds of miles downwind of a single nuclear explosion. An all-out nuclear war

Nuclear News: Winter to Fall?

Scientific debate on the climatic effects of nuclear war raged throughout the 1980s. By 1990, however, most studies were converging to the conclusion that significant climatic effects would follow nuclear war, although those effects would not be as severe or prolonged as the earliest models had predicted. So would there be a nuclear winter or not? Some scientists argue that climatic disturbance would be sufficiently mild to constitute a "nuclear autumn," whose global effects would be dwarfed by devastation of the warring nations. But others maintain that the original and frightening predictions of nuclear winter have been essentially vindicated, albeit in somewhat tempered form. If these predictions hold, casualties of a nuclear winter could number in the billions.

The nuclear-winter debate is far from academic. The world is armed with many more nuclear weapons than are necessary to trigger global climate change. Nuclear war may be undesirable, but it remains possible. In that context, might our species be more secure if nuclear arsenals were reduced below the triggering point for nuclear winter?

News sources: "Nuclear Winter Theorists Pull Back," *New York Times,* January 23, 1990; "Don't Relax About Nuclear Winter Just Yet," *New York Times,* March 5, 1990.

would leave survivors with few means of economic recovery, and could lead to a total breakdown of remaining social institutions. Fallout from an all-out war would expose most of the belligerent nations' surviving populations to radiation levels ranging from harmful to fatal. And the effects of nuclear war would extend well beyond the warring nations, possibly including climatic change severe enough to threaten much of the planet's human population.

Debate about national and global effects of nuclear war continues, and the issues are unlikely to be decided conclusively without the unfortunate experiment of an actual nuclear war. But enough is known about nuclear war's possible effects that there is near universal agreement on the need to avoid them. But how? And at what risk? Those are the real issues in the ongoing debates about the future course of nuclear weaponry.

Further Reading

Paul P. Craig and John A. Jungerman, *Nuclear Arms Race: Technology and Society* (McGraw-Hill, 1986). Chapters 13–18 detail the physical effects of nuclear weapons; chapters 19 and 22 describe the physical and psychological effects of nuclear war.

Lydia Dotto, *Planet Earth in Jeopardy: Environmental Consequences of Nuclear War* (Wiley, 1986). Despite its alarmist title, this short book provides a good qualitative introduction to the consequences of nuclear war. It does not include the results of more recent models of climatic effects.

Samuel Glasstone and Philip J. Dolan, *The Effects of Nuclear Weapons,* third edition (U.S. Department of Defense and Energy Research and Development Administration, 1977). The authoritative reference on nuclear weapons effects; comes complete with a "Nuclear Bomb Effects Computer" that calculates all manner of weapons as a function of explosive yield.

Mark A. Harwell, *Nuclear Winter: The Human and Environmental Consequences of Nuclear War* (Springer-Verlag, 1984). A thorough account of the many considerations necessary to understand the global effects of nuclear war. Although the book is useful and enlightening, it does not include the more modest predictions of recent computer models.

Nuclear Weapons: Report of the Secretary-General of the United Nations (Autumn Press, 1980). A report on international trends in nuclear weapons through 1980. Chapter 4 and appendix I detail the weapons' effects.

Michael Riordan, ed., *The Day After Midnight: The Effects of Nuclear War* (Cheshire Books, 1982). This revision of the Office of Technology Assessment's study *The Effects of Nuclear War* contains very nearly the same material.

Scientific Committee on Problems of the Environment (SCOPE), *Environmental Consequences of Nuclear War* (Wiley, 1986). Detailed technical analyses cover physical, atmospheric, ecological, and agricultural effects of nuclear war.

F. Solomon and R. Q. Marston, eds., *Medical Implications of Nuclear War* (National Academy Press, 1986). A compendium of authoritative papers covering not only medical but also physical, climatic, and psychological effects of nuclear warfare.

James Thompson, *Psychological Aspects of Nuclear War* (Wiley, 1985). This official statement of the British Psychological Society describes both psychological responses to nuclear war and psychological factors that could lead to it.

U.S. Congress, Office of Technology Assessment, *The Effects of Nuclear War* (Government Printing Office, 1979). Clear, straightforward discussions of the effects of nuclear weapons and nuclear warfare. Includes scenarios of attacks on Detroit and Leningrad, limited and all-out nuclear war, and long-term effects. A fictional account envisions life after a nuclear war.

Glossary

air burst A nuclear explosion detonated at an altitude—typically, thousands of feet—that maximizes blast damage. Because its fireball never touches the ground, an air burst produces less radioactive fallout than a ground burst.

blast wave An abrupt jump in air pressure that propagates outward from a nuclear explosion, damaging whatever it encounters.

direct radiation Nuclear radiation produced in the actual detonation of a nuclear weapon, and constituting the most immediate effect on the surrounding environment.

electromagnetic pulse (EMP) An intense burst of radio waves produced by a high-altitude nuclear explosion, and capable of damaging electronic equipment over thousands of miles.

fireball A mass of air surrounding a nuclear explosion and heated to luminous temperatures.

ground burst A nuclear explosion detonated at ground level, producing a crater and significant fallout but less widespread damage than an air burst.

nuclear winter A substantial reduction in global temperature that might result from soot injected into the atmosphere during a nuclear war.

overpressure Excess air pressure encountered in the blast wave of a nuclear explosion. Overpressure of a few pounds per square inch is sufficient to destroy typical wooden houses.

thermal flash An intense burst of heat radiation in the seconds following a nuclear explosion. The thermal flash of a large weapon can ignite fires and cause third-degree burns tens of miles from the explosion.

Delivering Nuclear Weapons

14

Nuclear weapons aren't of much military use without some means of delivering them to their targets. Such **nuclear delivery systems**, important to military strategists, are also key elements in the international balance of political and military power.

Nuclear delivery systems range from "suitcase bombs" carried by individual soldiers (and perhaps by nuclear terrorists) to submarines over 500 feet long. Here we will focus primarily on **strategic** delivery systems—those designed to carry nuclear weapons to an adversary's homeland.

Features of Nuclear Delivery Systems

What are some desirable features of nuclear delivery systems? That depends in part on what a nation wants to do with its weapons; in turn, delivery systems may give clues to a nation's nuclear intentions.

Range describes the distance a delivery vehicle can travel from its base to a target. Today's longest-range delivery vehicles have truly intercontinental range and can travel from the heartlands of the superpowers to virtually anywhere on the globe. **Flight time** is the time it takes from base to target. Shorter flight times would be especially useful in a surprise attack—a **first strike**, as it is called in nuclear parlance. The flight times of today's strategic weapons range from about 10 minutes for submarine-launched missiles to many hours for bombers. **Payload** is the maximum weight a vehicle can deliver to its target. There is a tradeoff between range and payload, and the payload of a given delivery vehicle can generally be increased at some cost in range. **Accuracy** describes how close to its target a weapon will actually be delivered. Accuracy is relatively unimportant for **soft targets** (such as cities) but becomes increasingly important to ensure destruction of **hard targets** (such as underground missile silos and military command centers). If a 1-megaton warhead lands a mile from its center-city target, the city will still be destroyed; on the other hand, only a direct hit will destroy a hardened underground concrete missile silo. Today's newest intercontinental missiles have accuracies of about 100 yards, meaning that a warhead has a 50-50 chance of landing within this distance of its target after a flight of 10,000 miles. There is a tradeoff between required accuracy and payload: With its huge destructive radius, a 9-megaton weapon needs less accurate targeting than one with a 300-kiloton yield. Extreme accuracy is often viewed as a first-strike capability, since destruction of an adversary's missiles would presumably be most important in a first strike.

A delivery system is no good if it is destroyed before reaching its target. **Penetration** characterizes the ability to reach targets deep within enemy territory in the face of defensive systems. Slow-moving aircraft have lower penetration capability than high-speed missiles, although even aircraft penetration is rapidly improving with technological advances.

A delivery system might also be destroyed before being launched, in a surprise attack. That possibility, a nightmare for nuclear strategists, demands that nuclear delivery systems have a measure of **invulnerability** in the face of an enemy's first strike. Invulnerability is the most important attribute of the weapons in today's nuclear equation. Why? Because the primary purpose of nuclear weapons is to deter an adversary's attack by threatening nuclear retaliation, and that threat is hollow if your adversary can destroy your retaliatory forces in a first strike. Invulnerable weapons keep the retaliatory threat credible, thus helping preserve the nuclear peace. The quest for invulnerability is a significant force driving the nuclear arms race, even with the end of the Cold War.

What about the danger of accidental nuclear war, or of war provoked inadvertently in time of crisis? Can the designs of nuclear

Nuclear News: Bringing Back the Big Ones

The general trend in the evolution of nuclear weapons has been toward lower explosive yields coupled with more accurate delivery systems. The United States' Titan II missile, retired in the mid-1980s, delivered a 9-megaton warhead with accuracy not much better than a mile. Minuteman II dropped the yield to 1.2 Mt, but with an accuracy of 700 yards. The warheads on the new MX missile pack only 0.3 Mt explosive yield each (although there are ten such warheads—a total of 3 Mt—as opposed to one each on Titan II and Minuteman II), but with an accuracy reported at 80 yards.

The trend toward lower yields has reversed in recent years, starting with the replacement of 170-kiloton warheads by 335-kt versions on some three-warhead Minuteman III missiles. Similarly, 475-kt Trident II submarine-based warheads are replacing earlier 100-kt devices. But a more dramatic and little-noticed turn occurred in 1987, when the U.S. Air Force redeployed a stock of 9-Mt bombs that had been mothballed. Built in the 1960s, these behemoths have more than 6 times the yield of any other U.S. nuclear weapon. And why were they reactivated? According to Air Force and Department of Defense officials, to provide "more flexibility in the planning process." Planning for what? For nuclear war: The big bombs would be used to "decapitate" Soviet leadership in deep underground bunkers. Critics charge that this redeployment indicates increased military interest in actually fighting a nuclear war, and argue that such preparation only makes war more likely. Who is right: the critics, with their heightened fear of nuclear war, or the Defense Department, with its need for nuclear flexibility?

News sources: "U.S. to Take Its Most Powerful Nuclear Bomb Out of Mothballs," *Boston Globe,* August 6, 1987; "The Dirtiest Bomb," *Boston Globe,* August 15, 1987.

weapons and their delivery systems affect these frightening possibilities? Yes. High-speed systems with short flight times are inherently more dangerous. They cannot be recalled in the event of an accidental or mutinous launch, and their short flight time gives the nation under attack little time to make rational decisions about how to respond. Other features, including the vulnerability of communications systems and the deployment of multiple-warhead missiles, also have significant bearing on the likelihood of nuclear conflict.

Today's nuclear world is in a complex balance that hinges, in large part, on the delivery systems the nuclear powers have chosen to deploy. In this chapter we will examine both the technical details of those delivery systems and their implications for international security. We will focus primarily on the United States' nuclear forces, which, along with those of the Soviet Union, dominate the global nuclear balance. As the nuclear rivalry of the Cold War fades, however, we face a world in which nuclear weapons and nuclear-capable delivery systems are spreading rapidly. The established nuclear arsenals provide a model for such emerging nuclear weapons states as Iraq, Israel, Pakistan, and India.

The Strategic Triad

Since about 1960, each superpower's nuclear forces have comprised a **triad** of delivery systems that differ in **basing schemes** (i.e., where the weapons are deployed) and in actual delivery vehicles. The triad includes bomber aircraft, land-based intercontinental missiles, and submarine-based missiles. One rationale for the triad is redundancy: If a surprise attack destroys one leg of the triad, the others remain capable of retaliation. The legs of the triad differ with respect to the delivery-system qualities we discussed above; since those qualities often involve tradeoffs, a diversified triad may give more security than would "putting all one's eggs in one basket." Another reason for the triad is surely political: Each major branch of the military would like its own share of nuclear forces, and the triad affords just such a sharing. Here we will examine separately the three legs of the triad, and then consider a new development—the cruise missile—that blurs somewhat the distinction among the three legs. We will focus primarily on the United States' nuclear triad, but will look briefly at other nations' nuclear forces. Broader questions of nuclear strategy will be addressed in the next chapter.

Bombers

The 2,000-mile-range B-29 bomber of World War II arrived just in time to carry the overweight first-generation nuclear bombs. The success of conventional and nuclear bombing in that war led to rapid postwar development of bombers capable of delivering nuclear weapons at long range. By 1952 the United States had begun deploying the B-52, with eight jet engines, intercontinental range, and in-flight refueling capability. Although the last B-52s were delivered in 1962, the United States retains nearly 200 of these venerable aircraft in its strategic forces; in 1990 they still made up about two-thirds of the bomber leg of the U.S. strategic triad. If you live anywhere near a strategic air base, you have probably seen B-52s pass overhead, and they may well have been carrying nuclear weapons.

The B-52s are old, but they are hardly obsolete. They fly at about 600 miles per hour, and they have undergone continual upgrading. Their newer capabilities include fuel-efficient engines for longer range, extensive electronic-warfare systems for foiling en-

Figure 14.1   A B-52 bomber. (U.S. Air Force)

emy defenses, hardening against EMP and blast effects so they can survive a nuclear attack on their base shortly after takeoff, and modifications to carry new air-launched nuclear missiles.

Today's most advanced B-52s are armed with a variety of nuclear weapons, including bombs, short-range attack missiles (SRAMs), and air-launched cruise missiles (ALCMs). SRAMs are small rocket-powered missiles launched up to 100 miles from a target. They fly supersonically, guided by an on-board navigation system that gives them good accuracy. Cruise missiles, which will be discussed separately in a later section, are subsonic jet-powered devices with a range of about 1,500 miles. They fly at treetop level, and they have extreme accuracy (as good as 30 feet) thanks to a system that compares radar images of the ground over which the missile is flying against a computerized map. Both SRAMs and ALCMs improve a bomber's penetration, by allowing it to "stand off" at a safe distance from a heavily defended target. A typical B-52 load might include four 1-megaton bombs; four SRAMs, each with a 200-kiloton warhead; and twelve ALCMs, each with a 300-kiloton warhead. That is a total of 8.4 megatons, or, in the measure we introduced in chapter 12, more than one World War II on a single airplane.

Since the 1950s there have been numerous attempts to replace the B-52. The B-70 project of the 1950s envisioned a supersonic bomber capable of three times the speed of sound, or over 2,000 mph. The project was canceled, over the objections of many in the defense establishment, when it became clear that intercontinental missiles provided superior high-speed delivery of nuclear warheads.

A bomber's ability to remain airborne is limited by its fuel capacity, a limitation partly overcome by in-flight refueling. In the early years of nuclear era, military technologists envisioned another solution: a nuclear-powered bomber that could remain aloft for weeks. Insoluble difficulties involving radiation shielding eventually halted this project in the early 1960s.

In the 1970s the United States developed the supersonic B-1, an advanced aircraft with "swing wings" that could be optimized for supersonic or subsonic flight. Controversy surrounded the B-1 from its inception. Did the United States really need a new bomber? Could the lighter, shorter-range B-1 really replace the B-52? What about the cost, which had reached $87 million per plane by 1976 and $280 million by the 1980s? In 1977 President Jimmy Carter canceled the B-1 program. Ronald Reagan revived it early in his presidency, and a modified version of the original B-1 went into service in 1986. Nearly 100 of these B-1Bs now make up about one-third of the United States' strategic bomber fleet.

On paper, the B-1 is an impressive airplane. It can fly at 42,000 feet at twice the speed of sound (about 1,400 mph), or at nearly treetop level at a subsonic 650 mph. Its visibility on radar is only 1/100 that of the B-52, giving the B-1 much better penetration. In a typical configuration, the B-1 might carry eight air-launched cruise missiles and fourteen short-range attack missiles, for a total of 5 megatons or nearly one World War II. In practice, the B-1's performance has been decidedly mixed, with a number of important design specifications unmet and an abnormally high number of crashes.

As envisioned in the early 1980s, the B-1 was to be a short-term supplement to the bomber fleet. Even its low radar visibility was considered too high to ensure penetration. Recent advances in materials technology permit a greater use of nonmetallic components in aircraft construction, further reducing radar visibility. In

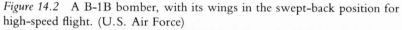

Figure 14.2 A B-1B bomber, with its wings in the swept-back position for high-speed flight. (U.S. Air Force)

Nuclear News: The B-1 Bombs

Under the B-1 program, the U.S. Air Force procured 100 planes at a cost of $280 million each. But three of them crashed in their first 2 years of operation. One struck a bird during a low-altitude training flight; another was damaged when its swing-wing mechanism punctured a fuel tank. Fuel leaks and other difficulties may reduce the B-1's payload and/or range. And the entire fleet has been plagued by problems in the defensive electronics that could compromise penetration capability. The U.S. General Accounting Office has estimated that improving the planes to meet their original design specifications could raise their cost to $400 million each. Do we need to pay that kind of money for security in the nuclear age? That nuclear question is coming under closer congressional scrutiny as the Cold War fades into history.

News source: "Stubborn Problems Plague B-1, Report Says," *New York Times,* February 5, 1989.

1988 a supersecret program exploiting such "stealth" technology came to fruition with the unveiling of the B-2 Stealth Bomber, a subsonic aircraft with the roughly the same weapons load as the B-1. With one in-air refueling, it can attack targets anywhere in the world. It flies as high as 50,000 feet, and fuel additives eliminate the contrails that form behind normal jet airplanes. It can also fly at near-treetop level, using sophisticated terrain-following systems. Most important, the B-2 is nearly invisible on radar: The 172-foot-wingspan plane has about the same radar profile as a pigeon. But at close to a billion dollars per plane, is it really something we can afford? And do we need a new, penetrating bomber in light of easing tensions with our global superpower neighbor, the Soviet Union? Those nuclear questions are matters of vigorous debate in the early 1990s.

How does the U.S. bomber fleet stack up against our criteria for nuclear delivery systems? The B-52, the B-1, and the B-2 all have intercontinental range with in-flight refueling. All carry substantial payloads—about one World War II's worth of destructive capability per airplane—and can strike with good to excellent accuracy. And features ranging from missiles for "standoff" attack to electronic-warfare capability to sophisticated stealth technology en-

Figure 14.3 A B-2 stealth bomber in flight. (Northrop Corporation)

sure penetration of enemy territory. Flight times of both subsonic and supersonic bombers are measured in hours, a disadvantage in a first strike but certainly a safeguard against accidental or inadvertent nuclear attack.

What about vulnerability? It takes 2–4 minutes for a bomber crew on alert to board its aircraft, and another minute or so to start the engines and take off. Missiles launched from submarines off the U.S. coast take 10–15 minutes to reach inland targets. That gives bomber crews a comfortable lead of 5–10 minutes. On the other hand, modifications of their trajectories could reduce the missiles' flight time to about 7 minutes, leaving the bombers only 1–2 minutes to spare. Whether bombers could survive a first strike depends on their getting off the ground in an absolute minimum of time, and on their being hardened enough to withstand blast and EMP effects from nuclear weapons exploding on the bases they had just left.

Their long flight times make bombers primarily second-strike weapons, intended for retaliatory nuclear attack. On the other hand, the virtual invisibility of the B-2 might make it useful in a first strike. How is a potential adversary to perceive the evolution of the U.S. bomber fleet? Is the introduction of stealth technology an indication of potential U.S. belligerence, or is even the B-2 deployed solely to deter others' nuclear aggression? In the nuclear era, an adversary's answers to those questions can be as important as the actual technical capabilities of a nuclear weapons system.

Intercontinental Ballistic Missiles

The development of intercontinental ballistic missiles (ICBMs) is closely related to that of rocket technology for space flight, and can be traced back to the work of the American rocket pioneer Robert Goddard and the Russian Konstantin Tsiolkovskii and to the German V-2 rocket of World War II. By the mid-1950s, rocket technology had advanced to intercontinental range, and the development of physically small but high-yield thermonuclear weapons made rocket delivery attractive. The launch of the Soviet *Sputnik* satellite in 1957 spurred rocket development, and within a few years both the United States and the Soviet Union had begun deploying intercontinental ballistic missiles aimed at each other's territory. Other nations have gradually followed suit, and today ballistic missiles are proliferating at an alarming rate, especially among nations that are believed nearly capable of producing nuclear weapons.

What is a ballistic missile? Throw a ball into the air. Once it leaves your hand, it moves (if we neglect air resistance) under the influence of gravity alone. Its path—called a ballistic trajectory—is fully determined by the ball's speed, direction, and position when it leaves your hand. A ballistic missile works the same way. It is boosted above the atmosphere by rocket motors, after which gravity alone governs the subsequent motion. The rocket motors on today's ICBMs burn for only about 3 minutes, leaving their payloads at just over 100 miles altitude and traveling at 15,000 miles per hour (over 4 miles per second). A precise **inertial guidance**

Nuclear News: Missile Proliferation

Considerable international attention has focused on the proliferation of nuclear weapons. The result has been at least a slowing in the spread of those weapons. Meanwhile, though, the proliferation of nuclear-capable ballistic missiles has continued virtually unabated. Those missiles threaten the peace in some of the world's most volatile regions, and when wed with nuclear weapons they may soon threaten even the superpowers.

Who has ballistic missiles? How did they acquire them? What are the missiles' capabilities? Are any armed with nuclear warheads? As superpower relations warm and the risk of a U.S.-Soviet nuclear war fades, answers to those questions play increasingly important roles in the equation of global security.

Many of the emerging ballistic-missile states are Third World countries with a history of involvement in regional hostilities. The Middle East, in particular, bristles with ballistic missiles: Iraq, Iran, Israel, Libya, Syria, Saudi Arabia, and Egypt all have ballistic missiles, with ranges from 200 to nearly 2,000 miles. India and Pakistan, potential nuclear rivals, have active missile programs; so do Argentina and Brazil.

Most of the Third World missile powers acquired their missiles from foreign sources. Early in the missile race, the United States and the Soviet Union often supplied short-range missiles to their allies in the world's trouble spots. In recent years China has emerged as a leading supplier; especially worrisome are its sales of 2,000-mile-range East Wind missiles to Saudi Arabia. India, Brazil, and Israel have well-developed missile programs of their own, and are themselves potential suppliers to other budding missile powers. For some, ballistic-missile development goes hand in hand with civilian space programs.

Although many of these missiles are capable of delivering nuclear warheads, few of the new missile owners are currently nuclear-armed. An exception is Israel, which almost certainly has a significant nuclear-missile arsenal. In most cases, Third World ballistic missiles today carry conventional explosive warheads or, increasingly, chemical weapons. Nevertheless, many of their owners are striving vigorously for nuclear capability. The turn of the century could see a half-dozen countries with nuclear missiles capable of striking targets thousands of miles distant.

What is to be done about this threatening situation? In 1987 seven countries—the United States, Canada, Japan, Germany, Italy, France, and Britain—established the Missile Technology Control Regime, a program aimed at slowing the spread of ballistic missiles. The Regime bans export of large rockets and cruise missiles and their manufacturing facilities. It also requires discretion in the sale of other technologies that might be used in missile development. Although the Soviet Union is not a party to the Regime, the United States and the Soviet Union agreed in 1988 to begin discussions on missile proliferation. And the United States has pressed China to halt ballistic-missile exports, although with little success.

News sources: "Missile Influx Raises Arab-Israeli Tension," *Chicago Tribune,* April 7, 1988; "A New Genie Emerging," *Christian Science Monitor,* April 25, 1988; "Space Research Fuels Arms Proliferation: India's First Intermediate-Range Ballistic Missile Suggests U.S., West German Parenthood," *Wall Street Journal,* July 6, 1989; "Greater Threats from Lesser Powers, *New York Times,* April 8, 1990.

system senses every change in the missile's motion, and at the end of the boost period small rockets fire to send one or more warheads on precise trajectories toward their targets. From that point on the warheads travel thousands of miles under the influence of gravity alone, with no further propulsion or guidance. That long flight without guidance makes the extreme accuracy of today's ICBMs all the more impressive. The total intercontinental flight time is only about 30 minutes.

Did that last paragraph say "one or more warheads"? Yes. In the late 1960s, American nuclear strategists worried that Soviet advances in missile-defense technology might prevent U.S. missiles from reaching their targets. The United States responded by developing multiple-warhead missiles, with the goal of overwhelming Soviet defenses. The current incarnation of the multiple-warhead concept is called **MIRV**, for **multiple independently targetable reentry vehicles.** (A reentry vehicle is a warhead-bearing structure that reenters the atmosphere and carries the warhead to its target.) The uppermost stage of a MIRVed missile is called the **bus**. It carries the individual warheads, and it maneuvers to release each one on a separate trajectory to its designated target. The warheads from a single missile can be targeted to land in a "footprint" measuring roughly 100 by 300 miles.

Are MIRVs a good thing? In retrospect, many nuclear strategists argue that they aren't, and that MIRV technology should have been banned by treaty. Its multiple warheads give one MIRVed

Figure 14.4 Warheads being mounted on an MX missile's bus. Fully loaded, the bus will carry ten warheads, each yielding 300 kt or nearly 30 Hiroshimas. After mounting, the warheads will be covered by the nose cone of the missile. (U.S. Air Force)

missile a very good chance of destroying one of an adversary's missiles before launch. And destruction of a single MIRVed missile means destruction of multiple warheads. For both of these reasons, a nation armed with MIRVs might be tempted to strike first at its adversary in an attempt to destroy the adversary's MIRVs. Combined with increased accuracy, MIRVing has greatly compromised the invulnerability of today's land-based missiles.

Since the late 1960s, the mainstay of the United States' land-based ICBM force has been the Minuteman, a solid-fuel rocket with an 8,000-mile range. The oldest of the three Minuteman models carries a single 1.2-Mt warhead; the other models are MIRVed, with the most recent carrying three 335-kt warheads. Accuracy is better than 300 yards. A total of 950 Minutemen are deployed in underground concrete silos scattered throughout the Great Plains. Together, they carry nearly 2,000 nuclear warheads, with a total yield of nearly 1,000 Mt—the equivalent of over 150 World War

Figure 14.5 Streaks mark the reentry of MX warheads in a test over the Pacific Ocean. (U.S. Air Force)

Figure 14.6 A Minuteman missile in its silo. (U.S. Air Force)

IIs on this single component of the United States' strategic forces. The number of Minutemen will drop somewhat as the Strategic Arms Reduction Treaty takes effect.

Like the B-52, the Minuteman missile has undergone continual upgrading. Yet military planners, seeking to "modernize" U.S. strategic forces, have long advocated new missiles to replace the Minuteman. As we have seen, the introduction of accurate, MIRVed missiles has made fixed, land-based ICBMs increasingly vulnerable to a first-strike attack. Although a treaty banning MIRVs might have been an effective response to that threat, nuclear strategists have instead sought technological advances to restore the invulnerability of ICBMs. An obvious solution is to deploy mobile missiles whose location changes frequently enough to foil any first-strike attack. In the 1970s the United States developed a new missile, the MX, intended originally to be shuttled among a large number of launching sites in the American west. Environmental and cost considerations eventually killed this basing scheme. Having missiles hidden among multiple launch sites also complicates verification of arms-control treaties. Controversy about MX basing raged through the 1980s, and eventually fifty MX missiles were deployed in existing Minuteman silos. Since that basing scheme leaves the vulnerability issue unresolved, the search for more secure MX basing continues.

However it is based, the MX is a formidable weapon. It carries ten 300-kt warheads, each with an accuracy reported at better than 100 yards. What is the point of that much destructive power and accuracy? High-yield, pinpoint-accuracy warheads are necessary for only one purpose: destroying hardened military targets, such as missile silos. In either its vulnerable silos or its semi-vulnerable rail garrison, the MX might well not survive a surprise attack. That fact, combined with its impressive yield and accuracy, leads many to argue that MX is really designed as a first-strike weapon. Its proponents claim the MX is not intended for a first strike, but in the nuclear age intentions are not all that counts; peace and stability are affected also by a potential adversary's *perception* of a weapon's capability. Does the MX deserve Ronald Reagan's designation, "Peacekeeper," or is it actually a threat to world peace? The answer to that nuclear question may directly influence your future survival.

Even as MX production advanced through the 1980s, a respected presidential commission headed by General Brent Scowcroft acknowledged that the United States' security interests might be better served by the deployment of a small, single-warhead mobile missile. Mobility would give the missile invulnerability, and its single warhead would make it a less tempting target for an adversary's first strike. The proposed missile, called "Midgetman," represents a very different weapons philosophy than the MX—a philosophy recognizing that highly accurate multiple-warhead weapons may endanger rather than enhance national security. Fully embracing that new philosophy would mean giving up the multiple-warhead MX in favor of the Midgetman. Debate over these two missile systems continues in Congress; meanwhile, the Bush administration has won funding to advance both programs.

Submarine-Based Missiles

Submarine-launched ballistic missiles (SLBMs) constitute the third leg of the nuclear triad. SLBMs are launched from submerged

Nuclear News: Where to Put the MX?

For years the MX has seemed a missile without a home. The demise of the multiple-launch-site plan led to "dense pack," a basing scheme whereby several hundred MX missiles would be deployed in closely spaced silos. Destruction of one silo by an attacking warhead would raise a cloud of debris that would disable other incoming warheads. Dense pack met with little enthusiasm, and attention turned to other ways of making MX invulnerable. One proposed scheme involves deep burial, with special machinery for digging to the surface prior to launch; this scheme could be foiled by a "burrowing" nuclear warhead capable of digging its way to an underground target before exploding.

Most proposed schemes involve making the MX mobile. Mounting the missiles on specially modified aircraft or submarines has been rejected in favor of the so-called rail garrison mode. Under this plan, the 50 MX missiles will be removed from their silos and mounted on special railroad cars, with two missiles on each of 25 trains. The trains would be stationed at Air Force bases throughout the United States, to be moved onto the nation's rail network in time of crisis. Rail garrison basing is expected to be completed by 1994, at a cost of about $6 billion.

Is rail garrison the right home for the MX? A true surprise attack would find the missiles sitting ducks at their Air Force base rail sidings. Moving them onto the rail network might escalate a crisis, making nuclear war more likely. On the other hand, rail garrison will give the United States a mobile missile, something the Soviet Union has had for years. But to what end? What is the real purpose of the MX, anyway? Is it well designed for that purpose? Those are substantive, and costly, nuclear questions.

News source: "50 MX Missiles to Be Shifted to Trains in 7 States," *New York Times,* November 30, 1989.

Figure 14.7 The MX missile is 71 feet long, weighs nearly 100 tons, and carries ten 300-kt warheads, upgradable to 475 kt each. (U.S. Air Force)

submarines stationed off an adversary's coast, and their guidance systems allow considerable accuracy despite the submarine's changeable location. Typical flight times of 10–15 minutes make defense or even warning highly impractical.

If any nuclear weapons system is invulnerable, the SLBM is. Despite continued research in undersea detection and tracking, ballistic-missile submarines remain virtually impossible to find once they are at sea, and nothing on the technological horizon is likely to change that in the foreseeable future. For some nuclear strategists, SLBMs' invulnerability makes them alone an entirely adequate retaliatory force. If that is the case, it renders moot many of the debates raging about ICBM and bomber forces, at least for a nation serious about not launching a nuclear first strike.

The invulnerability of SLBMs is made possible by the nuclear reactor. (As was noted in chapter 9, today's pressurized-water reactors for generating electric power are scaled-up versions of PWRs first developed in the 1950s for submarine propulsion.) The nuclear difference means that a submarine can travel a long way on a small quantity of nuclear fuel. A modern nuclear sub is refueled at 10-year intervals, and may travel half a million miles on less than 100 pounds of highly enriched uranium. Equally important is the fact that fission does not require oxygen as does the burning of chemical fuels; consequently, nuclear-powered submarines can remain submerged almost indefinitely. In fact, nuclear-generated electricity is used to dissociate seawater into hydrogen and oxygen, providing oxygen for the crew to breathe. In practice, a nuclear missile submarine carries a crew of about 150 on patrol missions lasting up to 3 months.

Two types of ballistic-missile submarines currently make up the U.S. SLBM fleet. Poseidon submarines, first launched in 1963, carry sixteen missiles each. Poseidon missiles are MIRVed, with up to fourteen 50-kiloton warheads per missile. That makes a maximum of 224 warheads per submarine, with a total yield of 11 megatons—almost 1,000 Hiroshimas or two World War IIs of destruction on a single boat. (The average warhead load is closer to ten per missile, leaving "only" about 640 Hiroshimas.)

In 1985 the United States began decommissioning its nineteen Poseidons to make way for the newer Trident submarine. "Make way" here has a legal meaning: Adding too many Tridents without removing Poseidons would have put the United States in violation of warhead ceilings set in the 1979 Strategic Arms Limitation Treaty (SALT II). By 1990 the U.S. SLBM fleet was a mix of Poseidons and Tridents, moving toward an all-Trident fleet by the turn of the century.

How is Trident different from Poseidon? Trident is a much

Figure 14.8 A technician prepares to load missiles into the submarine *Calhoun*. (Department of Defense)

Figure 14.9   A Trident submarine at sea. (U.S. Navy)

larger boat, carrying 24 missiles to Poseidon's 16. Advances in submarine technology make it a quieter and therefore even less vulnerable missile base. At some $2 billion per submarine, Trident is far more expensive. But the most significant difference, and one that has spurred a great deal of controversy, lies in the actual missiles on board. Poseidon missiles and the early Trident missiles (the so-called Trident I or C-4) lack the range and accuracy of land-based ICBMs. The Poseidon missile is accurate to about 1/3 mile over a 3,000-mile range. Its relatively poor accuracy hardly qualifies it as a "missile killer" or first-strike weapon, and its limited range requires submarines to operate fairly close to enemy territory. The Trident I missile has greater capabilities, but is still useful only against "soft" targets such as airfields, army bases, industrial areas, and cities.

Enter the Trident II or D-5 missile. With a range up to 7,000 miles, Trident II can reach its targets from anywhere in a substantial portion of the world's oceans. Each Trident II is reported capable of delivering from ten to fifteen half-megaton warheads. Those

Figure 14.10 A Trident II missile goes spectacularly out of control in a 1989 test launch from the submerged *U.S.S. Tennessee.* (U.S. Navy)

figures could give a fully loaded Trident submarine 360 warheads with a total destructive yield of nearly 30 World War IIs. The Trident II raises many of the same questions as the MX. Is this highly accurate, high-yield missile a first-strike weapon? Even if it isn't, how will potential adversaries perceive it? Why do we need the accuracy of Trident II if our SLBM fleet is truly a retaliatory force? Are we deploying Trident II because it is necessary, or because it is technologically possible? Why, if Poseidons remain undetectable, do we need Trident at all? Ultimately, does Trident enhance or reduce global security?

As we have seen, submarine-based missiles are the least vulnerable leg of the nuclear triad. Increasing accuracy and range give SLBMs high scores on the other criteria that were introduced early in this chapter. But what about safety in the event of a crisis or an accidental launch? Here SLBMs are distinctly inferior. Their extremely short flight time allows little time for warning or for last-minute diplomacy (which could prevent the launching of a few

SLBMs from escalating to an all-out nuclear war). And communication with submarines is a problem. The oceans' impenetrability to common signaling mechanisms is what gives submarines their invulnerability, but it also rules out easy communication with submerged submarines. Today, submarines must come very close to the ocean surface and drag long antennas to communicate with military authorities on land; in the process they compromise their invulnerability. Land-based antennas thousands of miles long would allow very slow communication with more deeply submerged subs, but public opposition has stalled proposed antenna projects. The difficulty of communication puts the submarine leg of the triad in a unique position: SLBMs are the only nuclear weapons that can be launched without first receiving special codes from high government authorities. That fact enhances SLBM's invulnerability, since no amount of disruption back home can disable them. But it also raises the specter that an accidental launch, a mutiny, or an act of insanity confined to a single submarine could initiate a nuclear war.

Cruise Missiles

We have now looked at the three legs of the U.S. nuclear triad. But a new missile deployed in the 1980s straddles all three legs, and requires a closer look. The **cruise missile** is not a new idea; the German "buzz bomb" of World War II was a primitive cruise missile. In contrast to a rocket-boosted ballistic missile, the cruise missile is a small, pilotless airplane with conventional wings and a jet engine that flies subsonically at low altitude. Two important developments led to the modern cruise missile: small, inexpensive jet engines, and sophisticated computerized guidance systems. Cruise-missile guidance involves terrain-contour matching (TERCOM), in which an on-board computer periodically compares radar images of the terrain over which it is flying against internal maps. The computer directs the missile to correct its course as necessary; as a result, it may be accurate to within 30 feet.

Cruise missiles are slow, so they are not an obvious first-strike choice. But they fly at nearly treetop level, making them undetectable by ground-based radar. And they can deliver high-yield nuclear warheads with missile-killing accuracy. Are cruise missiles needed

Figure 14.11 The U.S. air-launched cruise missile (ALCM) is only 21 feet long. It flies at 500 mph, 100 feet above the ground, and delivers a 1.2-Mt warhead with an accuracy of 30 feet. (U.S. Air Force)

for a credible retaliatory force? Or are they provocative weapons, capable of a first strike?

Cruise missiles confound the international nuclear balance in several ways. They are small and easily hidden, making it difficult to verify treaty limits on missiles. Their extreme accuracy makes them valuable delivery systems for conventional as well as nuclear explosives, and it is not externally obvious whether a cruise missile carries a nuclear or a conventional warhead. Proliferation of cruise missiles could render a comprehensive arms-control treaty unverifiable; indeed, the cruise-missile problem nearly derailed negotiations toward the Strategic Arms Reduction Treaty. Yet air-launched (ALCM), ground-launched (GLCM), and sea-launched (SLCM) cruise missiles have now been deployed. Is this an example of a "technological imperative," where a weapon gets developed not because it is needed but because its development is technologically possible? Or do cruise missiles play an essential role in preserving the nuclear balance?

The Nuclear Competition

In taking a detailed look at what makes up the nuclear triad, we have focused on the United States' strategic forces. What about other nuclear weapons nations? Only the Soviet Union has a fully developed nuclear triad comparable to that of the United States; indeed, to a great extent, the effort to maintain equivalent nuclear forces has driven the superpowers' nuclear arms race.

The American and Soviet nuclear triads are equivalent, but not identical. Accidents of geography and technology have given them different emphases. The Soviet Union's vast land area and relatively poor ocean access favor land-based ICBMs. Strength in high-pay-load rocketry and lagging guidance technology have further tipped the Soviets toward large, land-based missiles with high-yield war-heads. The Soviets have traditionally had less faith in bombers, and their submarine technology has generally been less advanced and less reliable than that of the United States. Figure 14.15 shows the results: The U.S. and Soviet triads, while roughly comparable in warhead numbers and yield, show distinctly different balances among the legs of the triad.

Who is ahead? The United States has more strategic warheads,

Figure 14.12 A Soviet Tu-16 bomber. (Department of Defense)

Figure 14.13 Soviet SS-X-15 ICBMs in their mobile launchers. (Department of Defense)

the Soviet Union more total megatonnage. In light of the huge destructive power of both arsenals, the question as to who is ahead is essentially meaningless. It is more important to ask whether the weapons systems making up the strategic arsenals tend to enhance or diminish international stability, especially in time of crisis. In the nuclear age, sheer destructive power does not buy security; circumstances as diverse as the first-strike temptation of multiple-warhead missiles or the prospect of nuclear winter may well mean that greater military might brings less security.

The United States and the Soviet Union are not the only nuclear-armed nations. Four others are known to have exploded nuclear devices, and three of those have well-developed nuclear arsenals. Still others are at or near nuclear capability, with still more acquiring missiles that could be used to deliver nuclear weapons.

France and China have developed largely independent nuclear capabilities, although France had some early help from the United

Figure 14.14 A Soviet submarine equipped with strategic cruise missiles (Department of Defense)

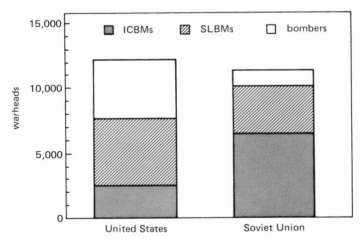

Figure 14.15 Compositions of the U.S. and Soviet nuclear triads in 1990, as measured by the distribution of warheads among the three legs of the triad. The Strategic Arms Reduction Treaty will reduce overall warheads by about 30 percent, but will have less effect on the distribution of warheads. (data source: *Bulletin of the Atomic Scientists,* January–February 1990, p. 49; March 1990, p. 49)

States, and China from the Soviet Union. Nuclear cooperation between the United States and Great Britain has been closer; the SLBM leg of the British force, for example, uses American submarine technology. Figure 14.16 compares the nuclear forces of Britain, China, and France. Although far smaller than the American and Soviet arsenals, all three of these forces have the capability to inflict unimaginable damage on any adversary.

Summary

We have looked here at the diverse components that make up the weapons-delivery systems of a modern nuclear arsenal. These include primarily the elements of the nuclear triad: bomber aircraft, land-based ICBMs, and submarine-based missiles. In addition, cruise missiles may appear in each leg of the triad.

Important features characterizing nuclear delivery systems include range, payload, accuracy, and penetration; all these relate to the ability to deliver weapons to their assigned targets. Flight times, which vary from minutes to hours, set the fundamental time scale for nuclear conflict. Invulnerability of delivery systems is strategically crucial, ensuring the retaliatory force needed to deter an ad-

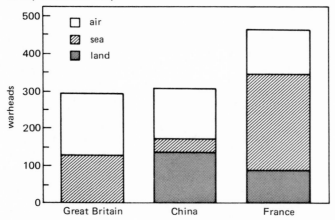

Figure 14.16 Great Britain, China, and France all have significant nuclear forces. Shown here are the numbers of nuclear warheads in each nation's arsenal, categorized as land, sea, or air based. (data source: *SIPRI Yearbook 1989*, tables 1.6–1.8)

versary's attack. The ideal nuclear weapons system affords stability during crisis and safety against accidental or mutinous launch.

Those features of weapons that seem important to military planners—destructive force, multiple warheads, warhead accuracy—may actually diminish national security by inviting preemptive attack or encouraging dangerous arms competition. Weighing perceived needs for military strength against the overwhelming necessity of avoiding nuclear war poses some of the most difficult and contested questions of the nuclear age.

Further Reading

Thomas Cochran et al., *Nuclear Weapons Databook* (Ballinger, 1984–1989). A four-volume compendium of data on nuclear weapons and delivery systems, and on the nuclear weapons industry's research, production, and testing facilities, prepared by the Natural Resources Defense Council. Volume I includes photographs and design specifications of airplanes, missiles, ships, reentry vehicles, and nuclear warheads. Volumes II and III detail U.S. nuclear-warhead facilities and production, and volume IV covers Soviet nuclear weapons.

Paul P. Craig and John A. Jungerman, *Nuclear Arms Race: Technology and Society* (McGraw-Hill, 1986). A general survey of nuclear weapons issues. Part 1 includes a description of U.S. and Soviet strategic weapons and delivery systems.

International Institute for Strategic Studies, *The Military Balance* (IISS, revised annually). Detailed statistics on world military power, including nuclear forces.

Jane's Weapon Systems (Jane's, revised annually). A detailed, illustrated catalog of *all* the world's military weapons systems, both nuclear and conventional. The advertisements constitute an eye-opening introduction to the magnitude of the international arms trade and to the world's infatuation with war preparation. Even the built-in bookmark champions international arms sales.

Janne E. Nolan and Albert D. Wheelon, "Third World Ballistic Missiles," *Scientific American* 263 (August 1990), p. 34. A disturbing look at the growing capabilities of ballistic missiles in Third World countries, including some with nuclear capability.

Dietrich Schroeer, *Science, Technology, and the Nuclear Arms Race* (Wiley, 1984), chapters 5–7. A detailed technical description of the components of the superpowers' strategic nuclear triads. Somewhat dated but still generally useful.

Stockholm International Peace Research Institute, *SIPRI Yearbook: World Armaments and Disarmament* (Oxford University Press, published yearly). Authoritative tabulations of weapons systems, military expenditures, weapons proliferation, and related topics, accompanied by summary essays.

Glossary

air-launched cruise missile (ALCM)   A cruise missile launched by an airplane.

ballistic missile   A missile that, after being boosted above the atmosphere, travels only under the influence of gravity.

basing scheme   The way nuclear weapons are deployed in readiness for wartime use. Examples include fixed, land-based deployment; mobile basing; and submarine basing.

bus   The uppermost stage of a multiple-warhead missile, responsible for carrying and aiming the individual warheads.

cruise missile   A subsonic, jet-powered missile that flies at treetop level to avoid radar detection.

delivery system   A means for getting nuclear weapons to their targets.

first strike   A nuclear surprise attack aimed at crippling an adversary's retaliatory capability.

ground-launched cruise missile (GLCM)   A cruise missile launched from a mobile ground-based launcher.

hard target   An underground missile silo, military command center, or other structure that has been fortified to resist a nuclear blast.

ICBM   An intercontinental ballistic missile.

inertial guidance system   A mechanism that senses changes in motion and from those changes determines accurately its position and motion. Used in accurate ballistic missiles and for submarine navigation.

invulnerability   A delivery system's ability to withstand an enemy's first strike.

multiple independently targetable reentry vehicles (MIRV)   Individual warhead-bearing reentry vehicles that can be independently targeted. Ten or more MIRVs may be carried on a single missile.

payload   The maximum weight a vehicle can deliver to its target.

penetration   A delivery system's ability to enter enemy territory in order to reach its target.

range   The distance a nuclear delivery vehicle can travel from its base to a target.

sea-launched cruise missile (SLCM)   A cruise missile launched from surface ships or submarines.

short-range attack missile (SRAM)   A rocket-powered supersonic missile launched from a bomber up to 100 miles from its target.

soft target   A target, such as a city or a factory, that is not fortified against nuclear blast.

stealth bomber A bomber aircraft employing nonmetallic components and other measures to reduce its radar visibility.

strategic In the context of nuclear weapons, *strategic* refers to weapons to be used against an adversary's home territory (in contrast to *tactical* weapons, designed for battlefield use).

submarine-launched ballistic missile (SLBM) A ballistic missile capable of being launched from a submerged submarine.

triad The three-part structure of the superpowers' nuclear forces, consisting of bomber aircraft, land-based intercontinental missiles, and submarine-launched missiles.

Nuclear Strategy

15

How should nuclear weapons be used? In view of the devastation that nuclear war would cause, an obvious answer is that they shouldn't be used. So let us broaden the concept of "use" to include the function of nuclear weapons as instruments of international politics. How, then, should these agents of mass destruction be "used"? How many should our country have, and how should they be deployed in order to enhance national and global security? What moral strictures attach to the possession of nuclear weapons and to the threat of their use? These deep and difficult questions were thrust on humanity with the dawn of the nuclear age, and they remain under vigorous debate. Here we will explore briefly some basic issues of nuclear strategy. Since this is a book emphasizing technology, the discussion will necessarily be sketchy; the reader is referred to the many published analyses of nuclear policy issues, some of which are listed at the end of this chapter.

The Vision of the Nuclear Scientists

On September 12, 1933, the Hungarian-born physicist Leo Szilard had a sudden, unprecedented vision of the possibility of a nuclear chain reaction. Szilard, a political idealist strongly influenced by the visionary novels of H. G. Wells, saw more than physics. He saw also nuclear bombs and a world order fundamentally altered by their presence.

Over the next decade, Szilard pushed hard to see his chain reaction brought to fruition. He believed the existence of nuclear bombs would mean an end to war and the advent of a new era of enlightened world community. During the Second World War, Szilard's vision was shared by the Danish physicist Niels Bohr, who directly influenced J. Robert Oppenheimer. After the war, Oppenheimer became an outspoken advocate of international control of nuclear weapons, and his urging helped the United States pursue that goal in the newly formed United Nations.

But the vision of the nuclear scientists was not to be realized. Growing mistrust between the United States and the Soviet Union derailed the UN negotiations, and led instead to an accelerating nuclear arms race. Disappointment with the course of events made many former bomb-makers outspoken and enduring critics of nuclear policy; Szilard founded the Council for a Livable World, a group that to this day supports disarmament-minded candidates for the U.S. Senate.

Nuclear Deterrence

Scientists were not alone in recognizing that their nuclear offspring had changed the world. In 1946 the strategic analyst Bernard Brodie stated: "Thus far the chief purpose of our military establishment has been to win wars. From now on its chief purpose must be to avert them."[1] In Brodie's statement lies the essence of nuclear **deterrence**, the idea that we possess nuclear weapons in order to prevent their use. In 1983 former U.S. Secretary of Defense Robert McNamara made it even clearer: ". . . nuclear weapons serve no military purpose whatsoever. They are totally useless—except only to deter one's opponent from using them."[2]

Deterrence requires the threat of nuclear retaliation: If you dare to attack me, I will use my nuclear weapons against you. For deterrence to work, that threat must be *credible*; a potential attacker has to believe its opponent *can* and *will* use its nuclear retaliatory force. What makes a nuclear deterrent credible?

First and foremost, a deterrent force must be capable of inflicting "unacceptable" damage on an opponent. "Unacceptability" depends on both technical and political factors—technical, because retaliatory weapons must be physically capable of sufficient destruction; political, because potential adversaries must believe that the level of damage suffered in a retaliatory strike would indeed be unacceptable to their country.

A retaliatory force is no good if it can be destroyed in a first strike; that is why there was so much emphasis on weapons systems' invulnerability in the preceding chapter. Invulnerability is primarily a technical issue, but questions of political will are also involved. For example, land-based ICBMs may be vulnerable to a first strike—*if* they are in their silos when the attacking warheads land. But a **launch-on-warning** strategy would have the missiles underway once warning of an impending attack was received, again giving them a measure of invulnerability. Launch-on-warning is risky business: It hangs the fateful decision to launch nuclear missiles on electronic warning systems, and reduces to almost zero the time for rational thinking. But the mere possibility of launch-on-warning should give a potential aggressor pause, so the possibility alone may contribute to deterrence.

It is not enough that a deterrent force be invulnerable and capable of ample destruction; it is also essential that its possessor show the political will to use its deterrent force if necessary. Therein lies part of the nuclear dilemma: Deterrent forces exist to deter war, but deterrence works only if a potential attacker believes they would actually be used. Paradoxically, a nation bent on deterring nuclear war must show willingness to engage in such war.

What constitutes a sufficient deterrent? That has been a central question of nuclear strategy since the 1950s. The ongoing debate between "hawks" and "doves" is not between those who favor nuclear war and those who oppose it; rather, it is a debate about the best way to prevent nuclear war. That "the best way" is not obvious is again related to the central dilemma of the nuclear age—that the principal use of nuclear weapons is to prevent their use. In that context, nuclear "hawks" would argue that we, and indeed the world, are safer with more and better nuclear weapons; nuclear "doves" counter that huge nuclear arsenals only increase the chance of inadvertent, accidental, or crisis-driven nuclear war, and ensure global catastrophe if deterrence should fail.

What? Deterrence fail? As long as deterrence holds—and with

it the necessity that nations be ready and willing to use their nuclear weapons—nuclear war remains a possibility. How should we plan for that possibility? Should we plan at all, or does planning only make nuclear war more likely? Should we, as some would argue, be prepared to prevail in a nuclear conflict, by maintaining nuclear superiority over potential adversaries? Should we engage in "damage limitation" measures, such as civil defense and "hardening" of key industries, to lower nuclear war's impact on our nation's health? Or would damage limitation only make us more likely to accept nuclear conflict? Again, a nuclear dilemma: Preparing for nuclear war might reduce our losses in the event of war, but might at the same time make war more likely.

Could we win a nuclear war? That question is important if we believe deterrence might fail, for an affirmative answer implies the desirability of nuclear arsenals far in excess of those needed purely for deterrence. Today the world's leaders are in consensus that nuclear war is not winnable, but that opinion has not always prevailed, even among those now in power. And many nuclear "doves" would argue that continuing technological advances and "modernization" of nuclear forces reveal a quest for nuclear superiority among military planners, despite what political leaders may say.

The possibility that deterrence could fail leads some "hawks" to advocate that we prepare to wage, and ultimately to prevail in, a protracted nuclear conflict. It leads some "doves" to reject deterrence altogether, arguing that its possible failure makes nuclear war an eventual inevitability. In between is a vast middle ground whose occupants generally support the concept of deterrence but disagree substantially about how best to achieve a stable and lasting deterrence. The evolution of nuclear strategy reflects the ebb and flow of these disagreements, the changing views of strategic thinkers, and the reality imposed by new technologies.

Nuclear Strategic Doctrines

The aftermath of World War II found the United States with a monopoly on nuclear weapons. But scientists realized there was no

Nuclear News: Our Leaders on Nuclear War

Today the leaders of the superpowers and their close advisors are united in the belief that nuclear war is a losing proposition for all parties. But that hasn't always been unambiguously the case. In the past, leaders on both sides have made statements of at least uncertain interpretation, and each side has at times accused the other of basing its policy on the belief that a nuclear war can be won. Lower government and military officials have sometimes been more overt in their acceptance of nuclear war. Some examples follow[3]:

"You have a survivability of command and control, survivability of industrial potential, protection of a percentage of your citizens, and you have a capability that inflicts more damage on the op-position than it can inflict on you. That's the way you can have a winner. . . ."—George Bush, candidate for 1980 Republican presidential nomination

"There are more important things than peace . . . In the nuclear age, the responsibilities are awesome, but there are things worth fighting for."—U.S. Secretary of State Designate Alexander Haig, January 9, 1981

"We will maintain sufficient strength to prevail if need be."—U.S. President Ronald Reagan, Inaugural Address, January 20, 1981

"I could see where you could have the exchange of tactical weapons against troops in the field without it bringing either side to pushing 'the button.'"—U.S. President Ronald Reagan, October 16, 1981

"The Soviets believe nuclear weapons are there to be used."—Edward Rowny, Special Representative for Arms Control and Disarmament Negotiations for the President of the United States, August 31, 1981

"Everything that has been said in their manuals indicates that, unlike us, the Soviet Union believes that a nuclear war is possible and they believe it is winnable."—U.S. President Ronald Reagan, October 16, 1981

"If there are enough shovels to go around, everybody's going to make it." "You can make very good sheltering by taking the doors off your house, digging a trench, stacking the doors about two deep over that, covering it with plastic . . . then pile dirt over it." "It's the dirt that does it."—Thomas K. Jones, Deputy Under Secretary of Defense for Research and Engineering, Strategic and Theater Nuclear Forces, autumn 1981

On the other hand, not all have been so frighteningly ambiguous; clear statements about the impossibility of nuclear war have sounded throughout the nuclear age:

"Nuclear war cannot be measured by the archaic standards of

'victory' or 'defeat.'"—U.S. President Jimmy Carter, October 4, 1977

"It is dangerous madness to try to defeat each other in the arms race and to count on victory in nuclear war."—Soviet President Leonid Brezhnev, October 21, 1981

"We rule out the possibility of 'limited' nuclear war. In a nuclear war, whether limited or unlimited, it will be difficult to tell the victor from the vanquished."—Leonid Zamyatin, Director, International Information Department, Soviet Communist Party Central Committee, November 2, 1981

"I believe thermonuclear weapons are good for two things—for deterrence or for suicide. Other suggested uses are pure fantasy."—Thomas J. Watson, former U.S. Ambassador to the Soviet Union, December 9, 1981

"A nuclear war cannot be won and must never be fought."—U.S. President Ronald Reagan, October 26, 1984

"We have something in common—our understanding that nuclear war is inadmissible, that it cannot be waged and that there will be no winner in a nuclear war."—Soviet President Mikhail Gorbachev, November 21, 1985

---

fundamental "nuclear secret," and that any industrialized country with the will to do so could build a nuclear bomb. Nuclear monopoly would not prevail, although estimates suggested it might take others from 5 to 20 years to duplicate the Manhattan Project's achievement. The United States tried briefly to flaunt its nuclear might in an effort to coerce others to do its bidding. That didn't work; in particular, the Soviet Union ignored the implicit nuclear threat and continued its occupation of Eastern Europe with an army that outnumbered U.S. forces in Europe by four to one. At the same time the Soviet Union began vigorous efforts to develop its own nuclear capability, culminating in its first fission explosion in 1949.

How was the United States to respond to the perceived threat from the Soviet Union's vastly greater conventional forces and its growing nuclear capability? In the early 1950s the United States still enjoyed an overwhelming nuclear advantage, and President Eisenhower's administration decided to base U.S. policy on that advantage. Instead of an expensive effort to match Soviet conventional forces, the United States would counter any Soviet aggression with devastating nuclear retaliation against the Soviet homeland. This policy earned the name **massive retaliation**.

Was massive retaliation an effective deterrent? When massive retaliation became official policy in 1953, the United States already had well over 1,000 nuclear weapons and a bomber fleet capable of delivering them to Soviet targets. Since 25 percent of the Soviet Union's population and 50 percent of its industrial capacity are concentrated in 100 cities, 1,000 weapons were far more than sufficient for massive destruction of the Soviet Union. And the American deterrent force was credible. In the early 1950s, the United States could have launched a massive nuclear attack without suffering "unacceptable" damage from the then puny Soviet nuclear arsenal.

The mid-1950s saw significant changes in the nuclear balance. Both the United States and the Soviet Union exploded their first megaton-range fusion weapons, and the Soviets developed intercontinental bombers capable of striking the United States. A U.S. nuclear strike in response to Soviet conventional aggression would no longer go unpunished; now the United States could expect massive Soviet nuclear retaliation against its homeland. Under those circumstances, would the United States really be willing to launch a nuclear attack on the Soviet Union in order to counter aggression in Europe or elsewhere? That question itself undermined the credibility of massive retaliation as a deterrent policy.

The eroding credibility of massive retaliation led to an emphasis on smaller, short-range nuclear weapons deployed in what was then a divided Europe. The notion of limited nuclear war developed, the idea being to restore deterrence with the more credible threat of using smaller nuclear weapons in a localized conflict that need not spread to the superpowers' homelands. Again, a nuclear dilemma: Is it really possible to make nuclear war less likely (i.e., deterred) by making it more acceptable (i.e., limited)? And would a limited war remain limited? Strategists envisioned a graduated sequence of nuclear war scenarios, ranging from a local battlefield conflict to all-out strategic bombing; with many possibilities in between, they argued, escalation to all-out war was unlikely. Nuclear "doves" rejected the strategists' assumption that rational decision making would prevail in the heat of conflict, and saw the risk of all-out war increasing with each new scenario.

By the early 1960s the development of intercontinental ballistic missiles made the threat of massive retaliation seem even more hollow; now, a swift, half-hour exchange would leave both sides in near total ruin. Although the United States' nuclear forces remained numerically superior at that time, it had become clear that the United States and the Soviet Union were already in a nuclear stalemate. Numerical advantage no longer translated into real military advantage, and the notion of nuclear superiority was fast becoming obsolete.

A strategy called **flexible response** evolved in light of the changing nuclear picture. Pioneered by the Kennedy administration in the early 1960s, flexible response called for deterrence of each possible form of aggression with the threat of a counterattack *in kind*. That is, a conventional attack would be countered with conventional weapons, a tactical nuclear strike with tactical nuclear weapons, and a strategic attack with a comparable level of strategic nuclear megatonnage. Only if deterrence failed at one level would the ante be upped; thus the United States would use nuclear weapons against a Soviet conventional attack in Europe only if Western conventional forces failed to halt the attack. Again, the idea was to make deterrence more credible; although the United States might not be willing to attack the Soviet homeland with thermonuclear weapons in response to an invasion of Western Europe, it would certainly have fewer qualms about unleashing conventional or even tactical nuclear forces.

Would flexible response work? Again, the same dilemma. Flexible response lowers the threshold for using nuclear weapons, making their use more likely. But it does so in the name of maintaining deterrence—that is, preventing the use of those weapons. And flexible response requires vastly expanded military capabilities, both conventional and nuclear. No longer are a few hundred strategic nuclear warheads enough to deter aggression; now it is necessary to have, in kind and in quantity, all the weapons possessed (or thought to be possessed, or possibly under development) by one's potential adversaries. Flexible response may or may not be a suitable deterrent, but it is surely a prescription for an escalating arms race.

What Are the Targets?

Where should we aim our nuclear weapons? At an opponent's population centers? Industries? Military facilities? Questions of targeting go hand in hand with decisions about nuclear strategy.

Most Americans imagine that our nuclear weapons are targeted on potential adversaries' major cities, and they envision nuclear war as an all-out assault on population centers. That impression is partly incorrect. Most of our weapons are aimed at specifically military targets. Some of those targets are, to be sure, in or near cities; many, though, are isolated missile silos and remote military installations. On the other hand, even the small fraction of weapons targeted against cities are sufficient to kill hundreds of millions of civilians.

In the cold parlance of nuclear strategy, targeting an adversary's population is known as a **countervalue** strategy; the name arises because live people are considered to have value, something a country would not want to lose. Targeting military facilities, on the other hand, is a **counterforce** strategy, aimed at destruction of an opponent's armed forces. The original massive-retaliation policy of the 1950s was a countervalue strategy, threatening destruction of Soviet population centers. The flexible-response strategy of the 1960s put more emphasis on counterforce capability, and in the 1970s and the 1980s counterforce emerged as the dominant theme in strategic nuclear planning. Today's counterforce strategy involves not only the destruction of an adversary's forces but also **decapitation**—the elimination of civilian and military leadership through precise nuclear targeting.

Which strategy is most likely to prevent nuclear war? Proponents of countervalue strategy, including many moderate nuclear "doves," argue that the horrible threat of population annihilation remains the most effective deterrent. Their argument rests on an adaptation of the massive-retaliation doctrine to an age in which the superpowers' nuclear arsenals are essentially equivalent. That doctrine goes by the acronym **MAD**, for **mutual assured destruction**. MAD holds nuclear adversaries' populations hostage against the threat of nuclear attack, maintaining the "balance of terror" which is often alleged to be responsible for keeping the nuclear peace. Under MAD, a nation contemplating nuclear attack is deterred by the knowledge that its own destruction in a nuclear counterattack is certain.

By assuring nuclear destruction, MAD advocates claim, their strategy guarantees that no rational leader would order a nuclear attack. Counterforce attacks, in contrast, might be viewed as acceptable, rational means to achieve national goals. MAD advocates see that view as illusory for two reasons: First, the presence of military targets in or near cities implies substantial collateral damage and civilian casualties, even with counterforce targeting. Second, a nuclear war initially involving only military targets could quickly escalate to an all-out conflict consuming population centers as well.

Proponents of counterforce strategy—including many nuclear "hawks"—reply with the same argument that derailed massive retaliation: that the threat of an all-out attack against population centers is too unbelievable to deter any aggression short of an initial all-out attack, so MAD cannot prevent more limited aggression. They argue instead that aggressive behavior—against the United States or elsewhere in the world—is best deterred if an adversary knows we can respond with a specific nuclear strike appropriate to the situation at hand, possibly including destruction of the adver-

Figure 15.1 A cartoonist's view of MAD. (Scott Willis; © 1983 *Dallas Times Herald*)

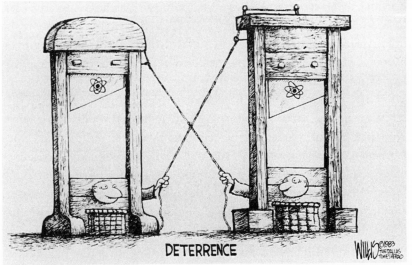

sary's own nuclear missiles. Counterforce proponents claim, too, that MAD is militarily ineffective, killing harmless civilians instead of destroying harmful weapons. Some of them add an ethical twist, arguing that it is immoral to hold civilians hostage to the threat of nuclear annihilation. Better, they say, to target weapons than human beings.

The double meaning of the acronym MAD is not lost on critics of assured destruction. But counterforce strategies have their own peculiar insanity, summarized in the acronym **NUTS**. Standing for **nuclear utilization target selection**, this term emphasizes that counterforce strategies may well result in actual use of nuclear weapons. But the distinctions between countervalue and counterforce and between the related but not quite synonymous MAD and NUTS are blurred, in part because the enormous destructive power of nuclear weapons means that even a limited counterforce attack would claim millions of civilian lives and in part because the superpowers' huge nuclear arsenals make the threat of assured destruction very real no matter what the targeting strategy. Some would argue that as long as sizable nuclear arsenals persist, some form of MAD will ultimately serve to keep the nuclear peace.

Technology and Strategy

How does nuclear technology affect strategy, and what demands do nuclear strategies place on technology? We have already seen how invulnerability of weapons—achieved largely through technology—is essential to deterrence. And our brief look at the evolution of strategic doctrines shows how technological advances have forced modifications in nuclear strategy (figure 15.2).

Counterforce and countervalue strategies have distinctly different technological implications. Destruction of population centers does not require particularly accurate delivery, although it does demand penetration of an adversary's defenses. Nor does countervalue require fast delivery, since cities don't move. And a relatively "soft" city can be destroyed by a single large warhead or several smaller ones. Counterforce strategies, in contrast, make quite different demands on weapons technology. Many military targets do move: submarines submerge, missiles get launched, aircraft take

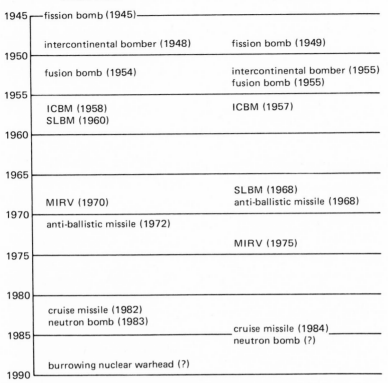

	United States	Soviet union
1945	fission bomb (1945)	
1950	intercontinental bomber (1948)	fission bomb (1949)
1955	fusion bomb (1954)	intercontinental bomber (1955) fusion bomb (1955)
1960	ICBM (1958) SLBM (1960)	ICBM (1957)
1965		
1970	MIRV (1970) anti-ballistic missile (1972)	SLBM (1968) anti-ballistic missile (1968)
1975		MIRV (1975)
1980		
1985	cruise missile (1982) neutron bomb (1983)	cruise missile (1984) neutron bomb (?)
1990	burrowing nuclear warhead (?)	

Figure 15.2 Technological advances made by the nuclear superpowers have done much to shape each other's responses and strategic doctrines. Note that the most recent developments incorporate counterforce accuracy, or warfighting capability.

off. Speed and surprise are essential in an effective counterforce attack. Destroying specific objects—especially hardened targets, such as missile silos—calls for high-yield, accurate nuclear weapons with multiple warheads. Accuracy, speed, and multiple-warhead capability are thus hallmarks of a counterforce arsenal.

Unfortunately, those features are precisely what is needed for first-strike capability. Opponents of counterforce claim the strategy is dangerous for that very reason: By deploying weapons that *could* be used effectively in a first strike, the nuclear powers make such a strike more likely. Even if weapons are intended only for counterforce deterrence, the fact that they double as first-strike weapons makes them so threatening to a potential adversary that they may invite a preemptive strike. Those policy considerations have a direct bearing on technological decisions about nuclear weapons.

War Fighting or Deterrence?

In the 1980s, counterforce strategy underwent a significant refinement. No longer was it enough to deter nuclear aggression; now, the United States had to be capable of waging a prolonged nuclear conflict and, in some sense, of ultimately "prevailing" over its adversary. No one quite said "winning," but Secretary of Defense Harold Brown of the Carter administration made clear that the United States must at least "deny victory" to its opponent. And Casper Weinberger, Ronald Reagan's Secretary of Defense, made it clearer: American nuclear capabilities "must prevail," with nuclear war ending "on terms favorable to the United States." This renewed emphasis on nuclear war fighting frightens many nuclear "doves," who fear that war-fighting strategies will make nuclear war less "unthinkable" and therefore more likely. Proponents of war-fighting capability respond that they are not advocating fighting a nuclear war. Instead, they claim that the clear-cut ability and will to fight and prevail in such a war are necessary to deter an adversary's attack. Today's multifaceted American weapons program is based on the perceived need for nuclear war-fighting capability, although the changing international climate and reductions in strategic arms may alter that situation. Is war-fighting capability a strategic necessity, or is it a dangerous whim of a military establishment out

Nuclear News: MX or Midgetman?

Technology and strategy come to a focus in the debate over modernization of the United States' ICBM forces. On the one hand is the MX missile, with its ten 300-kt warheads. Fifty MXs are already deployed, and they are soon to be mounted on railroad cars in a compromise between fixed and mobile basing. On the other hand is the Midgetman, a much smaller single-warhead missile designed for mobility. Which missile would best serve our future security needs?

The MX has all the hallmarks of a first-strike weapon: It is big and accurate, and it carries many warheads. Furthermore, even its "rail garrison" basing scheme hardly guarantees invulnerability. Its multiple warheads and its silo-killing capability make the MX a tempting target for an adversary's first strike. MX opponents argue that the missile is a dangerously destabilizing addition to our strategic arsenal. They see the Midgetman as a far safer deterrent; its single warhead makes it a less tempting target, and its true mobility (the Midgetmen would be mounted on truck-like launchers) enhances invulnerability. Many more Midgetmen than MXs would be needed to carry the same number of warheads; combined with the Midgetman's mobility, that would discourage an adversary from launching a first strike. MX proponents counter with cost arguments: The Midgetman would cost 3 times as much as the MX for the same number of deliverable warheads. And, MX proponents claim, their missile is available today, and for that reason the United States should continue to rely on its deterrent capability.

Which missile, MX or Midgetman? Faced with that choice, President George Bush took the easiest, but most expensive, route of all: His 1990 defense budget included funding for *both* missile programs.

News sources: "The MX: A Missile for All Reasons," *New York Times,* September 20, 1988; "Mobile Missiles? A Waste," *New York Times,* June 25, 1989; "Senate Gives Bush a Missiles Victory," *New York Times,* August 2, 1989; "Stocking the Atomic Arsenal: How Much Deterrence to Buy," *New York Times,* May 23, 1990.

of touch with a changing world? That is an important nuclear question with which governments must grapple through the coming years.

Minimum Deterrence

How many nuclear weapons do we really need? Today's arsenals, measured in the tens of thousands, surely represent "overkill"—the ability to destroy one's adversary many times over. Nuclear "doves" take that as indicative of the fundamental insanity of today's nuclear policies. That overkill capability is certainly real, if by destruction we mean the elimination of much of an adversary's population and industrial base. But counterforce advocates would argue that our huge arsenal is necessary because we have to be capable of destroying an adversary's equally huge arsenal. Don't count the cities or the people when figuring overkill, they would say; count instead an adversary's weapons. On that score our nuclear arsenal may never be excessive, no matter how big it gets.

In the early 1960s, Secretary of Defense Robert McNamara took a serious look at the question "How much is enough?" By analyzing destructive effects on the Soviet Union, he concluded that 400 one-megaton warheads would be a more than adequate deterrent. The "unacceptable damage" inflicted by those 400 warheads included destruction of nearly 80 percent of the Soviet Union's industrial capacity and half its population (figure 15.3). McNamara's **minimum deterrence** was quickly trampled in the proliferation of sophisticated counterforce strategies that called for more numerous and more varied nuclear weapons systems. But in the warming superpower relations of the late 1980s and the 1990s, minimum deterrence has emerged from the cloak of naiveté that, for two decades, kept it from serious consideration. The 1990 Strategic Arms Reduction Treaty (START) may represent a first step toward minimum deterrence, although that treaty still leaves each superpower with many thousands of warheads.

Would minimum deterrence be a good thing? Advocates claim that much smaller arsenals would reduce the danger of nuclear war and would reduce the chance of global catastrophe in the event that

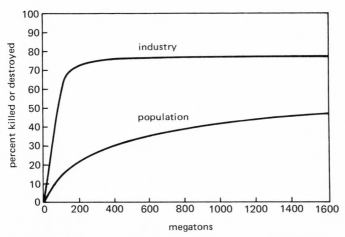

Figure 15.3 Robert McNamara's estimate of a 400-warhead minimum deterrence is based on these curves, suggesting a "saturation" in destructive effects at the equivalent of around 400 megatons. (data source: Enthoven and Smith, *How Much Is Enough?* [Harper & Row, 1971])

deterrence failed. Opponents claim that smaller arsenals could make a first strike more effective, increasing the risk of war. And would minimum deterrence deter all forms of aggression, or only an all-out attack? Again, opponents of minimum deterrence raise the specter of a deterrent force incapable of preventing smaller-scale aggression. And what about some secret technological advance on the part of an adversary that might that render a small deterrent force suddenly vulnerable? On the other hand, advocates argue that a smaller arsenal's more limited damage enhances the credibility of deterrence by strengthening its owner's will to use it. But again a paradox: Even the minimum deterrent must be capable of "unacceptable" damage; otherwise it is no deterrent at all. So are we better off with minimum deterrence? Or doesn't it matter?

Minimum deterrence could be to the world's advantage as much for political and economic reasons as for military reasons. Agreement on minimum deterrence would bring the costly nuclear arms race to a halt, allowing nations to turn economic resources to serious internal problems or to such rising global threats as environmental deterioration and climatic change. International cooperation on those issues would further reduce the mistrust that makes nuclear weapons seem necessary, as would the very act of cooperative negotiation leading to minimum deterrent forces. In short, minimum deterrence could reduce international tensions and therefore diminish the threat of war.

International trust may, in fact, be the best deterrent of all. After all, the United States does not seriously worry about deterring a British nuclear strike—even though Britain has the nuclear capability to devastate the entire East Coast. In the 1990s the world may be poised for the first time to achieve such a trust-based deterrence, at least between the traditional nuclear superpowers. Achievement of that state could replace decades of nuclear competition with cooperation that might enhance the security of all nations.

Could the world go below minimum deterrence? At 400 warheads, the United States and the Soviet Union would be in rough nuclear parity with Great Britain, China, and France—the other overtly nuclear nations. Minimum deterrence at McNamara's 400-warhead level would therefore require more than bilateral trust; it would demand that all five nuclear-armed nations halt further qualitative and quantitative development of their nuclear arsenals.

Today there are probably more than five nuclear-armed nations. India once exploded a nuclear device, Israel and South Africa probably have nuclear weapons, Iraq is striving for a nuclear force, and a number of other nations are at or close to nuclear capability. Had the 1990 confrontation between Iraq and the United States occurred a few years later, we might have been faced with a significant Iraqi nuclear threat. Ten warheads, effectively delivered, would be devastating even to the United States. The proliferation of nuclear weapons and related technology is already so far advanced that it is unlikely the existing nuclear powers could feel secure without something approaching McNamara's level of minimum deterrence. In degree and complexity, the international cooperation needed for nuclear trust among all the potential nuclear-weapon states poses a far greater challenge than did the improvement of bilateral relations between the United States and the Soviet Union.

Are we then faced with a best-case scenario in which five nations possess 400 warheads each while others build toward that level? Together, those warheads could cause nearly total devastation of modern civilization, and their numbers would be well above the threshold for the "nuclear winter" effects discussed in chapter 13. Can we conceive of a world where the nuclear threat is really gone, or must we learn to live indefinitely under that threat?

One thing is a near certainty: Humanity isn't going to forget about the atomic nucleus. Nuclear weapons will be with us, in concept if not in reality, as long as the human species survives. The nuclear critic Jonathan Schell argues that nuclear knowledge alone could be a sufficient deterrent in a world free of nuclear weapons themselves, freeing us from the "brink of doom" that characterizes deterrence with real weapons. Will that become the ultimate nuclear strategy—achieving peace through nuclear knowledge rather than nuclear might? Or would Schell's vision just encourage clandestine stockpiling of nuclear weapons?

Nuclear Morality

We have been considering nuclear strategies as means to further national goals, to enhance national and global security, and above all to avoid nuclear war. But what is the *right* way to "use" nuclear weapons? Is morality in nuclear matters synonymous with avoiding nuclear war, or are there conditions under which wartime use of nuclear weapons would be morally justified? And what about nuclear strategies that succeed in avoiding war; are they implicitly above moral reproach? These have been troubling questions since the beginning of the nuclear age.

Long before there were nuclear weapons, philosophers and religious leaders grappling with the morality of human aggression conceived the notion of the "just war." Not all agree with that concept, but it has helped shape moral thinking, particularly in the Judeo-Christian cultures of the West. Central to the "just war" concept are the requirements that the damaging effects of war must be *proportionate* to the good expected from the war and that the use of force must be *discriminating,* harming not innocent civilians but only those who actually commit aggression.

How do these moral criteria apply to nuclear war and to the

strategies for its prevention? Nuclear weapons are unique in the enormity of their destructive effects. From the millions of dead in the war-fighting strategists' most precise nuclear exchange to the extinction of humanity in the most extreme "nuclear winter" scenarios, conceivable nuclear wars fail miserably on the moral criterion of proportionate damage. And the indiscriminate nature of nuclear destruction makes their use only against aggressors essentially impossible. Innocent civilians within a target country would be slaughtered in even a limited attack, and fallout would spread nuclear death beyond the combatants' borders.

Suppose we agree that nuclear war is immoral. Then what about nuclear deterrence, based on the threat of mutual assured destruction? MAD-based deterrence can't be an empty threat, but requires the technical capability and political readiness to use nuclear weapons. Some would argue that threatening the immoral act of nuclear war is enough to make MAD an immoral posture. Others agree that MAD's threat is immoral, but argue that it is necessary in order to prevent the greater immorality of nuclear war itself. Even questioning the morality of nuclear war raises another moral dilemma: If we really believe nuclear war is immoral, doesn't that weaken deterrence by making us less willing to use our deterrent forces? That, in turn, could increase the likelihood of war itself.

Nuclear weapons pose deep and complex moral issues. Those issues are not just abstractions for philosophical debate; as we have seen, morality even enters the controversy between advocates of counterforce and countervalue strategies. How are we to weigh moral considerations in the face of a technology that radically expands our ability to harm our fellow humans and indeed our entire planet? This nuclear question engages our deepest moral and religious convictions.

Summary

Nuclear weapons have irrevocably altered the role of military forces in international relations. No longer can a nuclear-armed nation expect to win a major war; instead, its nuclear weapons must serve to prevent war. But how best to achieve that goal of nuclear deter-

Nuclear News: Clergy on Nuclear War

How does religion cope with the moral dilemmas of the nuclear age? People across the religious spectrum are involved in building nuclear weapons, in developing strategies for their use, and in the military forces that would fight a nuclear war. And nearly all of us, whatever our moral and religious views, pay the taxes that buy nuclear weapons. What do our religious leaders have to say about all this?

Many of the world's major religions have issued statements condemning, to one degree or another, the nuclear arms race and preparation for nuclear war. In the United States, the National Conference of Catholic Bishops has been especially visible for its delicately worded position on nuclear war and the morality of nuclear deterrence.[4] Other denominations have issued stronger statements, often supporting those who, in good conscience, violate the law in expressing their nuclear concern.

Issues of nuclear morality can be especially intense at the local level, and are particularly controversial when they impinge on economic well-being. In 1981 Bishop Leroy Matthieson of the Roman Catholic diocese of Amarillo, Texas, appealed to those involved in manufacturing nuclear weapons to "consider what they are doing, to resign from such activities, and to seek employment in peaceful pursuits." The bishop's appeal was no empty request: Amarillo is home to Pantex, the final assembly plant for U.S. nuclear weapons. When Amarillo's Catholic Family Services refused to stop counseling Pantex employees concerned about the morality of their work, the local United Way withdrew its support. Amarillo was locked in a nuclear dilemma: how to reconcile its own economic health with the global threat of nuclear war.

Not all religious leaders find involvement with nuclear weapons morally problematic. In the swirling nuclear debate of the early 1980s, New York's Roman Catholic archbishop Terrence Cardinal Cooke declared nuclear weapons "tolerable" for deterrence, and said "those who produce . . . the weapons . . . can do so in good conscience." Bob Nelson, an Episcopal priest from Nevada, goes even farther. Concurrent with his religious duties, Nelson has served as deputy manager of operations at the Nevada nuclear test site, where he has been personally responsible for approving the detonation of more than thirty nuclear explosives. While he would prefer a world without nuclear weapons, Nelson sees his work as helping ensure the security of a free and just society in the real, nuclear-armed world.

News sources: "Texas Catholic Bishops Reject Neutron Bomb," *New York Times,* September 13, 1981; "Cook Sees Atomic Weapons as 'Tolerable' for a Deterrent," *New York Times,* December

15, 1981; "U.S. Bishops Oppose Anti-Missile Plan," *New York Times,* April 15, 1988; "Churches Renew Morality Debate on Nuclear Arms," *Chicago Tribune,* June 26, 1988; "Presbyterians Adopt Statement on Nuclear Stance," *New York Times,* June 16, 1988; "Priest Tells of His Role in Nevada Bomb Tests," *New York Times,* December 8, 1989.

rence? For some, the destructive potential of the world's huge nuclear arsenals is enough to deter their use; mutual assured destruction, or MAD, has kept the nuclear peace for half a century, and, insane as its acronym implies, MAD nevertheless remains our best hope. Others argue that MAD-based deterrence will inevitably fail and that only drastic reductions in nuclear forces, leading eventually to total nuclear disarmament, can save us in the long run. Still others feel the threat of massive destruction lacks credibility as a deterrent; they argue that counterforce strategies and the ability to fight and prevail in a variety of nuclear-war scenarios will enhance deterrence. Opponents counter that preparation for nuclear war can only make such war more likely.

Questions of nuclear strategy have been the focus of political debate throughout the second half of the twentieth century. Although the world has moved toward a consensus that nuclear war is unacceptable, we are far from agreement on how to make the unacceptable also improbable. Debate on that issue will continue, and it is no exaggeration to say that our very survival hinges on the outcome.

Notes

1. Bernard Brodie, ed., *The Absolute Weapon* (Harcourt, Brace, 1946), p. 76.

2. Robert McNamara "The Military Role of Nuclear Weapons: Perceptions and Misperceptions," *Foreign Affairs* 62 (fall 1983), p. 79.

3. George Bush and Thomas Jones quotes from Robert Scheer, *With Enough Shovels: Reagan, Bush and Nuclear War* (Random House, 1982), pp. 261 and 18, 23, respectively. All other quotes are referenced in Pater and Pater, eds., *What They Said in 198_: The Yearbook of Public Opinion* (Monitor Book Co., published yearly).

4. The bishops' statement, along with opposing articles, can be found in Kegley and Wittkopf, *The Nuclear Reader* (St. Martin's 1985).

Further Reading

David P. Barash, *The Arms Race and Nuclear War* (Wadsworth, 1987). Chapter 6 describes the evolution of deterrence as a strategic doctrine and outlines policy questions associated with maintaining deterrence.

Peter R. Beckman et al., *The Nuclear Predicament* (Prentice-Hall, 1989). A thorough and well-illustrated look at many dilemmas of the nuclear weapons age.

Louis R. Beres, *Mimicking Sisyphus: America's Countervailing Nuclear Strategy* (Heath, 1982). A critique of nuclear strategies that threaten an adversary's military capability (counterforce strategies).

Freeman Dyson, *Weapons and Hope* (Harper & Row, 1984). A former weapons scientist looks at nuclear strategy and at the people responsible for its evolution.

Harvard Nuclear Study Group, *Living with Nuclear Weapons* (Harvard University Press, 1983). Written for the general public, this book is a thought-provoking analysis of the dilemmas faced by a nuclear-armed world. The book also presents detailed scenarios on how a nuclear war might start.

Fred Holroyd, ed., *Thinking About Nuclear Weapons* (Croom Helm, 1985). A collection of essays that includes European perspectives on nuclear arsenals.

Charles W. Kegley and Eugene R. Wittkopf, eds., *The Nuclear Reader: Strategy, Weapons, War* (St. Martin's 1985). Selected essays on nuclear strategy, the role of nuclear weapons, and the nature of nuclear war by writers covering a broad spectrum of conflicting views. Part I includes especially strong arguments on the moral dimensions of nuclear war and deterrence.

Michael Mandelbaum, *The Nuclear Revolution: International Politics Before and After Hiroshima* (Cambridge University Press, 1981). A political scientist describes how the nuclear difference affects international politics.

A. G. Mojtabai, *Blessed Assurance: At Home with the Bomb in Amarillo, Texas* (Houghton Mifflin, 1986). A city that is home to a bomb plant wrestles with its moral dilemmas.

Robert Scheer, *With Enough Shovels: Reagan, Bush and Nuclear War* (Random House, 1982). A frightening look at the nuclear war-fighting mood that prevailed in the United States in the early 1980s, largely through interviews with government officials from the president on down.

Jonathan Schell, *The Fate of the Earth* (Knopf, 1982). An eloquent statement of what the author sees as humanity's choice: extinction by nuclear war or a world free of nuclear weapons.

Glossary

counterforce   A nuclear strategy that targets an opponent's military forces.

countervalue   A nuclear strategy that targets an opponent's population centers.

decapitation   Elimination of an adversary's civilian and military leadership, usually with precisely targeted nuclear weapons.

deterrence   The use of weapons to deter an adversary's attack by threatening retaliation.

flexible response   A nuclear strategy calling for response in kind to an aggressor's actions: conventional attack with conventional counterattack, tactical nuclear strike with tactical nuclear weapons, strategic attack with strategic nuclear weapons.

launch-on-warning strategy   A strategy that calls for land-based missiles to be launched upon receipt of a warning that an enemy attack is underway.

massive retaliation   A nuclear strategy of the 1950s, declaring that the United States would respond to *any* Soviet aggression with a massive nuclear attack on the Soviet homeland.

minimum deterrence   A strategic deterrent force of the minimum size and capability necessary to deter an adversary's attack.

mutual assured destruction (MAD)   An international nuclear balance in which nuclear-armed adversaries are capable of inflicting certain and devastating destruction on each other.

NUTS   Acronym for **n**uclear **u**tilization **t**arget **s**election, describing counterforce strategies.

Defense in the Nuclear Age

16

Is there a way out of the MAD dilemma, a way to attain security that does not rest on the threat of annihilation? In a 1983 speech, President Ronald Reagan gave his answer:

After careful consultation with my advisers, including the Joint Chiefs of Staff, I believe there is a way. Let me share with you a vision of the future which offers hope. It is that we embark on a program to counter the awesome Soviet missile threat with measures that are defensive. Let us turn to the very strengths in technology that spawned our great industrial base and that have given us the quality of life we enjoy today.

What if a free people could live secure in the knowledge that their security did not rest upon the threat of instant U.S. retaliation to deter a Soviet attack, that we could intercept and destroy strategic ballistic missiles before they reached our own soil or that of our allies?[1]

Reagan's speech launched the Strategic Defense Initiative (SDI), a program whose ambitious aim was a complete defense against ballistic missiles. (Among Reagan's advisors, and a vigorous SDI proponent, was Edward Teller, who had helped make fusion weapons practical 30 years earlier.)

Nuclear Defense: The Early Years

Ronald Reagan's Strategic Defense Initiative was hardly the first defensive effort of the nuclear age. The bomber era of the 1950s saw extensive development of air-defense systems using small ground-based rockets to shoot down attacking planes. Updated versions of these systems continue in operation; the Soviet Union has a particularly elaborate network for air defense. Tracking and destroying relatively slow bomber aircraft is, in principle, well within the capability of present-day technology. However, electronic countermeasures and stealth technology have greatly offset advances in anti-aircraft defense.

With the advent of ballistic missiles in the late 1950s, anti-aircraft defenses were adapted for use against incoming ballistic-missile warheads. Defending against such warheads is far more difficult, since their small size and high speed (15,000 miles per hour, over 4 miles per second) make them difficult targets indeed. Early anti-ballistic-missile (**ABM**) systems included radar antennas for tracking incoming warheads and small nuclear-tipped rockets designed to intercept the warheads either in the atmosphere or in space. Atmospheric interceptors would detonate their nuclear explosives near an attacking warhead, destroying it with blast effects; space-detonated explosions would "kill" several warheads with intense bursts of x rays.

In the United States, ABM development proceeded through the 1960s amidst considerable debate over the utility and desirability of ABM systems. Technology ultimately converged on a two-phase system consisting of "Sprint" low-altitude atmospheric interceptor rockets and high-altitude "Spartan" rockets for interception above the atmosphere. By 1966, advances in Soviet ICBM and ABM technologies had moved the United States toward deployment of its own ABM system. Originally the high-altitude Spartan was to defend the entire United States, with low-altitude Sprint rockets grouped for additional defense of 52 major cities; the combined system would be called Sentinel.

But the American ABM effort faced a problem: Its effectiveness against current and future Soviet ICBMs was far from certain. By 1967 the government conceded that an ABM system could not counter a full-scale Soviet attack. Instead, the system would serve as insurance against an accidental missile launch or a future Chinese ICBM threat. Note the pattern: initial optimism for full-scale missile defense followed quickly by relegation of the proposed system to less significant roles.

Figure 16.1   A Spartan missile being launched from the Safeguard ABM base in North Dakota. (U.S. Army Air Defense Artillery Museum)

ABM advocates in the late 1960s faced another surprising disappointment: public opposition to nuclear-armed ABM sites near cities. In 1969 the Nixon administration reaffirmed Sentinel's impotence in the face of an all-out attack, and reshaped the U.S. ABM system toward protection of ICBM silos only. The new system, called Safeguard, barely won approval in Congress. Construction of a Safeguard system to protect ICBMs at Grand Forks, North Dakota, soon followed.

The ABM Treaty

Even as ABM development and debate continued through the late 1960s, the United States and the Soviet Union moved slowly toward the first negotiations aimed at limiting strategic nuclear arms. Both sides had come to recognize their essential nuclear stalemate, and both realized that maintaining the MAD balance was in their mutual interest. The Strategic Arms Limitation Talks (**SALT**) began formally in 1969, and led in 1972 to the **SALT I** agreements.

SALT negotiators recognized that missile defenses could have several adverse consequences. A truly effective system would give a defended country the ability to strike first with impunity, knowing its defenses would fend off a retaliatory strike. Even a partially effective system might encourage a first strike, since ABMs might then limit damage from weakened retaliatory forces to "acceptable" levels. And in any event, ABM development would surely lead to an expensive and potentially destabilizing offensive-arms race, as each side sought to maintain the nuclear balance with offensive weapons capable of overwhelming its adversary's defenses.

The SALT I agreements acknowledged the danger of defensive systems with the **ABM Treaty**, limiting each superpower to only two ABM sites, one to defend its capital city and the other at a selected ICBM field. A 1974 protocol further restricted each superpower's ABM deployment to one site. The Soviet Union chose ABM defense for Moscow, while the United States went with its Safeguard site in North Dakota. Safeguard was abandoned in 1976 as ineffective and costly, leaving the Soviet "Galosh" defenses ringing Moscow as the world's only operational ABM system.

The ABM Treaty embodies an essential paradox of nuclear

arms: that effective defense may be more likely to cause war than to prevent it. And the treaty formalizes the "hostage" roles of the superpowers' populations; by denying them any defense against nuclear attack, it upholds MAD and makes that attack less probable.

At the time the ABM Treaty was signed, missile defense meant interceptor rockets and associated radars for warning, tracking, and guidance. What about future technological developments? The treaty anticipated them with these words: ". . . in the event ABM systems based on other physical principles . . . are created in the future, specific limitations on such systems and their components would be subject to discussion. . . ." SALT negotiators put in place mechanisms for such discussion: a treaty review every 5 years, and a Standing Consultative Committee to meet as needed. The ABM Treaty was to hold indefinitely, withdrawal being allowed on 6 months' notice. The provisions on "other physical principles" and treaty withdrawal figured prominently in the SDI debate of the 1980s.

Defending against Ballistic Missiles: The Problem

Is defense against ballistic missiles possible? The ABM systems of the 1960s and the 1970s could not give a clear Yes answer. Meanwhile, advances in offensive technology have made the question even more challenging.

Understanding the problem of ballistic-missile defense (BMD) requires a detailed look at the flight of an intercontinental ballistic missile. As figure 16.2 shows, the trajectory of a typical ICBM comprises four distinct phases. Each phase offers separate challenges and opportunities for a BMD system.

During the **boost phase**, rocket motors lift the missile from its launch site to an altitude of about 125 miles, just outside the atmosphere. For today's missiles the boost phase lasts 3–5 minutes, although this time could be lowered significantly. Attacking missiles in the boost phase offers two significant advantages to a defensive system. First, the flame from the rocket boosters is tremendously bright, making it easy for space-based sensors to detect and track the missile during the boost phase. Second, the boost-phase missile constitutes a single relatively large target, with

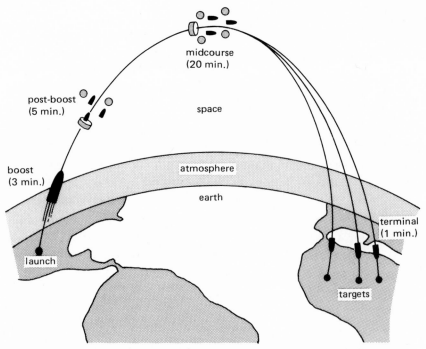

Figure 16.2   The phases of an ICBM's trajectory. Post-boost shows one warhead still on the bus; by midcourse up to 10 warheads and 100 or more decoys may have been released. Only the warheads survive atmospheric reentry. Maximum altitude and thickness of the atmosphere are exaggerated for clarity.

all its warheads riding on one fragile vehicle. Destruction of a single missile during the boost phase puts many warheads out of commission.

Boost-phase defense is decidedly advantageous, but it is not without difficulties. The boost phase is relatively short, giving the defense limited time for the decision to attack, for the actual destruction of rising missiles, and for assessment of defensive success. In an all-out attack involving thousands of simultaneous missile launches, those are formidable tasks to accomplish in a few minutes. The defense is further constrained because a missile is above the atmosphere only in the last part of its boost phase. Many of the proposed BMD weapons cannot penetrate the atmosphere, so a boost-phase defense may have even less time to do its work.

The several rocket motors that boost a missile into space fall away as their fuel is exhausted. The boost phase ends as the last rocket stage shuts down and separates, leaving the so-called bus on which the warheads are mounted. There follows a **post-boost phase** of about 5 minutes, during which the bus uses small rockets to orient itself in slightly different directions, releasing individual warheads on the trajectories that will take them to their targets.

It is unfortunate for the defense that what was one threatening object has now become ten or more individual warheads on separate trajectories. But things may get even worse. The bus can also release **decoys**, typically lightweight metalized balloons that superficially resemble warheads. In the vacuum of space, the absence of air resistance means that the decoys travel on the same ballistic paths as the much heavier warheads. Since decoys are light, inexpensive, and compact until they inflate, a typical missile could carry ten or more decoys for every warhead. A single missile thus becomes a **threat cloud** comprising 100 or more objects that all look like nuclear warheads. In addition to decoys, the bus might release clouds of aluminum-foil strips to confuse defensive radars, and aerosol mists to shield warheads from the prying eyes of infrared sensors. A nuclear strike involving thousands of missiles could easily put a million objects on ballistic trajectories toward the target country. An effective defense must destroy all this junk or must somehow single out the several thousand actual warheads.

The **midcourse phase** begins once the bus has finished releasing all its stuff. A system for midcourse defense faces stiff challenges. As we have already seen, it may have to deal with a million seemingly threatening objects. And with no more rockets firing, those small, cold objects are very hard to detect—all the more so because they are moving several miles each second. Despite formidable problems involving detection and discrimination of warheads from decoys, there are two advantages to midcourse defense. First, midcourse is the longest phase of missile flight, lasting about 20 minutes for an ICBM. Second, the entire threat cloud spends midcourse on ballistic trajectories, moving in accordance with simple physical laws under the influence of gravity alone. If an object can be tracked to determine its position and velocity, then its future trajectory can be predicted accurately. That predictability, coupled with the relatively long time available, might allow a midcourse defense to pick off its many targets one by one.

As it nears the target country, a missile's threat cloud reenters the atmosphere. Reentry marks the beginning of the **terminal phase**, which lasts only about a minute before the weapons detonate at their targets. Terminal-phase defense has little time to act, and there is no room for failure. Nuclear warheads, still traveling at several miles per second, *must* be destroyed or they will wreak their unimaginable damage. Terminal defense is a last-ditch effort, but it does have one advantage. Drop a rock and a piece of paper and you will see how air resistance dramatically alters the motion of the paper. The same thing happens to the lightweight decoys—they are stripped away by air resistance high in the atmosphere, leaving only the true warheads.

Whether terminal defense can work depends in part on the targets being defended. A missile silo would be saved if an attacking warhead were detonated a few miles up instead of in a direct-hit ground burst. Under the same circumstances a city would be destroyed. This is one reason why the ABM systems of the 1960s and the 1970s—exclusively terminal defenses—could not protect the United States' population.

Submarine-launched ballistic missiles also go through the four phases of flight described above. But SLBMs typically would be launched from just off the target country's coast, making the overall flight time much shorter. The constraints on a defense system are correspondingly tighter, and many experts believe that systems designed for ICBM defense could not function effectively against SLBMs.

SDI: Layered Defense

Ronald Reagan's 1983 speech gave rise to the Strategic Defense Initiative, a technological program initially intended to produce a 100 percent effective "shield" against ballistic missile attack. As originally envisioned, SDI was to rely heavily on advanced space-based weapons; thus the program came to be nicknamed "Star Wars." Although the emphasis has shifted in recent years toward more established technologies, research on advanced weapons continues.

The SDI plan differs radically from the relatively simple terminal defenses of the 1960s and 1970s. SDI favors a **layered defense**, with separate weapons engineered for effectiveness against the boost, midcourse, and terminal phases of the attacking missiles' flight. Ideally, boost-phase defenses would destroy most of the attacking missiles. The few that got through would be largely eliminated in midcourse. If any survived midcourse, terminal defenses would take care of them. The formidable problems faced by midcourse and terminal defense would be reduced by elimination of missiles in earlier phases.

Would layered defense work? It might, but it also might not. Failure of an early layer could leave a subsequent layer not just slightly weakened but overwhelmed. And the layers might not enjoy complete independence; for example, they might share tracking satellites or communication systems. Destruction of a crucial link could then cripple all the layers. Ultimately, the effectiveness of a layered strategic-defense system will remain uncertain without actual wartime experience.

The Technology of SDI

How would SDI weapons work? The high speed of ballistic missiles calls for defensive weapons that "kill" rapidly and over long distances. The most exotic weapons proposed for SDI are so-called **directed-energy weapons** that deliver lethal effects at or near the speed of light.

A **laser** is a source of light or other electromagnetic radiation—including infrared radiation and x rays—capable of producing a high-intensity beam that can travel long distances without spreading. A laser differs from a conventional light source not only in its intensity, but also because the individual waves making up a laser beam are all exactly in step with one another. This property, called *coherence,* is what allows a laser to maintain its intensity over long distances.

Lasers for ballistic-missile defense would use several "kill mechanisms" to disable their targets. Visible-light or infrared lasers would shine on the target long enough (typically from seconds to minutes) to cause melting and structural failure. X-ray lasers would deliver energy in a sudden burst, driving a destructive shock wave into the target. Either of these kill mechanisms requires enormous laser power.

The most technologically mature of the available high-power lasers use the rapid reaction of hydrogen with other gases to produce energetic molecules that emit the laser light. Several other designs involve electrical excitation, and would require power sources of some tens of megawatts. The x-ray laser uses a different power source: At its heart is a nuclear weapon surrounded by rods, whose vaporized remains channel x rays from the nuclear explosion in the general direction of attacking missiles. None of these laser technologies has yet achieved the full requirements for ballistic-missile defense.

How would laser weapons be deployed? Visible-light lasers and some infrared lasers could be based on the ground. These units would fire their beams toward large mirrors in geosynchronous orbit, hanging motionless over fixed points on the Earth's equator as communication and TV satellites do. Geosynchronous mirrors would reflect the beams to so-called battle mirrors in lower orbits, and the battle mirrors in turn would reflect the light to its targets (figure 16.3). Beams from ground-based lasers suffer from atmospheric distortion, and compensating mechanisms (perhaps involving computer-controlled flexible mirrors) might prove necessary.

An alternative to ground-based lasers is deployment in space (figure 16.4). Indeed, some forms of laser radiation do not penetrate the atmosphere and thus would require space deployment. A space-based laser defense system would involve a "constellation" of laser battle stations orbiting the Earth at altitudes of several hundred miles. Numerous battle stations would be necessary, because not all would be in appropriate orbital positions when needed. Space-based lasers might fire directly at their targets, eliminating the complicated mirror systems required with ground-based lasers.

How would space-based lasers be powered? Although a variety of power sources are under investigation, it appears likely that some components of a space-based defense system would require nuclear reactors in space. That route would reopen many of the controversies discussed in part II of this book, but in the new context of outer space.

One weapon that *must* be space-based is the x-ray laser, since

Nuclear News: Reactors in Space?

How should large satellites, including SDI weapons, be powered? Most U.S. satellites in Earth orbit use solar photovoltaic cells to produce electricity from sunlight. But the Soviet Union's largest satellites use compact nuclear reactors. Should U.S. satellites, especially power-hungry SDI battle stations, be nuclear-powered? What are the implications of nuclear power in space?

Proponents of a U.S. space-reactor program worry about a "reactor gap" caused by the Soviet lead in space-based nuclear reactors. Opponents counter that Soviet reliance on nuclear power reflects the Soviet Union's less advanced photovoltaic capability and is no cause for U.S. technological alarm. Opponents further point to the dangers of launch accidents or radioactive contamination from reactors falling out of orbit, citing the 1978 contamination of Northern Canada by a fallen Soviet space reactor. Proponents argue that nuclear reactions would not start until reactors achieved orbit, leaving only mildly radioactive uranium fuel at launch. They claim further that safety systems developed by the Soviet Union boost a reactor's core to a higher orbit when its host satellite nears the upper atmosphere.

Nuclear reactors also present environmental hazards in space. Already, astronomical research satellites have been affected by gamma radiation from orbiting reactors, in much the same way that city lights interfere with earthbound telescopes. This "celestial pollution" would only worsen with proliferation of nuclear-powered satellites.

Space reactors are expensive—so much so that the U.S. Department of Energy recently sought to cancel its $1.8 billion program to develop a 100-kW space reactor. But the Department of Defense and the National Aeronautics and Space Administration, whose ambitious space plans call for new power sources, both oppose cancellation of the space-reactor program. The U.S. Office of Management and Budget is left to adjudicate this cosmic nuclear argument.

Do we want to live on a planet enveloped with orbiting nuclear reactors? That is a question that must be asked as the United States moves ahead with ambitious military and civilian space programs.

News sources: "New Plans for Space Reactors Raise Fears of Nuclear Debris," *New York Times,* October 18, 1988; "Soviet Space Radiation Hampering U.S. Satellite," *New York Times,* November 17, 1988; "A Debate Over Putting Nuclear Power in Orbit," *New York Times,* January 22, 1989; "Energy Dept. Aims to Stop Spending on Space Reactor," *New York Times,* September 11, 1989.

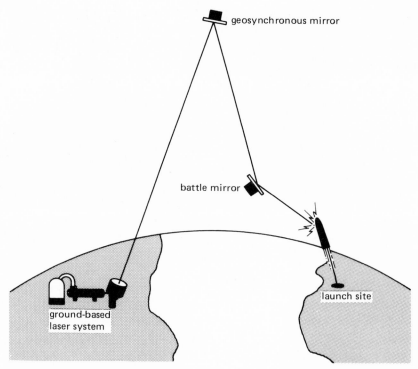

Figure 16.3 A ground-based ballistic-missile-defense laser might use two or more mirrors to reach its target.

x rays do not penetrate far in air. But this device would be powered by a nuclear weapon, and a 1967 treaty prohibits nuclear weapons in space. Advocates of x-ray lasers therefore envision "pop-up" basing, with the weapons mounted on submarine-launched missiles off an adversary's coast. On warning of a missile launch, x-ray lasers would be quickly "popped up" into space to get a direct line on attacking missiles as they emerged from the atmosphere (figure 16.5).

All laser systems face major technical and logistical hurdles. Even under laboratory conditions, none of the lasers under consideration has yet achieved performance close to what would be needed for long-range missile defense. All but the short-burst x-ray laser require guidance systems capable of holding the beam for several seconds on a spot only a few feet in diameter on a missile moving at several miles per second—this from a distance of thousands of

Figure 16.4 An artist's conception of an orbiting laser weapon, based on the Alpha system, a chemical laser now under development. (TRW, Inc.)

miles. Orbiting mirrors are exceptionally vulnerable to space debris or to hostile attacks, and must be specially treated to handle the extremely powerful laser beams. Mirrors and lasers must be capable of rapid and accurate redirection, to destroy many targets in quick succession. Finally, laser systems must withstand long periods of inactivity and be ready to fire on on a few seconds' notice in time of attack.

Particle-beam weapons constitute a second class of directed-energy weapons. These devices would use accelerators to bring protons, electrons, or other particles to very high energy. Traveling at close to the speed of light, particles could penetrate several inches into missile warheads, destroying electronic circuits, guidance systems, and other delicate mechanisms. Technology to achieve the requisite particle energies is widely used in scientific research, but producing intense beams of high-energy particles and maintaining them over thousands of miles is an altogether different matter. One complication comes from the Earth's magnetic field, which bends

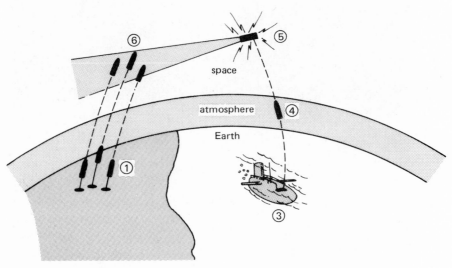

Figure 16.5 A scheme for "pop-up" x-ray lasers. A missile launch (1) is detected by a warning satellite (2), which communicates with a submarine (3). The submarine launches a rocket (4) carrying a nuclear-weapon-powered x-ray laser. Above the atmosphere, the weapon (5) detonates, sending an intense beam of x rays to destroy several missiles (6) as they rise into space.

electrically charged particles out of straight-line paths. The technology of particle-beam weapons is far from the practical stage, and even if perfected it will require space basing because particle beams do not penetrate the atmosphere.

Although most people think of SDI in terms of exotic "Star Wars" technologies, simpler weapons would complement these technologies in a complete SDI system (figure 16.6). Because ballistic missiles travel at high speeds, collision with *anything* would be fatal to a missile. So-called **kinetic-kill** weapons would utilize this fact by attacking missiles with projectiles moving at up to 20 miles per second. The projectiles would be accelerated by small rockets or electromagnetic launchers. They might be as simple as inert plastic spheres, or they might be "smart rocks" or "brilliant

Figure 16.6 An artist's conception of a space-based midcourse defense system. Ten or more small rocket interceptors are mounted on satellite platforms like that shown in right foreground, and are launched on command from separate tracking satellites. (Department of Defense)

pebbles" equipped with homing devices to ensure collision with the target.

Finally, old-fashioned rockets would play the major role in terminal-phase defense. Actually, these would be modern rockets; they would be capable of very high acceleration and would carry sophisticated guidance and homing systems. They might kill with low-yield nuclear blasts or by direct collision. Swarms of small, inexpensive rockets might be used instead of fewer but "smarter" weapons. Although rocket-based defense was hardly part of the initial SDI concept, difficulties with more exotic technologies have moved rockets to the forefront of SDI efforts. At the same time, the overall goal for the SDI program has grown considerably less ambitious; now defense planners speak of guarding missile silos or making an adversary uncertain of the outcome of an attack, not of wholesale population defense.

SDI: The System

Weapons alone do not make a ballistic-missile defense. Also needed are a host of sensors to acquire information about attacking missiles and about the course of defensive action; communication networks linking sensors, weapons, and human command; and computer systems for rapidly processing an overwhelming amount of information. How will the elements of this SDI system fit together? What will the system have to do? Figure 16.7 suggests some of the answers.

Consider the sensors. Obviously, they have to detect missiles and track them. But they also have to provide information sufficient to distinguish warheads from decoys. And what happens after an SDI weapon is fired? How does the defense know it has been successful? Laser and particle-beam weapons may render attacking warheads useless without obvious damage. Across thousands of miles of dark, empty space, something has to determine whether a defensive shot has done its job, and if it hasn't to order another try. How does one build a sensor that can see a 3-foot-long warhead 2,000 miles away, and find out enough about it to know that it is

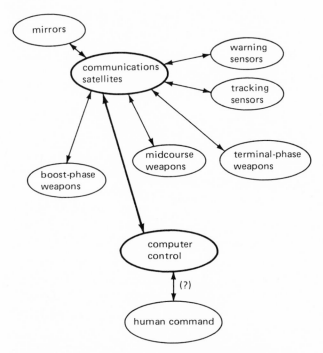

Figure 16.7 A complete missile-defense system includes not only weapons but also sensors, communications, and computers. The system's tasks include identifying and tracking targets, allocating weapons to targets, aiming and firing, and verifying kills. An anti-missile battle would take place so rapidly there would be little role for human intervention.

not a decoy and that it has suffered sufficient damage to be rendered inoperable? This problem of "kill verification" is among the most difficult but most important tasks of the SDI system.

Consider communications. You can't stretch wires between orbiting battle stations, mirrors, and related hardware. And because at least some SDI weapons would be useful only over enemy territory, radio communication requires satellite relays. How do you design a satellite communication system that will function reliably during a nuclear war, and that will be secure enough so a few key explosions will not render it useless? And what about the electromagnetic pulse that, with a single nuclear explosion, might destroy electronic devices for thousands of miles?

Finally, what about the computer systems that will coordinate the entire defensive effort? An anti-missile battle would take place too fast for human intervention, so computers must run the show. Computers will have to process information from thousands of sensors, compute trajectories of perhaps a million threatening objects, distinguish warheads from decoys, allocate available weapons to targets, aim and fire weapons, verify "kills," and, in the event that the defense proves inadequate, decide which targets in the country under attack are most valuable. (Your city? Mine? ICBM silos?) All this in a matter of minutes! Computers may even have to initiate defensive warfare, especially with "pop-up" weapons that would allow essentially no time for human decisions.

Many computer specialists regard software development as the most serious technical problem facing a layered ballistic-missile defense. A centralized SDI control program could involve tens of millions of lines of computer code, and would have to be free of serious errors even though it could not be tested under actual wartime conditions. Some have proposed that the software be written not by humans but by other computer programs endowed with "artificial intelligence." SDI opponents view that prospect with alarm, suggesting that it may further remove war-related decisions from human beings. A more sensible approach might be to distribute computational tasks throughout the system, perhaps giving individual orbiting battle stations the "intelligence" to handle their own tracking, firing, and kill verification. But even then some computerized overseer would need to know which defensive weapons had been destroyed in battle, how many warheads had eluded the defense, and what their targets were.

Can It Work?

Is ballistic-missile defense possible? Even the most staunch SDI advocates have backed away from Ronald Reagan's 1983 vision of an impregnable shield protecting the entire United States from nuclear attack. Debate continues about whether the more exotic directed-energy weapons could ever work and whether, even with technically successful weapons, a complete and coordinated system could operate effectively.

In many respects, the SDI program has followed the same

pattern as the ABM systems of two decades ago. Originally touted as a complete population defense, SDI is now regarded even by many advocates as strictly a defense for our ICBMs, or as a means of protecting our population from the small but growing missile forces of China, Iraq, and other potential adversaries of the post-Cold War era, or as insurance against an accidental launch. Those may be worthy goals, but do they justify the huge expense and the potentially destabilizing effects of the SDI program?

SDI also faces serious logistical questions, foremost among them the need to launch hundreds or thousands of components into orbit. Some of those objects would be very heavy, requiring far more launch capacity than that of today's Space Shuttle. Alternatively, large battle stations could be assembled in orbit by astronauts operating out of a space station. What would be the impact of so much SDI-related space activity on the rest of our space program? On our national budget? Or on our planet's atmosphere, whose ozone layer could be endangered by the enormous amounts of rocket exhaust? How do the answers to those questions weigh against SDI's purported enhancement of our security?

Countermeasures

How might the United States' potential nuclear adversaries respond to this country's deployment of a ballistic-missile-defense system? SDI enthusiasts often assume that a technically successful BMD would remain invincible. But a nuclear adversary would undoubtedly strive for the capability to overcome that defense. How difficult would that be? Many indications suggest that it may be easier, both technically and economically, to foil a defensive system than to build one.

Numerous fairly simple technical measures might make the defense's task much more difficult if not impossible. We have seen the importance of boost-phase defense in light of multiple-warhead missiles and their prolific decoys. But the boost phase is short— only 3 minutes for a modern MX missile—and it could be shortened further with the introduction of "fast-burn boosters." These high-powered rockets would achieve ballistic speeds in just 40 or 50

Nuclear News: The Costs of SDI

How much would the SDI system cost? That is really anyone's guess, although figures from $100 billion to $1 trillion have been mentioned. Through 1990, SDI activities have cost the United States about $20 billion.[2] The 1990 budget, however, marked a milestone: That was the first year in which congressional opponents forced a decrease in SDI funding.

SDI's cost is measured not only in dollars. Opponents claim that the huge SDI research budgets have diverted U.S. scientific efforts from more pressing problems. Even with the decline of the SDI program, the SDI budget in the early 1990s is still nearly double that of the entire National Science Foundation, which funds much of the United States' civilian research. Indeed, federal research funding in the 1980s shifted dramatically toward the military sector, in large part because of SDI. SDI proponents counter that national-security considerations justify SDI expenditures, and that the civilian economy will benefit from technological spinoffs. But a substantial number of physicists, mathematicians, and engineers, including more than half of the physicists at leading U.S. universities, have pledged not to accept SDI funding. And studies by respected scientific panels confirm many scientists' views that proposed SDI technologies cannot provide meaningful ballistic-missile defense. Although many scientists oppose a large-scale SDI program, most feel that continuing anti-missile research at a modest level makes good sense.

News sources: "Physicists Express 'Star Wars' Doubt," *New York Times,* April 23, 1987; "Lawmakers Agree on 'Star Wars' Cut," *New York Times,* November 3, 1989.

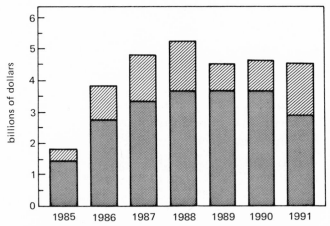

Figure 16.8 Funding for the SDI program. Darker shading represents congressional allocations; lighter shading is the president's budget request. Figure shows Department of Defense activities only; the Department of Energy spends an additional $200–300 million on SDI. (data source: Strategic Defense Initiative Organization)

seconds, leaving their payloads well within the atmosphere when the rocket flames burned out. That new time frame and the atmospheric shielding could rule out the "pop-up" x-ray laser, and would greatly complicate other boost-phase defense weapons. Fast-burn boosters would be expensive, but there is no technological barrier to their deployment. An even simpler countermeasure would be to set missiles spinning as they rise; then laser beams could not dwell continuously on a single spot and would therefore have to be much more powerful to work their harm. It is also possible to "harden" missiles with special coatings, or to cover them with protective shrouds through which laser beams might burn harmlessly. Hardening presents a cost in money and in payload, but that cost might be far less than the increased cost of a laser capable of damaging a hardened missile.

We have already seen how midcourse defense is complicated by such countermeasures as decoys, radar reflectors, and infrared-emitting aerosols. It would also be possible to shield the bus with an umbrella-like structure so the defense can't watch it deploy warheads. Maybe a clever defense system can still discriminate the warheads, perhaps by probing all objects in a threat cloud with laser or particle beams to see how they respond. But then the countermeasures get more clever still: How about putting some real warheads *inside* decoy balloons? Again a simple move, but one that could foil elaborate and expensive warhead-discrimination systems.

What about countermeasures against terminal defense? In the terminal phase, air resistance strips away today's lightweight decoys, while cone-shaped warheads proceed to their targets with minimal aerodynamic disturbance. But put a tiny rocket on a decoy, and use it to make the decoy mimic a warhead's behavior well into the atmosphere; then high-altitude interceptor systems won't be able to distinguish warheads from decoys. Or put tiny wings on the warheads, making them **MaRV**s (**Ma**neuverable **R**eentry **Ve**hicles), and have them suddenly swoop out of their ballistic trajectories in a way that makes them very difficult targets for terminal-defense rockets. (MaRVs were under development in the late 1980s, and were reportedly an option for the Trident II missile.)

The entire SDI system would be vulnerable to simple countermeasures. Sensors could be blinded with laser beams of modest power. The electromagnetic pulses from high-altitude nuclear explosions could disable computers and electronic communication systems essential for the SDI system's operation. And all space-based SDI components would be vulnerable to attack. Space mines—simple nuclear or conventional explosive devices placed in orbits adjacent to an adversary's defensive weapons—could be detonated at any time to cripple the defense. A space mine costing a few million dollars could destroy a defensive battle station worth billions. Even cheaper, how about putting a truckload of rocks in the same orbit as an SDI weapon, but going in the opposite direction? The resulting 34,000-mile-per-hour collision would destroy *anything*. These destructive countermeasures raise a difficult question: Is disabling an enemy's defense to be considered an act of war? Should we launch our ICBMs if an adversary's lasers blind our warning sensors? What about deployment of space mines? That alone does no harm, but it means our defense can be wiped out whenever an adversary chooses. Would the proliferation of space-based weapons and countermeasures make nuclear war less likely, or would it only provide new channels for initiating war?

Other Uses for SDI

We have been considering the use of SDI technology in defense against land-based intercontinental missiles. But ICBMs are only one leg of the strategic triad. Can SDI defend against the other legs, or against novel means of nuclear delivery? And is SDI capable of offensive roles?

SDI is proposed as a defense against ballistic missiles. It is not intended for use against aircraft or cruise missiles, and for that reason even a technologically perfect SDI system would not provide a complete defense against existing strategic weapons. Some have argued that space-based SDI weapons could be adapted for use against low-flying bombers, but that prospect remains even more visionary than the goal of disabling missiles.

SDI might be effective against submarine-launched ballistic missiles, which share with their land-based cousins the flight phases that make possible a layered defense. But SLBMs' short flight times exacerbate the already difficult technical challenges to the defense. Because their trajectories are short, SLBMs spend particularly little time in space. If their trajectories are "depressed," submarine-launched missiles can be made to remain inside the atmosphere. Although depressed-trajectory flight requires more fuel, it rules out attack by any SDI weapons that cannot penetrate the atmosphere.

The elaborate defensive systems now under consideration are impotent in the face of novel delivery systems that might be used by nuclear-armed terrorists or small countries. What good is a billion-dollar space laser against a nuclear bomb delivered to its target city by car, truck, or boat? How about "suitcase" bombs? Or maybe a nuclear adversary has already implanted clandestine bombs within our cities, ready for detonation at a moment's notice. Can SDI protect against that grim possibility?

Is the proposed SDI system really defensive, or does it have offensive applications as well? Space-based weapons whose effects penetrate the atmosphere could conceivably be used against targets on the ground, perhaps to set fires or to pinpoint damaging effects with surgical precision. It would be easy to harden ground targets against SDI effects, and in any event SDI weapons represent an inefficient means of attacking large ground-based targets. Still, the

Nuclear News: An ASAT Role for SDI?

Even as the SDI program slipped from popularity in the late 1980s, SDI officials suggested adapting at least one anti-missile weapon for use against satellites. MIRACL, a high-power ground-based laser at White Sands, New Mexico, had been developed originally for naval use and was later adapted for SDI. In 1989 officials proposed upgrading MIRACL's beam-directing mirrors to test the laser's effects on orbiting satellites. Congressional opposition to the program was mollified by assurances that tests involving objects in space would not be carried out for several years. More vigorous debate on the strategic wisdom of ASAT, especially in the context of futuristic "Star Wars" weaponry, is to be expected.

News sources: "Military to Ready Laser for Testing as Space Weapon; Anti-Satellite Role Seen," *New York Times,* January 1, 1989; "A MIRACL Weapon Better Left Untried," *New York Times,* July 26, 1989.

near-instantaneous speed with which directed-energy weapons work means that space-based lasers, in particular, might see offensive use in a surprise attack.

There is one offensive task for which SDI is ideal: destroying satellites. Any SDI system designed to find, track, and destroy warheads minutes after their unexpected launch can easily eliminate satellites in predictable orbits. Satellite targets might include an adversary's communication links, warning sensors, defensive weapons, or intelligence-gathering devices. Anti-satellite weapons (ASATs), some space-based and some in the form of small rockets launched from aircraft, are in various stages of development. Should we encourage this latest step toward a battleground in space? ASAT advocates argue that the increasing use of space for military purposes and the Soviet Union's advances in ASAT capability make a U.S. ASAT program imperative. Opponents see ASAT technology as profoundly destabilizing. An ASAT attack on space-based defenses would give the suddenly defenseless country a strong incentive to launch its nuclear missiles. Destruction of communication satellites would eliminate civilian and military controls that might otherwise contain an escalating crisis. And even in terrestrial peacetime, ASAT attacks in space could disable the spy satellites that verify treaty compliance and warn of military buildups. Again, crisis stability would suffer.

Strategic Implications

The 1972 ABM Treaty codified the then-prevalent view that ballistic-missile defense is strategically unwise because it undermines deterrence. Have technological and political developments superseded that wisdom? Much of the debate over SDI hinges on that question; so do efforts to "reinterpret" the ABM Treaty in a light more favorable to the testing of space-based defenses.

SDI proponents argue that advancing technology makes ballistic-missile defense not only possible but also desirable. They see an enhanced deterrence based not on the threat of massive retaliation but on the knowledge that defensive systems could prevent an effective first strike. Some envision a redirection of the arms race toward a competition in defensive weapons, and find that a safer alternative to offensive-arms competition. What about the fact that ballistic-missile defense could not be 100 percent effective? That is not a problem for many SDI advocates, who argue that even a limited defense of ICBM silos would enhance deterrence by making an attacker unsure of the outcome of a missile strike.

Critics counter that SDI would weaken deterrence. The first side to deploy a system might be tempted to launch a first strike, knowing that its defensive weapons would dent its opponent's retaliatory effort. On the other hand, a country that found itself behind in the defensive-arms race might be tempted to launch a nuclear strike before its more advanced adversary had fully deployed defenses.

Critics also argue that even a defensive-arms race is dangerous and costly, with the emergence of space as a battleground providing new avenues to nuclear conflict. They see no end to a cycle of countermeasures, counter-countermeasures, and so forth, with no permanent gain in global security. And critics are convinced that SDI deployment would also spur a buildup in offensive arms, with adversaries striving for offenses so strong they could overwhelm any defense.

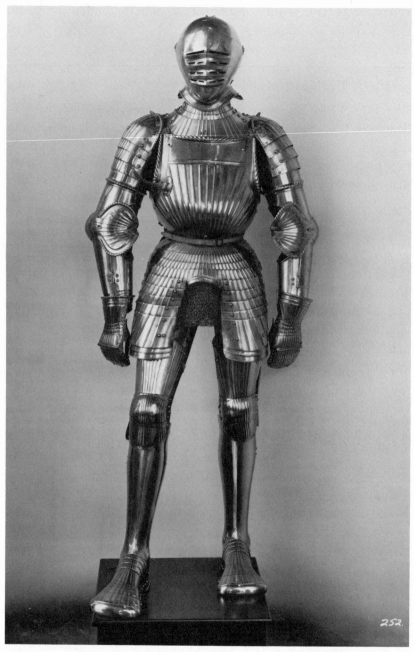

Figure 16.9 SDI of the Middle Ages? SDI critics argue that defensive innovations have always led to more deadly offensive weapons. (Maximilian ar-
mor, sixteenth century, German; courtesy of Higgins Armory Museum,
Worcester, Massachusetts)

Summary

Defense in the nuclear age is a complex issue. Concepts for defensive systems have evolved from relatively simple rockets for terminal interception to elaborate space-based weapons providing multilayered protection against ballistic missiles in their boost, midcourse, and terminal phases. Technical hurdles and the existence of simple countermeasures leave the effectiveness of even high-tech ballistic-missile defense in considerable doubt. The early-1980s goal of a complete defensive shield based on lasers and other directed-energy weapons has faded in favor of more limited defense emphasizing rocket interceptors.

Despite the changing emphasis, proponents of ballistic-missile defense champion their systems as providing a safer and more ethical alternative to deterrence based on massive retaliation. Opponents see defensive efforts spurring a costly, never-ending competition in both defensive and offensive arms, bringing with it strategic instability and new ways to nuclear war.

Who is right? That is a nuclear-age question with major implications for your pocketbook and your future security.

Notes

1. From "Defense Spending and Defensive Technology," a speech televised on March 23, 1983.

2. Although the SDI program began formally only in fiscal year 1985, research on exotic ballistic-missile-defense systems had been funded at lower levels for many years. The budget jumped dramatically with the onset of SDI.

Further Reading

Zbigniew Brzezinski et al., eds., *Promise or Peril: The Strategic Defensive Initiative* (Ethics and Public Policy Center, 1986). Edited by a former U.S. National Security Advisor, this volume includes 35 essays by statespeople, scholars, and strategic analysts with differing views on strategic defense. It includes a historical overview going back as far as Winston Churchill.

Robert Jastrow, *How to Make Nuclear Weapons Obsolete* (Little, Brown, 1985). An astronomer and vigorous supporter of SDI champions the SDI program.

Franklin A. Long et al., eds., *Weapons in Space* (Norton, 1986). A collection of essays by leading authorities on all sides of the SDI controversy. Topics include political, strategic, economic, and legal implications as well as technology.

Rosy Nimroody et al., *Star Wars: The Economic Fallout* (Ballinger, 1988). A look at the economic implications of SDI by the Council on Economic Priorities. Useful explications of SDI funding and "pork barrel" politics, and of the connection between SDI and academia.

Keith B. Payne, *Strategic Defense: "Star Wars" in Perspective* (Hamilton 1986). An analysis of strategic-defense issues from a leading supporter of the SDI program.

Union of Concerned Scientists, *Empty Promise: The Growing Case Against Star Wars* (Beacon, 1986). Essays by leading SDI critics.

U.S. Congress, Office of Technology Assessment, *Ballistic Missile Defense Technologies* (Government Printing Office, 1985). A thorough and cautious analysis of missile-defense technologies and their implications for deterrence and for international stability.

Glossary

ABM Treaty   Part of the SALT I agreements, limiting missile-defense systems.

anti-ballistic-missile (ABM) system   A system designed to destroy attacking ballistic missiles.

ASAT   Anti-satellite weapon; a weapon designed to seek out and destroy an adversary's militarily significant satellites, including those for warning, communication, intelligence gathering, and missile defense.

BMD   Ballistic-missile defense; a defensive system aimed at destroying ballistic missiles.

boost phase   The first part of a missile's flight, in which rocket motors fire to boost the missile above the atmosphere.

decoys   Lightweight objects carried by an ICBM and released along with warheads to confuse defensive systems.

directed-energy weapon   A weapon that delivers intense bursts of energy in the form of electromagnetic radiation or subatomic particles.

kinetic-kill weapon   A weapon using high-speed projectiles to destroy its targets.

laser   A source of intense and coherent light, x rays, infrared, or other electromagnetic radiation.

layered defense   An anti-missile system with weapons tailored for use against each phase of a missile's flight.

maneuverable reentry vehicle (MARV) A warhead capable of changing course once it reenters the atmosphere; proposed as a countermeasure to terminal defenses.

midcourse phase The longest part of a ballistic missile's flight, during which warheads and other elements of the threat cloud travel on ballistic trajectories under the influence of gravity alone.

particle-beam weapon A device using beams of high-energy atomic particles to destroy its targets.

post-boost phase The second part of an ICBM's flight, during which individual warheads and decoys are released.

SALT Strategic Arms Limitation Talks.

Strategic Defense Initiative (SDI) A U.S. program begun in the 1980s with the aim of providing complete defense against ballistic missiles.

terminal phase The final phase of a ballistic missile's flight, as warheads plunge through the atmosphere.

threat cloud The totality of objects—warheads, decoys, chaff, etc.—released once a ballistic missile is above the atmosphere.

Controlling Nuclear Weapons

17

The years from 1945 to 1990 saw an essentially steady increase in the world's nuclear arsenals. The two nuclear superpowers—the United States and the Soviet Union—accounted for much of that buildup; other nations showed smaller but still significant increases. The **Strategic Arms Reduction Treaty (START)**, now in its final negotiations, would mark the first substantial reversal of the nuclear buildup, with the United States and the Soviet Union agreeing to cut their strategic nuclear forces by some 30 percent.

Will the 1990s mark the beginning of a new age of nuclear sanity? Or are we seeing only a temporary lull in the growth of nuclear arsenals? Will the nuclear competition shift to the Mideast, South Asia, and other potential trouble spots? Are we doomed to a future in which nuclear weapons proliferate in quantity, variety, and national origin? Or will restraint prevail, guided and enforced by international agreements?

These and related questions will be addressed in this final chapter of part III. Our nuclear future will never be certain, but we can assess its prospects with a look at the history of international agreements aimed at controlling nuclear weapons.

The Nuclear Weapons Club

The United States' nuclear monopoly ended in 1949, when the Soviet Union exploded its first fission bomb. Today six nations are known to have tested nuclear explosives, and at least five of those have well-established nuclear arsenals and delivery systems. Possession of nuclear weapons is a near certainty for several more countries, and others are rapidly approaching nuclear capability. And as overall technological sophistication increases, nuclear weapons come within the grasp of an ever greater fraction of Earth's sovereign states.

Table 17.1 lists the six countries known to have nuclear weapons capability, along with the year each first tested a nuclear device, the estimated size of its strategic nuclear arsenal, and its status with regard to two treaties that will be discussed shortly. This group constitutes the **nuclear-weapon states** (**NWS**), in contrast to the **non-nuclear-weapon states** (**NNWS**).

How did the NWS acquire their nuclear capabilities? For the United States the answer is the Manhattan project, reviewed at

Table 17.1
Nuclear-weapon states.

	Year of first nuclear explosion	Number of strategic warheads (estimated)*	Delivery systems†	Signed 1963 test-ban treaty?	Signed nonproliferation treaty?
United States	1945	12,100	ICBMs, SLBMs, bombers, CMs	yes	yes
Soviet Union	1949	11,320	ICBMs, SLBMs, aircraft, CMs	yes	yes
Britain	1952	300	SLBMs, aircraft	yes	yes
France	1960	500	IRBMs, SLBMs, aircraft	no	no
China	1964	300	ICBMs, IRBMs, SLBMs, aircraft	no	no
India	1974	0–20?	aircraft, IRBMs?	yes	no

* As of 1990; estimates from *SIPRI Yearbook 1989*. U.S. and Soviet figures do not include START reductions.
† CM=cruise missile, IRBM=intermediate-range ballistic missile.

length in chapter 12. The Soviet Union launched its nuclear weapons research in 1939, but only after World War II did the Soviet project accelerate. The Soviets were aided by Klaus Fuchs, a physicist who had joined the Los Alamos effort in 1944 with a British team and who had been passing nuclear secrets to the Soviets for several years. Fuchs' efforts may have shaved a year or two off the Soviet bomb program, but even without any espionage the Soviets were certainly capable of producing their own bomb. Both the United States and the Soviet Union followed their fission devices with successful fusion-weapon designs in the mid-1950s.

The British had been active in nuclear weapons research from the beginning, and were partners with the United States in the Manhattan Project through World War II. After the war the United States withdrew from weapons collaboration, and in 1947 the British resolved to develop their own nuclear weapons. The first British fission explosion occurred in 1952, by which time Britain had well-developed plutonium-production facilities and other essentials of a nuclear weapons industry. A British fusion test followed in 1957. Today's British nuclear force is largely home-grown, although some of the missiles (but not their nuclear warheads) are procured from the United States.

France, fiercely independent in its military posture, began in 1954 to develop a fully indigenous nuclear force. A French fission test occurred in 1960, and a fusion explosion in 1968. Although its nuclear independence was compromised somewhat by recent revelations of clandestine U.S. assistance, France today remains insistent on charting its own nuclear course, and has steadfastly refused to enter most arms-control or test-ban negotiations.

Next to join the nuclear club was China, with its 1964 fission explosion. During the 1950s the Chinese had received considerable nuclear help from the Soviet Union under a program designed to bring the benefits of nuclear energy to the socialist world. That help included uranium-enrichment facilities and plutonium-production reactors. With the Sino-Soviet split in 1960, China chose to defer plutonium production in favor of its better-developed uranium capability. As a result, China remains the only nation whose first nuclear test involved a uranium device. By the mid-1960s the general principles of thermonuclear fusion weapons were widely known, and Chinese scientists had little trouble assembling their

Nuclear News: Rewriting Nuclear History

The history of nuclear weapons continues to spark controversy, as government secrecy makes it difficult to know just who helped whom do what. Recent revelations shed new light on the Soviet and French nuclear efforts.

A long-standing view holds that the spy Klaus Fuchs aided the Soviet fission and fusion programs with the secrets he passed from Los Alamos. But it now appears that Fuchs' fusion information, gleaned before he left the U.S. bomb program in 1946, was less than useless. In the late 1940s Edward Teller advocated a "superbomb" design that later proved unworkable. It was the rough idea of Teller's early "super" that Fuchs gave the Soviets, although neither Fuchs nor anyone else knew at the time that the idea was flawed. Only in the early 1950s, with the Teller-Ulam invention, did the fusion bomb become a reality. A full-fledged Soviet fusion weapon followed the United States' first true fusion explosion by 3 years.

How did the Soviets learn the fusion secret? New revelations suggest they did it by analyzing fallout from U.S. fusion tests. And Britain's first fusion explosion followed the Soviets' by the same 3-year interval that separated the U.S. and Soviet tests. Why? Perhaps because of Soviet fallout samples supplied to the British by their American friends!

And what about the French? It took them 8 years between fission and fusion weapons. But after the French fusion test of 1968, a clandestine U.S. assistance effort began, helping France to miniaturize its fusion weapons and ultimately to develop small enough warheads for MIRVed missiles. That secret collaboration came to light only in 1989. Apparently it lay just barely within the law, as French nuclear specialists asked their American counterparts a series of yes-or-no questions whose answers guided their work; that way, the Americans didn't actually pass restricted information. In 1985 U.S.-French nuclear collaboration gained legitimacy with a little-known pact permitting the two nations to share restricted nuclear data.

News sources: "Spy's Role in Soviet H-Bomb Is Played Down," *New York Times,* January 3, 1990; "U.S. Secretly Helped France Develop Nuclear Weapons, An Expert Writes," *New York Times,* May 28, 1989.

own fusion weapon. A successful thermonuclear test followed in 1967.

The most recent nation to demonstrate nuclear weapons capability was India, whose sole nuclear test occurred in 1974. Ostensibly an experiment in the peaceful use of nuclear explosives, the Indian test involved a plutonium fission device. And where did India get the plutonium? From a Canadian-supplied research reactor of the CANDU type (figure 17.1; see also chapter 8), moderated with heavy water from the United States. India has not tested again, and claims not to possess a nuclear arsenal. Circumstantial evidence, however, suggests that India is at least stockpiling plutonium as a hedge against Pakistan's nuclear weapons program, and that by 1990 that stockpile contained enough for about fifty weapons.

What can we learn from the stories of the six nuclear-weapon states? First, that nations of even modest technological advancement can build nuclear weapons. That task appears relatively straightforward; as far as anyone knows, all six NWS were successful on the first nuclear try. And what about their routes to nuclear development? Although exchanges of information and technology—both licit and illicit—played some role, the first four NWS built bombs largely on their own, each having first established indigenous industries for producing enriched uranium and/or fissile plutonium. The last two entries are a little different. Neither China nor India was at the forefront of world technological ranks, and both relied heavily on technology and expertise from abroad. That technology was not explicitly for weapons production; nor, on the other hand, did either country use nuclear power plants in its weapons effort. Instead, both had acquired a generalized capability for research and development in all areas of nuclear technology. The Chinese and Indian models may be particularly appropriate as we consider the further spread of nuclear weapons.

Who Is Next?

After India's 1974 test, nuclear experts estimated that a dozen more countries could go nuclear by 1980, and another dozen by 1985. Yet here we are in the 1990s, and no one has yet overtly followed

Figure 17.1   This small CANDU-type research reactor at the Bhabba Atomic Research Center in Trombay, India, was used to produce plutonium for India's sole nuclear test. (United Nations)

India's entry into the NWS group. Does that mean the world has been spared the further spread of nuclear weapons? Probably not.

The universal success of the six NWS in first-time nuclear testing, coupled with increasingly widespread knowledge of weapons design and the availability of supercomputer simulations, means that a potential nuclear nation can build weapons in considerable confidence that they will work without being tested. It need not even assemble the weapons; fabricating the pieces is enough for an incipient nuclear arsenal that could be completed on a few days' notice. Although none are known to have tested nuclear weapons, a number of countries beyond the six NWS are today at various stages on the road to nuclear weapons, from fledgling weapons programs to substantial nuclear arsenals. Here we look briefly at some of the more serious contenders for membership in the "nuclear club"; a summary is provided in figure 17.2.

Israel

It is believed that Israel secretly developed nuclear weapons in the late 1960s, in part with uranium diverted or stolen from the United States, Britain, and France and in part with plutonium bred in a French-supplied research reactor. The Israeli nuclear arsenal today may number 50–200 weapons, including fusion-boosted devices with at least 10 times the yield of the Hiroshima bomb. Delivery systems include aircraft and missiles, the latter with increasing range and accuracy. Already the purported Israeli nuclear arsenal could wipe out every Middle Eastern city with a population over 100,000. Although Israel's nuclear capability is widely acknowledged, the Israeli government maintains a purposeful nuclear ambiguity, neither denying nor confirming its nuclear might.

South Africa

Well endowed with uranium, South Africa has had active nuclear programs. In 1988 the South African government openly declared itself capable of building nuclear weapons. Some proliferation experts believe that South Africa today has a slowly growing nuclear arsenal of some 10–20 weapons. Although other countries in sub-Saharan Africa have little nuclear experience, South Africa's nuclear initiative has led several others to consider advancing on the nuclear

---

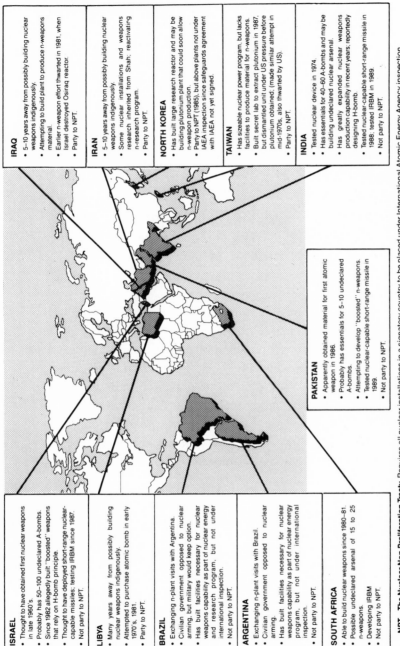

ISRAEL
- Thought to have obtained first nuclear weapons in late 1960's.
- Probably has 50–100 undeclared A-bombs.
- Since 1982 allegedly built "boosted" weapons that rely on H-bomb principle.
- Thought to have deployed short-range nuclear-capable missiles; testing IRBM since 1987.
- Not party to NPT.

LIBYA
- Many years away from possibly building nuclear weapons indigenously.
- Attempted to purchase atomic bomb in early 1970's, 1981.
- Party to NPT.

BRAZIL
- Exchanging n-plant visits with Argentina.
- Civilian government opposed to nuclear arming, but military would keep option.
- Has built facilities necessary for nuclear weapons capability as part of nuclear energy and research program, but not under international inspection.
- Not party to NPT.

ARGENTINA
- Exchanging n-plant visits with Brazil.
- Civilian government opposed to nuclear arming.
- Has built facilities necessary for nuclear weapons capability as part of nuclear energy program, but not under international inspection.
- Not party to NPT.

SOUTH AFRICA
- Able to build nuclear weapons since 1980–81.
- Possible undeclared arsenal of 15 to 25 n-weapons.
- Developing IRBM
- Not party to NPT.

PAKISTAN
- Apparently obtained material for first atomic weapon in 1986.
- Probably has essentials for 5–10 undeclared A-bombs.
- Attempting to develop "boosted" n-weapons.
- Tested nuclear-capable short-range missile in 1989.
- Not party to NPT.

IRAQ
- 5–10 years away from possibly building nuclear weapons indigenously.
- Attempting to build plant to produce n-weapons material.
- Earlier n-weapon effort thwarted in 1981, when Israel destroyed Osiraq reactor.
- Party to NPT.

IRAN
- 5–10 years away from possibly building nuclear weapons indigenously.
- Some nuclear installations and weapons research inherited from Shah; reactivating n-research program.
- Party to NPT.

NORTH KOREA
- Has built large research reactor and may be building plutonium plant that could soon allow n-weapon production.
- Party to NPT (1985), but above plants not under IAEA inspection since safeguards agreement with IAEA not yet signed.

TAIWAN
- Has sizeable nuclear power program, but lacks facilities to produce material for n-weapons.
- Built secret lab to extract plutonium in 1987, but dismantled unit under US pressure before plutonium obtained; (made similar attempt in mid-1970s, also thwarted by US).
- Party to NPT.

INDIA
- Tested nuclear device in 1974.
- Has essentials for 40–60 A-bombs and may be building undeclared nuclear arsenal.
- Has greatly expanded nuclear weapons production capability in recent years; reportedly designing H-bomb.
- Tested nuclear-capable short-range missile in 1988; tested IRBM in 1989.
- Not party to NPT.

NPT—The Nuclear Non-Proliferation Treaty. Requires all nuclear installations in a signatory country to be placed under International Atomic Energy Agency inspection.

Figure 17.2 Nations believed at or approaching nuclear-weapons capability in 1990. (Carnegie Endowment for International Peace)

Carnegie Endowment for International Peace

path. And political upheaval within South Africa could lead to rival groups' gaining control of nuclear weapons.

Pakistan

Even before India's 1974 nuclear test, Pakistan began a nuclear weapons program to preclude another defeat such as it had suffered in the 1971 Indo-Pakistani war. Throughout the 1970s and the 1980s, Pakistan managed to acquire sufficient sensitive technology to permit construction of its own uranium-enrichment plant. Although the United States sought to discourage Pakistani nuclear weapons development in the mid-1980s, U.S. efforts were less than wholehearted because of Pakistan's proximity to Afghanistan, then occupied by the Soviet Union. Today Pakistan is believed to be at or past the nuclear weapons threshold, although its leaders consistently deny any intention to produce nuclear weapons.

Israel, South Africa, and Pakistan are today probably the only nuclear-capable countries beyond the six certain nuclear-weapon states. But others are vigorously pursuing nuclear weapons. Despite the setback of a 1981 Israeli raid on its nuclear facility, Iraq appears to be within 5 or 10 years of building a nuclear bomb. Iran is reviving a nuclear program left over from the days of the Shah, and may also be within a decade of nuclear weapons capability. Libya, although far from producing its own nuclear bomb, has sought nuclear cooperation from a number of countries, and in the early 1980s responded seriously to a former CIA agent's fraudulent offer to provide Libya with a nuclear weapon. Elsewhere in the world, Argentina and Brazil have significant nuclear technology, including uranium-enrichment and plutonium-reprocessing plants. Although these countries' desire for nuclear weapons has abated recently, at times South America has seemed on the verge of a nuclear arms race. In Asia, Taiwan has backed away from an earlier interest in nuclear weapons, in part because of pressure from the United States. The North Korean nuclear program remains worrisome, although Soviet pressure in the mid-1980s brought an official North Korean commitment to nuclear nonproliferation.

Most of the technologically advanced countries could develop nuclear weapons if they chose. Germany and Japan are particularly

Nuclear News: Potential Bomb, Preemptive Strike

On June 7, 1981, eight Israeli fighter-bombers left their base on what appeared to be a training mission. But in fact the planes carried 2,000-pound conventional bombs, and their target was a nuclear reactor at the Osiraq nuclear complex outside Baghdad, Iraq. The surprise attack destroyed the reactor, then nearing completion under the supervision of French and Italian technicians. The Israelis timed their raid carefully, bombing on a Sunday when few foreign workers would be present. They also knew the reactor would begin operation within a few months, and that a later attack could blanket Baghdad with radioactive material.

Was Israel's preemptive strike justified? Amid strong condemnation from the international community, Israeli Prime Minister Menachem Begin called the raid essential to prevent a nuclear first strike by Iraq. By the time of the raid Israel almost certainly possessed a significant nuclear arsenal, but Israeli strategists had evidently concluded that even the threat of nuclear destruction might not deter an Iraqi nuclear strike. So they struck first, setting back a nuclear program that might have produced an Iraqi bomb by the mid-1980s. The French protested that an International Atomic Energy Agency inspection just months earlier had found the Osiraq reactor free of military use; on the other hand, Iraq's president had earlier indicated that the reactor was being built for use against Israel. Did Israel act in self-protection, or was its raid an act of unjustified aggression? Was the Iraqi nuclear program halted, or did the raid only increase Iraq's resolve to go nuclear?

A partial answer to these questions emerged in 1990, when six people were arrested in Britain on charges of smuggling from the United States to Iraq electronic devices that can be used in triggering nuclear explosions. Most experts in nuclear proliferation agreed that the devices were indeed intended for use in weapons. Some felt that Iraq was still as much as a decade from actually producing a nuclear bomb, but others expressed alarm at the advanced state of the Iraqi nuclear program. Coupled with gains in Iraqi missile technology, nuclear weapons capability could make Iraq a formidable power. Iraq's 1990 invasion of Kuwait further heightened fears of a nuclear-armed Iraq; indeed, some commentators believe that the 1981 Israeli raid prevented the crisis over Kuwait from becoming a nuclear confrontation.

News sources: "Israeli Jets Destroy Iraqi Atomic Reactor," *New York Times,* June 9, 1981; "6 Held in Britain in Scheme to Send Atom Gear to Iraq," *New York Times,* March 29, 1990.

strong in nuclear technology, although the legacy of World War II would still inhibit moves toward nuclear arms. Canada has long been capable of producing nuclear weapons and in the 1950s Sweden was well along the nuclear road. Today there is little worry that these countries will go nuclear; at the same time, exports of nuclear technology from Germany, Canada, Belgium, and elsewhere in the industrialized world have figured prominently in others' attempts to acquire nuclear weapons.

Nuclear Motivations

Why would a country want to go nuclear? For many reasons. Nuclear weapons seem to have ended war among the nuclear-armed powers and their major allies, but dozens of minor wars continue each year. For a nation perceiving belligerent neighbors, what better deterrent to aggression than nuclear weapons? And nuclear weapons are powerful status symbols in the community of nations, with good reason: Even the "modest" nuclear forces of Britain, China, France, and probably Israel are sufficient for massive devastation of any country on Earth. A nuclear-armed country is freer to chart its own international course, because it need not rely on outsiders for its defense. And nuclear weapons could give developing nations First World status in the international arena. Nuclear weapons might also bolster domestic pride, perhaps even distracting a population from poverty and other ills. Finally, in an age of nascent nuclear proliferation, a nation may choose to develop nuclear weapons as a hedge against its neighbors' doing so.

Were these reasons alone compelling, technological considerations suggest that nuclear weapons could spread rapidly. The general technological level throughout the world continues to advance, and with that advance often go nuclear expertise and facilities. As the study of nuclear power in part II made clear, there are many technical links between nuclear reactors and weapons capability; we have also seen some nations taking illicit steps to ensure access to weapons-grade uranium and plutonium. Although nuclear power is temporarily stalled in the United States, its worldwide growth and the related international plutonium trade open new channels for acquisition of fissile materials. So why haven't we seen more proliferation of nuclear weapons?

Politics is part of the answer. Although India's nuclear test garnered surprisingly mild rebukes from the international community, adverse international reaction remains a serious concern for a budding nuclear-weapon state. Emergence as an NWS—or even evidence of efforts to acquire nuclear weapons—could invite a preemptive strike such as Israel dealt Iraq in 1981. Although the development of nuclear weapons is within many nations' technical capabilities, it is economically burdensome and may divert resources and brainpower from other national priorities. Many potential nuclear-weapon states are strong advocates of nuclear arms control, and embarking on the nuclear path would tarnish their credibility. Some who could easily achieve nuclear capability are already well protected by the superpowers' nuclear arsenals; they see no reason to undertake the burdens of nuclear weaponry or to lose an ally's "nuclear umbrella." Finally, many nations have willingly incurred a legal obligation to forswear nuclear weapons, as we will see next.

The Nonproliferation Treaty

It is generally but not universally agreed that the spread of nuclear weapons would be detrimental to global security. In the late 1950s and early 1960s Ireland, Sweden, and India first broached the idea of an international agreement to halt the proliferation of nuclear weapons. Initially opposed, the nuclear superpowers soon realized that nonproliferation was in their best interest. The result was the 1968 Treaty on the Non-Proliferation of Nuclear Weapons (the **Nonproliferation Treaty**, or **NPT**).

The NPT overtly establishes a two-class structure in the society of nations. On the one hand are the nuclear-weapon states (NWS); on the other, the non–nuclear-weapon states (NNWS). Each group has its own specific set of treaty obligations.

NNWS signatories to the NPT agree not to acquire nuclear weapons or "peaceful" nuclear explosives, and to subject their nuclear power plants and other nuclear facilities to inspections by the **International Atomic Energy Agency (IAEA)**, an autonomous

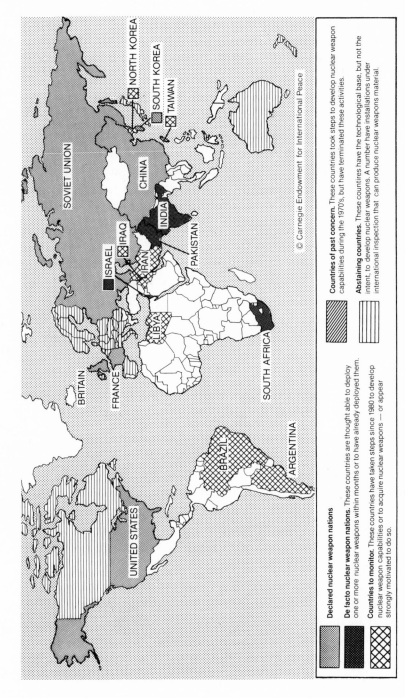

Declared nuclear weapon nations

De facto nuclear weapon nations. These countries are thought able to deploy one or more nuclear weapons within months or to have already deployed them.

Countries to monitor. These countries have taken steps since 1980 to develop nuclear weapon capabilities or to acquire nuclear weapons — or appear strongly motivated to do so.

Countries of past concern. These countries took steps to develop nuclear weapon capabilities during the 1970's, but have terminated these activities.

Abstaining countries. These countires have the technological base, but not the intent, to develop nuclear weapons. A number have installations under international inspection that can produce nuclear weapons material.

© Carnegie Endowment for International Peace

Figure 17.3 Worldwide nuclear capability and intentions as of 1990. (Carnegie Endowment for International Peace)

body loosely affiliated with the United Nations. NWS signatories agree not to transfer nuclear-weapons technology to NNWS, and not to assist NNWS in developing weapons; however, the NWS incur an explicit obligation to help NNWS with the development of peaceful nuclear technologies. The NWS further agree to engage in negotiations on nuclear arms control and disarmament.

The NPT is a remarkable international agreement, and since its inception only India—which did not sign the treaty—has overtly joined the ranks of nuclear-weapon states. But it is not clear whether the NPT has been a major barrier to weapons acquisition. Although Iraq is an NPT signatory, many regard the 1981 Israeli raid, rather than the NPT, as the reason Iraq did not have nuclear weapons to use in the 1990 confrontation over Kuwait. Although more than 130 nations have signed the NPT, some have not—including three of the six NWS and others at or close to nuclear-weapons capability. Among the most important nonsignatories are India, Pakistan, Israel, South Africa, Argentina, and Brazil—all of whom were mentioned above as actual or potential nuclear-weapon states. Of the two NWS nonsignatories with known nuclear arsenals, France has declared that it will act as though it had signed the treaty and China, while often critical of the NPT, has indicated some unwillingness to encourage proliferation. Indeed, both China and France agreed for the first time in 1990 to send representatives to an IAEA review of the nonproliferation treaty.

Even among its signatories, the NPT is hardly an iron-clad guarantee against the development of weapons. By encouraging the transfer of peaceful nuclear technology, the treaty virtually ensures advancement in the general level of nuclear capability. As we have seen, the line between peaceful and weapons applications is less clear than the NPT would indicate. Research reactors, breeder reactors, and plutonium-reprocessing plants are considered "peaceful" under the NPT, but they can easily be turned to weapons use. IAEA safeguards are intended to spot illegal activities, such as production of weapons-grade fissile material or diversion of plutonium from power-reactor fuel cycles. But the small critical mass needed to make a nuclear weapon, coupled with technical and political complications of on-site inspection, mean that safeguards remain far from perfect. Today, nearly 1,000 nuclear facilities are under IAEA safeguards worldwide, with sensitive materials that include well over 100 tons of plutonium, over 20,000 tons of mildly enriched uranium, and more than 1,000 tons of heavy water. By 1990 the total number of inspectors for this global nuclear enterprise had reached only about 200.

The NPT is narrowly precise about what constitutes weapons proliferation: The only banned activities are "acquisition" and "testing" of nuclear explosives. This means that a would-be nuclear nation can legally produce and stockpile weapons-grade plutonium without violating its treaty obligations. And the NPT permits withdrawal on 3 months' notice, giving an incipient nuclear nation plenty of time to take the last steps to a full-fledged nuclear arsenal. Finally, the treaty itself has a limited lifetime. In 1995 an international conference is to decide whether and how long it should be extended. That will require a truly global nuclear choice.

Although the Nonproliferation Treaty is the dominant international instrument to prevent the spread of nuclear weapons, other multilateral agreements have also helped. Among the earliest is the 1959 Antarctic Treaty limiting that continent to peaceful uses; nuclear explosions and the disposal of radioactive waste are explicitly banned. Placement of nuclear weapons in space and on or beneath the ocean floor is prohibited by the 1967 Outer Space Treaty and the 1971 Seabed Treaty, respectively. In addition the 1967 Treaty of Tlatelolco and the 1985 Treaty of Rarotonga establish nuclear-weapons-free zones in Latin America and the South Pacific, respectively.

Acquiring Nuclear Weapons

How would a nation with nuclear ambitions obtain its arsenal? We have seen that the bottleneck is in the acquisition of fissile plutonium or enriched uranium, and throughout this book we have noted avenues to these materials. Four approaches to nuclear-weapon status are summarized below.

1. Mine and enrich your own uranium. This requires a uranium-enrichment plant, the construction of which is expensive and technologically challenging. Pakistan, nevertheless, has managed to build one, despite supplier nations' export controls on many of the needed components.

2. Make plutonium from natural or mildly enriched uranium in a nuclear reactor. This requires a reactor and a plutonium-reprocessing plant; heavy water is also needed if unenriched natural uranium is to be used. Reactors are widely accessible, with nearly 1,000 power and research units throughout the world. Plutonium-reprocessing plants may be purchased commercially and are far less expensive than uranium-enrichment facilities. The amount of plutonium in unreprocessed fuel now amounts to perhaps 20,000 bombs' worth, and is growing rapidly. India used the plutonium approach, which is probably the choice of most emerging nuclear nations.

3. Steal fissile materials. There is strong evidence that Israeli agents conspired with a U.S. nuclear-materials corporation in the 1960s to divert 200 pounds of enriched uranium to Israel's nuclear weapons program. Israeli commandos have reportedly hijacked British and French uranium trucks, and possibly a German ship carrying uranium. Significant quantities of plutonium are unaccounted for, and that amount is likely to increase as plutonium reprocessing becomes widespread. Theft may be a particularly attractive way to initiate a nuclear weapons program while a nation is building its own plutonium-production facilities.

4. Steal complete nuclear weapons. Practices followed in today's nuclear-weapon states make this a difficult approach; as modern weapons have evolved, so have technical and personnel-related safety mechanisms to prevent accidental or unauthorized detonation. On the other hand, newcomers to the NWS could hardly be expected to emphasize safety in the rush toward nuclear capability, so their weapons might be more vulnerable to theft. And even among the established nuclear powers, accidents and political upheaval may make weapons vulnerable.

Nuclear Testing

So far we have considered primarily **horizontal proliferation**, the spread of nuclear weapons to nations that don't already have them. But what about the proliferation of weapons—in number, yield, and variety—within the existing nuclear-weapon states? In the remainder of this chapter we will look at measures aimed at control-

Nuclear News: Bomb Security

Nuclear weapons are supposed to remain under competent military control, with authority for their use vested in high military and civilian authorities. How likely is it that these weapons could fall into other hands?

At least fifty U.S. and Soviet nuclear warheads have reportedly been lost at sea as a result of aircraft and naval accidents. Recovery of these weapons by either their owners or by would-be nuclear nations is unlikely.

Even the established nuclear powers are subject to political instability that could render weapons vulnerable. In 1961 mutinous French generals in Algeria may have briefly gained control over France's desert test site and with it a French nuclear weapon. The weapon was tested somewhat hastily near the end of the four-day revolt, perhaps to make it unavailable to the rebels. Had the insurgents maintained control of the bomb, the outcome of the revolt might have been very different.

The Soviet Union, which exercises tighter central control over its nuclear weapons than the United States, entered a period of unprecedented political turmoil in the late 1980s. How likely is it that the possible disintegration of the Soviet Union could put nuclear weapons in the hands of political extremists or splinter groups? Might independent former Soviet republics emerge as nuclear powers? Some experts dismiss these possibilities as unlikely, but others point to worrisome trends. For example, Azerbaijan, the site of widespread ethnic unrest, hosts one of the Soviet Union's major nuclear weapons stockpiles. What can be done to guarantee the integrity of Soviet nuclear forces in light of revolutionary change? That is an important nuclear question for a Soviet leadership preoccupied with other matters, and a question that is important to the entire planet.

News sources: "France Explodes Nuclear Bomb at Sahara Test Site in Algeria," *New York Times,* April 25, 1961; "Specter is Raised of Nuclear Theft," *New York Times,* January 28, 1990.

ling that so-called **vertical proliferation**. We start with an important key to the continuing development of weapons: the ability to test one's nuclear explosives. Testing helps in designing new weapons, in verifying the reliability of existing weapons, in analyzing weapons' effects, and in developing security systems.

Through the mid-1950s the nuclear-weapon states tested vigorously, with little thought given to environmental or security implications. By the late 1950s public concern began to rise and serious proposals to ban nuclear tests emerged. In 1958 the Soviet Union declared a voluntary moratorium on testing, and the United States soon followed suit. A technical conference on test-ban verification produced agreement that all but underground tests could be detected with near certainty using existing technology. Verification of underground tests would also be possible, but would require intrusive, on-site inspections. Soviet refusal to allow such inspections, coupled with U.S. insistence on perfectly reliable verification techniques, led to a breakdown in the test moratorium. Renewed atmospheric testing of high-yield weapons soon raised fears of global radioactive contamination, and the Cuban Missile Crisis again focused international attention on the desirability of a test ban. The result was the Partial Test Ban Treaty of 1963, prohibiting all but underground nuclear tests. Although the Test Ban Treaty affected primarily the Soviet Union, the United States, and Great Britain, it has been signed by over 100 other nations. (Although China and France have not signed, they eventually moved their tests underground.) The Partial Test Ban Treaty was complemented in 1974 with the bilateral U.S.-Soviet Threshold Test Ban Treaty limiting underground tests to a maximum of 150 kilotons. Disagreement over verification procedures left the Threshold Treaty unratified until 1990, although both sides continued to honor it.

Have the Partial Test Ban Treaty and the Threshold Test Ban Treaty been successful? That depends on your criteria for success. The Partial Test Ban Treaty immediately forced U.S., Soviet, and British nuclear testing underground, resulting in a dramatic reduction in radioactive fallout (figure 17.5). But nuclear testing has continued unabated, with more than two-thirds of the world's nearly 2,000 nuclear tests occurring since the 1963 Test Ban Treaty (figure 17.6). And despite the Threshold Treaty's limitation to 150-

Figure 17.4 Since 1963, all U.S., Soviet, and British nuclear tests have been conducted underground. Here, preparations for an underground test proceed at the Nevada Test Site. Craters from previous tests dot the landscape; they formed as earth subsided over cavities created by underground blasts. (U.S. Department of Energy)

kt tests, the nuclear nations have confidently developed larger, untested devices, such as the United States' 330-kt Minuteman missile warhead.

Would a comprehensive test ban be a good thing? For years the United States' answer was Yes, with every U.S. president before 1980 advocating such a ban; indeed, the Partial Test Ban Treaty requires that nuclear-weapon states strive for "the discontinuance of all test explosions of nuclear weapons for all time." Through the 1970s Soviet reluctance to permit on-site inspections remained a major stumbling block in negotiations on a comprehensive test ban, and in the 1980s the Reagan administration chose to discontinue test-ban negotiations. The tables turned as a liberalized Soviet Union emerged under Mikhail Gorbachev. In 1985, on the fortieth anniversary of the Hiroshima bombing, Gorbachev announced a unilateral Soviet test moratorium, and invited the United

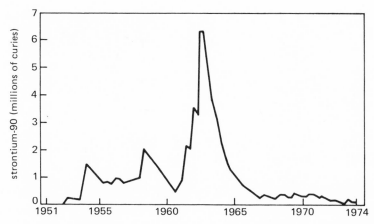

Figure 17.5 Atmospheric levels of strontium-90 rose dramatically with the nuclear tests of the 1950s and early 1960s, then fell abruptly when the 1963 Partial Test Ban Treaty went into effect. (From Glasstone and Dolan, *The Effects of Nuclear Weapons,* third edition, p. 449)

States to join. Soviet pleas fell on deaf American ears as the Reagan administration, with support from military leaders and weapons-lab scientists, rejected calls for the United States to halt its testing. The Soviets resumed testing in 1987, and, as figure 17.6 shows, nuclear testing continues today at moderate levels.

Opponents of a comprehensive test ban argue that it would be impossible to verify, so an adversary could cheat and eventually develop new weapons that would upset the strategic balance. Proponents counter with increasingly sophisticated seismic detection methods that, coupled with a new Soviet willingness to permit on-site monitoring and inspection, would permit near-certain detection of tests down to the kiloton level or even lower. Test-ban opponents claim a ban would weaken deterrence by making the success of a retaliatory strike less certain. Proponents use the same reasoning to argue that a ban would reduce the likelihood of a first strike. Who is right? The answer to that nuclear question will have a lasting effect on global security.

Arms Control and Disarmament

How well have the nuclear nations stood by their Nonproliferation Treaty obligation to pursue arms control and nuclear disarmament?

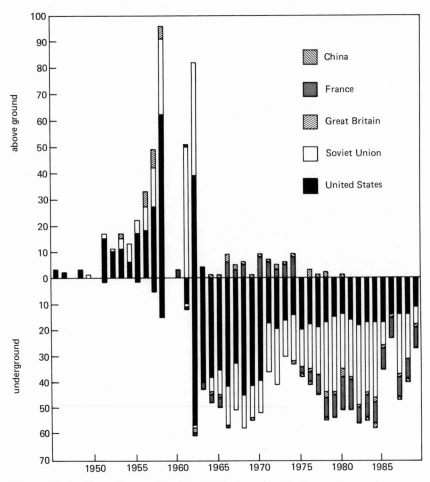

Figure 17.6 Chronology of nuclear explosions by the five major nuclear-weapon states. (data source: *SIPRI Yearbook 1989*)

Nuclear News: Verifying Nuclear Tests

Verification has been a major stumbling block in test-ban negotiations. Initially the United States argued for a verification technology called CORRTEX, which involved placing a cable in a hole adjacent to an underground nuclear blast. The Soviet Union countered with a less intrusive plan for seismic monitoring stations placed as far as several hundred miles from the test site. The stalemate began to break in 1986 when a private U.S. environmental group, the Natural Resources Defense Council, arranged to set up seismic monitoring equipment on Soviet territory. The results were encouraging, showing that even small chemical explosions could be detected hundreds of miles away. By 1988, American and Soviet scientists had jointly observed both nations' nuclear tests, and had gathered evidence that the seismic-monitoring approach might actually be more effective than the on-site method. The success of these joint verification experiments led to agreement on verification procedures involving both seismic and on-site methods, and ratification of the 1974 Threshold Test Ban Treaty followed in 1990.

Not everyone looks favorably on the TTBT. Much of the world sees the treaty's 150-kiloton threshold as being much higher than is needed for meaningful constraints on the development of nuclear arms. In late 1989, the UN General Assembly voted 127 to 2 in favor of a conference to amend the Partial Test Ban Treaty to make it a comprehensive treaty banning *all* nuclear tests. Of the five known nuclear powers, only the Soviet Union supported the resolution; the United States and Britain cast the sole opposing votes, while China and France abstained.

News sources: "U.S. Team Joins Russians in Monitoring Explosion," *New York Times,* September 3, 1987; "In Remotest Nevada, a Joint U.S. and Soviet Test," *New York Times,* August 18, 1988; "Atomic Test Data Weaken U.S. View," *New York Times,* September 11, 1988; "Nonaligned Nations to Seek Total Nuclear Test Ban," *New York Times,* November 5, 1989.

Until recently the answer has been decidedly mixed. Here we look briefly at the most significant arms-control agreements between the two nuclear superpowers.

SALT Agreements

We saw in the preceding chapter how the first Strategic Arms Limitations Talks (SALT I) resulted in a treaty limiting anti-missile defenses. SALT I also led in 1972 to the first limits on offensive nuclear weapons, with the U.S.-Soviet signing of the so-called Interim Agreement (intended to provide temporary limits until the SALT II negotiations were completed later in the decade).

The Interim Agreement set limits on the total number of strategic-missile launchers—ICBM silos and SLBM launch tubes. However, it set those limits at the existing number of missiles and therefore did nothing to reduce strategic weapons. Complications involved in counting bombers and shorter-range delivery vehicles stationed in Europe led SALT I negotiators to leave these out of the agreement. But the most glaring omission in the Interim Agreement was that it said nothing about MIRVs, which the United States had just begun deploying. With fixed limits on missiles, the agreement just channeled more warheads into multiple-warhead missiles. The result was a rapid increase in the number of strategic nuclear warheads on both sides (figure 17.7).

Two years after signing the Interim Agreement, the United States and the Soviet Union agreed on a modified strategic-arms-limitation pact known as the Vladivostok Accords. This agreement equalized the number of strategic launch vehicles at 2,400 for each side, and it counted the long-range strategic bombers left out of the Interim Agreement. The Accords were significant in being the first international agreement requiring dismantling of existing weapons systems; in particular, the Soviets were forced to scrap 99 strategic launchers. However, the Accords did nothing to limit MIRVs, as figure 17.7 makes quite obvious.

Through the 1970s the SALT process continued, complicated by new technological developments such as the cruise missile. In 1979 the United States and the Soviet Union signed the SALT II treaty, which continued limits on strategic launchers at slightly

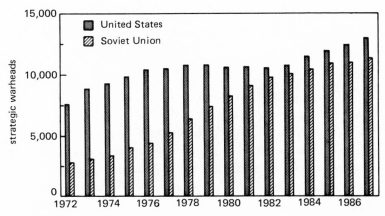

Figure 17.7 Evolution of the U.S. and Soviet strategic arsenals. The rapid growth in the 1970s is associated with the MIRVing of missiles, first by the United States and later by the Soviet Union. (data source: *Bulletin of the Atomic Scientists,* May 1988, p. 56)

below the Vladivostok level. Although each nation could set its own balance among ICBMs, SLBMs, and bombers, SALT II did spell out some additional specific limitations. For example, each side was allowed to develop at most one entirely new missile, with only limited modifications permitted to existing missiles. SALT II also set a limit of 1,200 MIRVed missiles, both ICBMs and SLBMs, and it required that no missile carry more warheads than the maximum number with which it had ever been tested. To aid in verification, SALT II banned the use of secret codes in the information broadcast from missile test flights; that way, each side could monitor the other's tests to check for treaty compliance.

SALT II established arms limitation as a basic premise of U.S.-Soviet relations. But it, too, left much to be desired. Opponents felt that SALT locked the United States into a position of strategic vulnerability at the hands of large, accurate Soviet ICBMs. Disarmament advocates bemoaned the lack of controls on cruise missiles, and envisioned a proliferation of these small, easily hidden, and highly accurate missiles that could elude future arms-control agreements. Whatever its virtues or failings, the SALT II treaty was signed in 1979 amidst deteriorating U.S.-Soviet relations. The Soviet invasion of Afghanistan swung the U.S. Senate solidly against SALT II, and the treaty was never ratified. Despite the treaty's legal impotence, both sides have remained essentially in compliance with its limits.

Verification

A major hurdle in arms-control negotiations has been verification. How is each side to be sure the other isn't cheating? Arms-control opponents have often used the verification problem, and their lack of trust in the Soviet Union, to argue against negotiated limits on nuclear weapons. Proponents claim that verification procedures are adequate, and that trusting a potential adversary is not necessary to ensure treaty compliance. Who is right?

Verification is one area where technological advancement appears unambiguously beneficial. Today, so-called *national technical means* of verification are based largely on surveillance by orbiting satellites. The capabilities of these satellites are highly classified, but the most advanced are believed to distinguish objects as small as 6 inches from altitudes of about 100 miles. A wide range of satellites now patrol space, providing detailed and wide-angle photographs, warning of missile launches, detecting atmospheric nuclear tests, and intercepting radio signals from missile flight tests. Most of the secret military Space Shuttle missions of recent years have involved launches of these spy satellites.

Although the verification capability of satellites is impressive and continually improving, it is still not perfect. Coupled with subtle questions of treaty language and national intentions, ambiguous satellite images have sometimes led to charges of treaty violations. And limitations on verification capability may indirectly drive the arms competition in undesirable directions; for example, the SALT I Interim Agreement's emphasis on missiles instead of warheads reflects the ease of counting missiles from space and the near impossibility of counting individual MIRV warheads; that limitation on MIRV verifiability is at least partly responsible for the proliferation of MIRVs in the 1970s.

Spy satellites, although less than 100 percent reliable, are probably the best verification method available in a climate of hostility and mistrust. We will soon see, though, how changing international relations in the late 1980s strengthened verification procedures in a way that would have been unimaginable a decade earlier.

Shorter-Range Weapons and the INF Treaty

The SALT agreements dealt only with strategic nuclear weapons—those each superpower aims against its adversary's homeland. But before 1990, military planners envisioned conflict in Europe as the most likely way for a U.S.-Soviet nuclear war to begin. For 45 years after World War II, Europe bristled with nuclear and conventional weapons as the United States and its NATO allies faced the forces of the Soviet Union and Warsaw Pact. Those weapons included short- and medium-range nuclear missiles, and NATO policy called for first use of nuclear weapons against aggression by the larger conventional forces of the Warsaw Pact. Particularly worrisome to the Soviet Union were medium-range missiles capable of striking well inside Soviet territory from bases in Germany; NATO forces, in turn, were alarmed at Soviet nuclear modernization programs that brought accurate MIRVed missiles to bear on Western Europe.

Negotiations on medium-range missiles began in 1981 but were suspended in 1983 in a Soviet protest against new NATO missile deployments. Both sides returned to the bargaining table in 1985 and, helped by the remarkable changes beginning in the Soviet Union, proceeded rapidly to fruition. On December 8, 1987, the United States and the Soviet Union signed the Treaty on Intermediate-Range Nuclear Forces, boldly eliminating all U.S. and Soviet missiles with ranges between 300 and 3,400 miles.

The INF treaty did not significantly reduce the number of nuclear warheads in the superpowers' arsenals. But it nevertheless made major advances in arms control, eliminating an entire class of nuclear weapon systems and requiring physical destruction of existing missiles (figure 17.8). (INF did not, however, require destruction of nuclear warheads; in fact, U.S. INF warheads were recycled for use in long-range missiles.) And the treaty established an unprecedented level of international cooperation and openness in the verification process, with American and Soviet observers stationed at each other's missile-production facilities. For optimistic arms-control advocates, INF marked the beginning of a serious superpower effort to bring nuclear weapons under control.

Figure 17.8   U.S. inspectors at Sayrozek, in the Soviet Union, observe the destruction of a missile as required by the INF treaty. (On-Site Inspection Agency)

Reducing Strategic Arms

The 1980s began in what was for many a frightening climate of superpower belligerence and nuclear threats. In 1982, there arose in the United States a popular movement advocating a bilateral freeze on the development and deployment of additional nuclear weapons; numerous communities voted resolutions in support of the freeze, and nearly a million Americans rallied in New York City to express their concern at the rising danger of nuclear war (figure 17.9). The Reagan administration responded by proposing that the United States and the Soviet Union join in **Strategic Arms Reduction Talks** (**START**) aimed at achieving the first significant cuts in strategic nuclear weapons. Prospects for START seemed dim in the confrontational climate of the early 1980s, and indeed negotiations broke off in 1983 in the Soviet protest over NATO missiles. When they resumed in 1985, the climate had begun to change. Driven by rapidly thawing U.S.-Soviet relations and by severe economic stresses both sides experienced from ongoing strategic-arms build-ups, the START negotiations proceeded quietly through the late 1980s. In early 1990 the START discussions bogged down (in part over the difficult issue of cruise missiles), but a compromise was

Figure 17.9 On June 12, 1982, nearly a million people from all walks of life joined in a massive demonstration of concern over the rising threat of nuclear war. Shown here are delegations from trade unions. (Steve Cagan/Impact Visuals)

reached in time for U.S. President George Bush and Soviet President Mikhail Gorbachev, at their 1990 summit meeting, to issue a joint statement committing both countries to START reductions. After ratification of START, reductions in strategic arms would take place over a seven-year period. And both sides would begin to work toward deeper cuts through a pact to be known as START II.

START is a complex web of detailed agreements reflecting the superpowers' mutual desire for arms reductions in the face of the differing makeup of their strategic triads (recall figure 14.15). The original goal of a 50 percent cut in strategic warheads was compromised by the difficulty of comparing nuclear weapons on different delivery vehicles. An airplane carrying nuclear bombs, for example, counts in the treaty as a single warhead no matter how many bombs it actually carries. And START limits on cruise missiles were set at levels higher than the existing arsenals, virtually ensuring growth in the number of cruise missiles. But START does make substantial reductions in warheads, especially in the case of ballistic missiles.

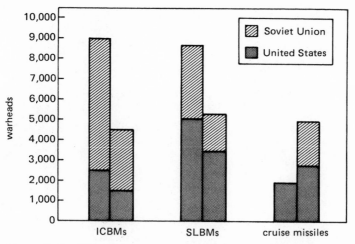

Figure 17.10 The START Treaty's effects on ballistic-missile and cruise-missile warheads. The left column in each category shows existing arsenals at the time the treaty was signed; the right column is the number of warheads allowed under START. The cruise-missile category includes air-launched and submarine-launched cruise missiles. Note how the treaty preserves the Soviet emphasis on land-based ICBMs and the U.S. emphasis on submarine-launched missiles. (data source: *New York Times*, June 2, 1990)

As figure 17.10 shows, Soviet land-based ICBM warheads will drop by more than 50 percent, and other classes of ballistic-missile warheads on both sides will be reduced by nearly that percentage. Treaty details further proscribe "heavy" ICBMs, such as the MX and the SS–18, although the treaty makes no effort to ban multiple-warhead ballistic missiles. An overall limit caps the total number of long-range delivery vehicles—land-based and submarine-based ballistic missiles and bombers—at 1,600 for each side. Although START permits modernization of weapon systems, the treaty specifically bans new missiles with more than ten warheads, multiple-warhead air-launched cruise missiles, new "heavy" ICBMs and SLBMs, mobile launchers for "heavy" ICBMs, and the development of rapidly reloadable ICBM launch silos.

Is START a good agreement? Nurtured by one of the United States' most conservative, anti-Soviet administrations, the treaty is nevertheless a major step in the direction nuclear "doves" have advocated for years. Although START enjoys wide support across the U.S. political spectrum, a few diehard Cold Warriors continue to view *any* arms-control measure as detrimental to U.S. interests. And "dovish" critics are quick to point out the treaty's failure to achieve its goal of 50 percent reductions in nuclear weapons, and the lack of significant progress toward the elimination of destabilizing multiple-warhead missiles. Nevertheless, nearly everyone agrees that START represents a major and welcome turning point in the nuclear age. The treaty's effects should extend beyond the superpowers; the fact that they have moved toward their Nonproliferation Treaty obligation to curb their nuclear arms race will give more credibility to their efforts to limit the spread of nuclear weapons. But START is only a beginning. The superpowers retain more than enough nuclear firepower to obliterate civilization. And no arms-reduction scheme can be complete without addressing the still-growing arsenals of the other nuclear nations.

Summary

Horrifyingly destructive nuclear weapons have been with us for half a century. During that time they have increased in numbers, explosive yield, accuracy, and means of delivery. Originated in the United States, they have spread to at least five other sovereign nations. Several more countries almost certainly have nuclear weapons, while others strive to acquire them. Can we bring nuclear weapons under control, or are we doomed to a future in which nuclear weaponry is commonplace and increasingly likely to be used?

We have seen in this chapter that international controls on nuclear weapons are possible, and that some have been in place for decades. Those controls are imperfect, as difficult international negotiations and domestic politics lead often to compromises that weaken arms-control agreements. But the incentive to control nuclear weapons is increasingly shared throughout the world, and that incentive comes from the nearly universal recognition that nuclear war is unacceptable. Arms-control agreements complement deterrent strategies in seeking to prevent nuclear war.

The best-known of the international arms-control agreements are multilateral treaties, such as the Partial Test Ban Treaty and the

Nonproliferation Treaty, and the major bilateral superpower pacts, including the various SALT agreements, the INF Treaty, and START. Lesser agreements help, too, in reducing the threat of nuclear war; the 1963 Hot Line communication agreement, for example, was the first of many U.S.-Soviet measures aimed at improving communications in time of crisis. Table 17.2 summarizes the significant international arms-control agreements, including those discussed earlier in this chapter.

Have arms-control agreements helped move the world away from nuclear war? Perhaps, although their exact role remains controversial. The nuclear sword-rattling of the early 1980s came on the heels of the SALT II treaty, and SALT I was followed by a rapid increase in nuclear warheads. The Partial Test Ban Treaty drove nuclear testing underground, but weapons development and testing continued at a brisk pace. The Nonproliferation Treaty may have helped curb the spread of nuclear weapons; on the other hand, it may have encouraged potential nuclear nations to go as far they can without actually producing completed weapons. The INF treaty eliminated an entire class of nuclear weapons, but hardly changed the magnitude of the nuclear threat. Even START, with its wholesale reduction in strategic warheads, still leaves the world awash in several thousand World War IIs of nuclear destructive power.

Perhaps what is important about these treaties is not just the technical limits they place on nuclear weapons, but the international cooperation and trust they engender. If we can negotiate with our global neighbors toward the goal of mutual security, then maybe in the process we will achieve enough understanding to prevent the nuclear threat from becoming reality.

Further Reading

David P. Barash, *The Arms Race and Nuclear War* (Wadsworth, 1987). Written by a psychologist, this book is especially good at posing questions on both sides of arms-control and nuclear-technology issues. Part 4 deals with the arms race and with controls on nuclear weapons.

Ashton Carter, John D. Steinbruner, and Charles A. Zraket, eds., *Managing Nuclear Operations* (Brookings Institution, 1987). Comprehensive analyses of the security and control of nuclear weapons, including both technical and policy issues.

Table 17.2
Agreements limiting nuclear weapons.

Year*	Agreement	Description	Signatories
1959	Antarctic Treaty	Limits Antarctica to peaceful activities; explicitly bans nuclear explosions and disposal of radioactive waste.	30
1963	Partial Test Ban Treaty	Bans testing of nuclear weapons in the atmosphere, in space, or underwater.	111
1963	Hot Line Agreement	Establishes direct communications links between Moscow and Washington so heads of state may confer in event of crisis.	US, USSR
1967	Outer Space Treaty	Bans nuclear weapons from space.	81
1967	Treaty of Tlatelolco	Bans production, acquisition, testing, storage, or deployment of nuclear weapons by Latin American nations; additional protocols require NWS to respect denuclearized Latin America. Requires IAEA incpection of nuclear activities.	25
1968	Nonproliferation Treaty	Bans acquisition of nuclear weapons by NNWS; bans nuclear-weapons technology transfers from NWS to NNWS; encourages non-weapons nuclear-technology assistance to NNWS; commits NWS to pursue arms control.	140
1971	Accidents Measures Agreement	Upgrades safeguards against unauthorized or accidental use of nuclear weapons.	US, USSR
1971	Seabed Treaty	Bans nuclear weapons on or beneath the ocean floor beyond 12 miles from coastline.	74
1972	ABM Treaty	Limits signatories to two anti-ballistic-missile sites each; later protocol reduces this to one site each.	US, USSR
1972	SALT I Interim Agreement	Places limits on strategic-missile launchers	US, USSR
1973	Prevention of Nuclear War Agreement	Requires US-USSR consultation when there is danger of nuclear war.	US, USSR
1974†	Threshold Test Ban	Limits underground tests to 150-kiloton maximum yield.	US, USSR

Table 17.2 (*continued*)

Year*	Agreement	Description	Signatories
1974	Vladivostok Accords	Limits strategic launch vehicles to 2,400 on each side. Bombers included for first time.	US, USSR
1979†	SALT II	Limits strategic launch vehicles, number of MIRVed missiles, bombers with cruise missiles. Allows each side to deploy only one new ICBM model. Prohibits secret coding of signals from missile test flights.	US, USSR
1985	Treaty of Rarotonga	Establishes nuclear-weapons-free zone in South Pacific; requires IAEA safeguards on other nuclear technology.	8
1987	INF Treaty	Eliminates all medium-range nuclear missiles (range: 300–3,400 miles) in U.S. and Soviet arsenals; establishes on-site inspection of missile-production facilities.	US, USSR
	START Treaty	30% reduction in strategic nuclear warheads, including 50% reduction in ballistic-missile warheads. Cap of 1,600 long-range delivery vehicles, including missiles and aircraft. First limits on cruise missiles. Detailed limitations on specific weapons systems.	US, USSR

* Year treaty was first signed. Unless noted, treaty became effective by the following year.
† SALT II was never ratified; Threshold Test Ban was not ratified until 1990.

John W. Lewis and Xued Litai, *China Builds the Bomb* (Stanford University Press, 1988). A nicely written history of the Chinese bomb project, in the spirit of Richard Rhodes' *Making of the Atomic Bomb*.

Amory B. Lovins and L. Hunter Lovins, *Energy/War: Breaking the Nuclear Link* (Friends of the Earth, 1980). A well-documented and articulately argued case against nuclear power based on its potential for weapons proliferation.

Morris McCain, *Understanding Arms Control* (Norton, 1989). A brief and readable introduction to nuclear weapons issues in the context of arms control. A series of "controversies" summarize the arguments on both sides of selected issues.

Progress in Arms Control? (Freeman, 1979); *Arms Control and the Arms Race* (Freeman, 1985). These sometimes overlapping collections from *Scientific American* provides incisive analyses of nuclear arms control and related issues.

Peter Pry, *Israel's Nuclear Arsenal* (Westview, 1984). This brief but thorough history of the Israeli nuclear weapons program elucidates many technical and political aspects of weapons proliferation.

Lawrence Scheinman, *The International Atomic Energy Agency and World Nuclear Order* (Resources for the Future, 1987). An up-to-date analysis of the IAEA's work in relation to nuclear nonproliferation.

Dietrich Schroeer, *Science, Technology, and the Nuclear Arms Race* (Wiley, 1984), section 4. An outline of arms-control issues, with emphasis on technological considerations.

Leonard S. Spector, with Jacqueline R. Smith, *Nuclear Ambitions* (Westview, 1990). The latest in a series of proliferation studies sponsored by the Carnegie Endowment for International Peace, this book details nuclear developments in aspiring nuclear-weapon nations. Two earlier titles include Spector's *The Undeclared Bomb* (Ballinger, 1988) and *Going Nuclear* (Ballinger, 1987).

Glossary

horizontal proliferation   The spread of nuclear weapons to nations that do not already have them.

non-nuclear-weapon states (NNWS)   Those nations not possessing nuclear weapons.

Nonproliferation Treaty (NPT)   A multilateral agreement signed in 1968. It bans transfers of nuclear weapons technology to non-nuclear-weapon states but encourages transfer of nuclear technology for peaceful purposes. It also requires the nuclear-weapon states to pursue arms control.

nuclear-weapon states (NWS)   Those nations possessing nuclear weapons, given special status by the Nonproliferation Treaty.

Partial Test Ban Treaty   A multilateral 1963 treaty banning all but underground nuclear tests.

Strategic Arms Reduction Treaty (START)   A 1991 U.S.-Soviet treaty calling for 30 percent reductions in strategic nuclear forces by 1997.

Threshold Test Ban Treaty   A 1974 U.S.-Soviet agreement limiting nuclear test yields to a maximum of 150 kilotons.

vertical proliferation   Proliferation of nuclear weapons within a nuclear-weapon state, including increases in the number and in the variety of weapons.

Nuclear Power, Nuclear Weapons, and Nuclear Futures

18

Parts II and III of this book dealt separately with nuclear power and nuclear weapons. How valid is the separation of those topics? What will be the role of nuclear technology in the future? Can we continue to expand our use of nuclear technology without increasing the threat of nuclear annihilation?

These are not easy questions, although experts of widely differing opinions claim to have answers. They range from assertions that nuclear power is "not a source for weaponry"[1] to the notion that nuclear power plants are "bomb factories,"[2] and from belief that the nuclear age is already past to visions of a prosperous nuclear future. Some see the risk of nuclear war rapidly fading, while others are certain of a nuclear Armageddon.

Nuclear Power and Nuclear Weapons: The Connection

Even those who believe nuclear power to be safer and cleaner than many of its alternatives are often troubled by possibility that the development of nuclear power may lead to the proliferation of nuclear weapons. All arguments in favor of nuclear power are vacuous if civilization is destroyed in a nuclear war made possible by the spread of nuclear power plants. *That* is one possibility we didn't weigh into chapter 11's comparison of coal and nuclear power. Should the potential for weapons proliferation and nuclear

Nuclear News: Nuclear Optimism?

How likely is nuclear war? Unless war occurs, that is a question with no firm answer. But it is a question with substantial implications for your future: Even a mere 2 percent per year chance of nuclear war means that today's young Americans are more likely to die nuclear deaths than to die of old age. Despite the uncertainties, nuclear experts have attempted to quantify at least the relative danger of war; nearly all would agree, for example, that the nuclear risk fell dramatically in the late 1980s. The *Bulletin of the Atomic Scientists'* famous "doomsday clock" has, since the 1940s, displayed that journal's assessment of the world's approach to the brink of nuclear war; figure 18.1 traces the *Bulletin*'s responses to major events of the nuclear age. The public, too, makes an instinctive judgment of the nuclear threat, with increasing political activism marking the periods of greatest concern. Figure 18.2 shows how public perception of the nuclear threat has declined through the 1980s. Still, on the eve of the START reductions, 18 percent of Americans believe a nuclear war likely in the next 10 years. If their assessment is correct, nuclear war remains more likely than not to occur in the lifetimes of today's young people. Nuclear experts might dismiss the public fear as exaggerated, but there can be no sure bets in our changing world of emerging nuclear nations. And many believe that an increasing global dependence on nuclear power may bring with it a growing risk of nuclear war.

News sources: *Bulletin of the Atomic Scientists,* April 1990, back cover; "American Fear of Soviets Declines, Survey Finds," *New York Times,* May 30, 1990.

war change our assessment of nuclear power or, for that matter, of other "peaceful" nuclear technologies?

This book doesn't need a whole new part on the relation between nuclear power and nuclear weapons. The fact that there *is* a connection has been stressed throughout the book. Nuclear power and nuclear weapons use the same basic fissile materials. Both require the mining and often the enrichment of uranium. Reactors produce plutonium, and a bomb made from a few pounds of plutonium can destroy a city.

You now know enough about nuclear technology to understand the several routes to nuclear weaponry, some of which might make use of nuclear power reactors. On the other hand, you have also seen some technical subtleties: the need for heavy water if one

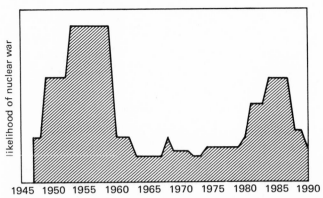

Figure 18.1 Relative likelihood of nuclear war, as measured by the *Bulletin of the Atomic Scientists*' "doomsday clock." Changes reflect major events in the history of nuclear weapons and superpower relations. What is actually plotted is the inverse of the clock's reading, so higher points on the graph correspond to greater risk. The doomsday clock ranged from 2 minutes before midnight at the highest point on the graph to 12 minutes during the low in the mid-1960s. (data source: *Bulletin of the Atomic Scientists,* April 1990, back cover)

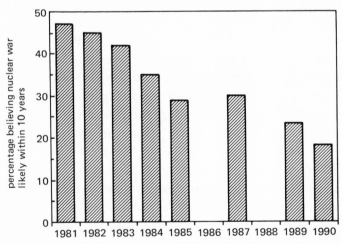

Figure 18.2 Public perception of the nuclear threat, as measured by the percentage of Americans believing nuclear war likely in the next 10 years. No polls were taken in 1986 and 1988. (data source: *New York Times,* May 30, 1990)

doesn't have enriched uranium; the problem of Pu-240 contamination in fuel that stays a long time in a reactor; the technical difficulty of rapidly configuring an explosive critical mass. So what about the connection between nuclear power and nuclear weapons? It is real, all right, but there is still room for controversy.

That controversy is compounded by the fact that no one has yet used civilian nuclear power reactors to produce weapons. India came closest, but its reactor was a research device that made clandestine plutonium production easier than it is in a power plant. And the power-weapons connection has different implications on national and international scales: Even if the international spread of nuclear power plants raises the risk of the proliferation of nuclear weapons, should that affect the development of nuclear power within an already nuclear nation such as the United States? And there are degrees of risk. Breeder reactors and international plutonium trade clearly carry a greater proliferation risk than light-water reactors. So, does nuclear power itself increase the danger of the proliferation of nuclear weapons? Perhaps. On the other hand, there are plenty of avenues to the acquisition of nuclear weapons that do not involve nuclear power plants.

The most significant role of nuclear power in the proliferation of nuclear weapons may be more subtle than direct production of weapons materials: Nuclear plants require nuclear technicians and engineers, so they create a cadre of people with expertise in nuclear technology. Facilities for handling nuclear fuel and waste add further to a country's nuclear infrastructure. Things might stop there, but on the other hand an interest in nuclear weapons could push the nuclear establishment in that less benign direction.

There is another link between nuclear power and warfare. Power plants could be tempting targets in a war using only *conventional* weapons. Blasted by an accurate conventional explosive, a single nuclear plant could release over 30 times the radioactivity of the Chernobyl accident. Would a conventional war on an opponent's nuclear power plants be a nuclear war? Would it call for the use of nuclear weapons in response? Is the spread of nuclear power plants lowering the threshold for all-out nuclear war?

On the other hand, alternatives to nuclear power may carry equally catastrophic risks. What about major climatic upheaval due to continued expansion of coal burning? Would it be better if we supplied nuclear power plants to developing countries to reduce their dependence on coal? Or does the risk of proliferation of weapons outweigh the uncertain hazards of climatic change? What about a nuclear war that starts with a conflict over Mideast oil supplies? Do we weigh that *nuclear* danger among the risks of *fossil* fuels? Those are complex, international nuclear questions.

The Future of Nuclear Technology

Questions of nuclear technology and the proliferation of nuclear weapons hinge on more than just the power-weapons connection, although that must certainly remain a central concern. Even a seemingly benign process such as radiation preservation of food, as we saw in chapter 4, may have a hidden weapons connection. And many of the other seemingly benevolent nuclear applications discussed in part I often make use of nuclear materials produced in research reactors such as the one India used to make plutonium for its nuclear explosive or the Iraqi reactor bombed by Israel. Should the developed nuclear nations withhold nuclear technology from their less prosperous global neighbors because of the weapons connection? Denying technological benefits where they might be most needed hardly seems just. For the United States, the Soviet Union, and Britain it is downright illegal; these Nonproliferation Treaty signatories are treaty-bound to help non-nuclear-weapon states develop peaceful nuclear applications.

The general level of nuclear technology throughout the world is almost certain to advance, and as it does more and more nations will come within reach of nuclear weaponry. Controls on sensitive technology and nuclear materials might slow that trend, but it is still easy to imagine that a few decades from now almost any nation could build nuclear weapons if it wanted. Turning back the technological clock, or eliminating nuclear technology altogether, might prevent that state of affairs. Either of those options is unlikely. Instead, the world community must develop mechanisms that diminish not only the technological potential but also the political desire for nuclear weapons.

Nuclear technology can do great harm; no one doubts that all-out nuclear war would be an unprecedented and unimaginable di-

Figure 18.3 This Philippine research reactor symbolizes the spread of nuclear technology to developing countries. This particular reactor aids research in biology, agriculture, and medicine; at the same time it breeds nuclear materials and creates nuclear expertise that could be turned to weapons activities. (United Nations)

saster. Nuclear technology can also do great good; few would argue against the life-saving techniques of nuclear medicine. And nuclear technology is certainly capable of supplying energy to run today's industrial civilizations, although the technology of nuclear power remains mired in controversy. The potential of nuclear technologies—whether for good or for ill—rests ultimately in the nature of the atomic nucleus and the forces that bind it together. But the realization of that potential—good or bad—rests with us. Humankind is not going to forget the nucleus, but must learn to live safely with nuclear knowledge. That learning can't be left to the experts. Responsible citizens of Planet Earth need to make nuclear choices, and they need to make them after analyzing complex and difficult nuclear questions in an informed and critical way.

Notes

1. This phrase appears in the headline of a letter to the editor of the *New York Times* by Richard Wilson, Mallinckrodt Professor of Physics at Harvard University (January 14, 1990).

2. Amory and Hunter Lovins, *Energy/War: Breaking the Nuclear Link* (Friends of the Earth, 1980), p. 19.

Further Reading

Listed below are some newspapers and magazines that can help carry your nuclear awareness into the future.

Aviation Week and Space Technology (McGraw-Hill, Inc., 1221 Avenue of the Americas, New York, NY 10020). Authoritative news on the aerospace industry, including the latest on delivery systems for nuclear weapons.

Bulletin of the Atomic Scientists (Educational Foundation for Nuclear Science, 6042 S. Kimbark Avenue, Chicago, IL 60637). Established in 1945 by scientists who had worked on the Manhattan Project, this monthly journal is, despite its technical-sounding title, actually a popular magazine dealing with political and technical issues of arms control. Its famous Doomsday Clock reflects the *Bulletin*'s assessment of the risk of nuclear war. Although its writing is authoritative, the *Bulletin* is decidedly with the anti-nuclear "doves."

National Review (National Review, Inc., 150 East 35th Street, New York, NY 10016). This articulate bastion of conservative thought frequently addresses nuclear issues, and is an outspoken proponent of all things nuclear.

Nuclear News (American Nuclear Society, 555 N. Kensington Ave., La Grange Park, IL 60525). Intended for nuclear professionals, this magazine features up-to-date articles on nuclear power, nuclear waste management, and related issues. The viewpoint is solidly with the nuclear industry, although nuclear accidents and other mishaps are reported.

Nuclear Times (Nuclear Times, Inc., 1601 Connecticut Avenue NW, Washington, DC 20009). A bimonthly magazine of nuclear news, from a decidedly anti-nuclear viewpoint.

Scientific American (Scientific American, Inc., 415 Madison Avenue, New York, NY 10017). This respected and authoritative monthly features frequent articles on nuclear technology and related issues, by experts in their fields. The articles are objective and informative, although a range of nuclear opinions is clearly evident.

The New York Times (or any other national or local newspaper). As the Nuclear News items throughout this book indicate, newspapers are an excellent source of up-to-date information on nuclear issues.

Index